Handbook of Plum Fruit

In the last ten years there has been an exponential increase in the adoption of high-density farming, which leads to better yield and higher-quality fruits, thus improving the economic return. *Handbook of Plum Fruit: Production, Postharvest Science, and Processing Technology* covers all the recent advances in plum production, harvesting, handling and processing.

Divided into two main parts, the first eight chapters provide insight about preharvest processing of plums, whereas the later chapters discuss the postharvest processing of plums. This book also includes vital chapters on varietal improvement and rootstock breeding, high-density planting, and pollination. After harvesting, plum quality quickly diminishes, mainly due to weight loss, total acidity, loss of firmness, and decay.

Key Features:

- In-depth information on the pre- and postharvest processing of plums
- Coverage on plum harvesting, handling, and storage practices
- Plum by-product utilization and potential health benefits

Handbook of Plum Fruit provides comprehensive information on recent advances in postharvesting technologies of plum. The health benefits of plum and its products are also addressed. This book will assist horticulturists, agriculturists, pomologists, food scientists and others working in various fruit-processing industries.

Handbook of Plum Fruit
Production, Postharvest Science, and Processing Technology

Edited by
Amir Gull
Gulzar Ahmad Nayik
Sajad Mohd Wani
Vikas Nanda

CRC Press
Taylor & Francis Group
Boca Raton London

CRC Press is an imprint of the
Taylor & Francis Group, an **informa** business

First edition published 2023
by CRC Press
6000 Broken Sound Parkway NW, Suite 300, Boca Raton, FL 33487–2742

and by CRC Press
4 Park Square, Milton Park, Abingdon, Oxon, OX14 4RN

CRC Press is an imprint of Taylor & Francis Group, LLC

Library of Congress Cataloging-in-Publication Data
Names: Gull, Amir, editor.
Title: Handbook of plum fruit : production, postharvest science, and processing technology / edited by Amir Gull, Gulzar Ahmad Nayik, Sajad Mohd Wani, Vikas Nanda.
Description: First edition. | Boca Raton, FL : CRC Press, 2023. | Includes bibliographical references and index.
Identifiers: LCCN 2022009539 (print) | LCCN 2022009540 (ebook) | ISBN 9781032062426 (hbk) | ISBN 9781032071176 (pbk) | ISBN 978100320544(ebk).
Subjects: LCSH: Plum—Handbooks, manuals, etc. | Plum—Postharvest technology—Handbooks, manuals, etc.
Classification: LCC SB377 .H36 2023 (print) | LCC SB377 (ebook) | DDC 634/.22—dc23/eng/20220625
LC record available at https://lccn.loc.gov/2022009539
LC ebook record available at https://lccn.loc.gov/2022009540

ISBN: 978-1-032-06242-6 (hbk)
ISBN: 978-1-032-07117-6 (pbk)
ISBN: 978-1-003-20544-9 (ebk)

DOI: 10.1201/9781003205449

Typeset in Times
by Apex CoVantage, LLC

Dr. Amir Gull dedicated this book to his beloved daughter, Abeeha Amir

Contents

Preface

Extensive growing technology, low yields and low-quality fruit, multitude cultivars and plum pox virus (PPV)–induced problems are main distinctions in plum production. In addition, the structure of the assortment of plums is unfavorable. About half of the total number of trees are native cultivars having small fruits, followed by poor quality. On the other hand, over 75% of the plum orchards are located in mountain areas, where a main limiting factor for plum intensive production is acidic soils, having deficiency of organic matter and inadequate availability of major nutrients. The appearance of nutrient deficiencies and responses to added nutrients indicated the prevalence of nutritional disorders of macronutrients and therefore, limited vegetative growth, low productivity and poor fruit quality.

The prunus genus (*Rosaceae* family) consists of more than 200 species, including some of the most common fruit or ornamental species that are of great economic importance. The plant is of medium height, has deciduous leaves and is ovate, dentate and glossy. White flowers with small green drupaceous fruits (3 cm), characterize the blossoming and ripening stages. It is commonly cultivated for its fruits, as a rootstock and hedge shrub. A recent study showed the potential of kernels recovered from fruit pits of this species as biodiesel feedstock. Plums are adapted to a broad range of geographic conditions, having great potential for production and consumption worldwide and therefore particularly suitable to use for selecting new cultivars.

Being rich in fibers and antioxidants, the fruits constitute an important nutritional source in the human diet. The aroma is one of the most important parameters of determining fruit quality and its perception and acceptability by consumers. The knowledge of volatiles emitted by different plant organs during different ontogenetic stages of growth evidence metabolites that are synthesized by the plant as possible intrinsic mechanisms of defense and interaction with other plants and animals, something of importance for further agronomical studies.

Plum fruits are small and oval in shape, with different skin color and flesh, which is consumed fresh, dried or used to prepare juice, jam or liquor. Plums are classified as climacteric fruits with ethylene controlling changes during ripening. However, some of them show a suppressed-climacteric pattern. Plums are appreciated by consumers due to their high-quality properties, such as sweetness and firmness, and their bioactive compound content, mainly phenolics, including anthocyanins, carotenoids, ascorbic acid and fiber. These bioactive compounds have been reported to have antioxidant properties with health benefits such as reducing the risk of cardiovascular illness.

After harvesting, plum quality quickly diminishes, mainly due to weight loss, total acidity, loss of firmness and decay. Thus, it is necessary to adopt proper postharvest technologies to maintain and extend the postharvest quality of plums for longer periods. Therefore, it is important to understand and document the modern production methods implemented in recent times for harvesting the maximum number of plums. This book will therefore bridge the gap for old and modern methods of production and will help students and researchers understand the complete pre- and postharvest handling of plum fruit. This book will also help fruit growers understand the best possible practice for modern plum production to harvest maximum yield, which in turn will increase their returns.

The book is divided into two main parts: the first provides insights about the pre-harvest processing of plum, which is discussed in the first eight chapters. Chapter 1 deals with the history, distribution, production and taxonomic classification of plums. Chapter 2 highlights the orchard planning, establishment and soil management of plums, followed by Chapters 3 and 4, which discuss recent advances in rootstock breeding and plum prorogation. Chapter 5 covers the flowering, fruit set and pollination of plum, while Chapter 6 deals with orchard manuring practices of plum trees. Chapter 7 focuses on recent developments in plum harvesting and handling, and Chapter 8 deals with diseases, pests and disorders in plum.

The second part gives insights about the postharvest processing of plum. Chapter 9 discusses the emerging packaging and storage technologies of plum, while Chapters 10 and 11 focus on recent plum-processing technologies and the utilization of plum by-products. Chapter 12 deals with the health benefits of plum fruit its products, and Chapters 13 and 14 deal with the effects of pre- and postharvest technologies/factors on the quality and bioactive compounds of plum. The last chapter emphasizes the extraction of anthocyanins from plum fruit and its products.

Lastly, this book assists horticulturists, agriculturists, pomologists, researchers, food scientists and other members working in various fruit-processing industries. It could be used by university libraries and institutes all around the world as a handbook and/or ancillary reading for students pursuing bachelor's and master's degrees in agriculture, horticulture or fruit and vegetable processing. We would like to thank and acknowledge one by one all authors for their fruitful contribution and for their dedication to editorial guidelines and timelines. We are fortunate to have had the opportunity to collaborate with many international experts from the United States, Australia, Romania, Spain, Colombia, México, South Korea, China and Pakistan. We would like to thank colleagues from the production team at Taylor & Francis for their constant help during the editing and production process. Finally, we as editors have a message to all readers that this book may contain minor errors or gaps. Suggestions, criticism and comments are always welcome, so please do not hesitate to contact us for any relevant issue.

Dr. Amir Gull,
Dr. Sajad Mohd Wani,
Dr. Gulzar Ahmad Nayik,
Dr. Vikas Nanda

About the Editors

Dr. Amir Gull completed his master's degree in Food Technology from Islamic University of Science & Technology, Awantipora, Jammu and Kashmir, India, and his PhD from Sant Longowal Institute of Engineering & Technology, Longowal, Sangrur, Punjab, India. Dr. Gull has published more than 35 peer-reviewed research and review papers in reputable journals. He has also published two books in Springer, more than ten book chapters and delivered a number of presentations at many national and international conferences. Dr. Gull's main research activities include developing functional food products from millets. He also serves as an editorial board member and reviewer of several journals. He is also an active member of the Association of Food Scientists and Technologists India and is a recipient of the Maulana Azad National Fellowship from the University Grants Commission of India.

Dr. Gulzar Ahmad Nayik completed his master's degree in food technology from Islamic University of Science & Technology, Awantipora, Jammu and Kashmir, India, and his PhD from Sant Longowal Institute of Engineering & Technology, Sangrur, Punjab, India. He has published more than 60 peer-reviewed research and review papers and 31 book chapters and has edited eight books with Springer, Elsevier and Taylor & Francis. Dr. Nayik has also published a textbook on Food Chemistry and Nutrition and has delivered a number of presentations at various national and international conferences, seminars, workshops and webinars. Dr. Nayik was shortlisted twice for the prestigious Inspire-Faculty Award, in 2017 and 2018, from Indian National Science Academy, New Delhi, India. He was nominated for India's prestigious National Award (Indian National Science Academy Medal for Young Scientists 2019–20). Dr. Nayik also fills the roles of editor, associate editor, assistant editor and reviewer for many food science and technology journals. He has received many awards, appreciations and recognitions and holds membership in various international societies and organizations. Dr. Nayik is currently editing several book projects with Elsevier, Taylor & Francis, Springer Nature, the Royal Society of Chemistry and more.

Dr. Sajad Mohd Wani is presently working as Associate Professor-cum-Senior Scientist in the Division of Food Science and Technology, SKUAST-Kashmir, Jammu and Kashmir, India. He has served as Assistant Professor in the Department of Food Science and Technology, University of Kashmir for more than a decade. He completed his BSc in agriculture and MSc in postharvest technology from SKUAST-Kashmir and his PhD in food technology from University of Kashmir. He has passed the CSIR-NET in life sciences and ICAR-ASRB NET in horticulture and food science and technology exams. He has acted as the principal investigator of four projects and co-principal investigator of ten major research projects. Two PhD scholars have received ICMR Senior Research Fellowship (SRF) under his mentorship. He has guided 40 MSc students, and 7 PhD scholars are currently working under his supervision. He has been awarded an International Travel Grant by SERB-DST to present a paper in EFFoST, held in Greece in 2015. He has presented 15 papers at several national and international conferences. He organized one national-level conference on “Recent advances in understanding the role of phytochemicals in human heath” in 2018. He has published more than 80 research and review articles in nationally and internationally reputable journals, a with some papers having an NAAS rating of more than 17. He has more than 1,800 citations to his credit, with an H-Index of 25. He has published one book and ten book chapters. He received the best paper award from AFST(I) in 2016.

Prof. Vikas Nanda completed his master’s degree and PhD in food technology from Punjab Agricultural University, Ludhiana, Punjab, India. He is currently working as Professor in the Department of Food Engineering & Technology, Sant Longowal Institute of Engineering & Technology, Sangrur, Punjab, India. Prof. Nanda has published more than 80 peer-reviewed research and review papers and many book chapters, has authored two books and has delivered a number of presentations at various national and international conferences, seminars, workshops, and webinars. Prof. Nanda has supervised 5 PhD scholars and 24 M.Tech students. Prof. Nanda has visited Israel, the Czech Republic, France, Greece, Argentina, the Netherlands, Thailand and the United Arab Emirates to attend international conferences. Prof. Nanda is also working as Co-Chairman for the International Honey Commission and Co-chairman/Mentor council formed by the Ministry of Labour and Employment, Government of India.

Contributors

Laura Natali Afanador-Barajas
Universidad Central
Bogotá, Colombia

Muhammad Haseeb Ahmad
Government College University
Faisalabad, Pakistan

Mukhtar Ahmed
Department of Higher Education
Government of Azad Jammu and Kashmir, Pakistan

Kashif Ameer
University of Sargodha
Sargodha, Pakistan

Mohammad Javed Ansari
Hindu College Moradabad
(Mahatma Jyotiba Phule Rohilkhand University)
Bareilly, UP, India

Jobil J. Arackal
Saintgits College of Engineering
Pathamuttom, Kottayam, Kerala, India

Muhammad Umair Arshad
Government College University
Faisalabad, Pakistan

Ioana M. Bodea
University of Agricultural Sciences and Veterinary Medicine Cluj-Napoca
Calea Mănăştur, Cluj-Napoca, Romania

Tridip Boruah
Madhab Choudhury College
Barpeta, Assam, India

Giorgiana M. Cătunescu
University of Agricultural Sciences and Veterinary Medicine Cluj-Napoca
Calea Mănăştur, Cluj-Napoca, Romania

A.J. Cenobio-Galindo
University of Hidalgo State
Tulancingo, Hidalgo, México

Ana B. Cerezo
Universidad de Sevilla
Sevilla, Spain

Debasish Das
Madhab Choudhury College
Barpeta, Assam, India

G. De Vega-Luttmann
University of Hidalgo State
Tulancingo, Hidalgo, México

Gargee Dey
Madhab Choudhury College
Barpeta, Assam, India

O. Fernández-Fernández
Chapingo Autonomous University
Chapingo, State of México, México

M.J. Franco-Fernández
University of Hidalgo State
Tulancingo, Hidalgo, México

M. Carmen Garcia-Parrilla
Universidad de Sevilla
Sevilla, Spain

Rajadurai Gothandaraman
Tamil Nadu Agricultural University
Coimbatore, Tamil Nadu, India

Thiribhuvanamala Gurudevan
Tamil Nadu Agricultural University
Coimbatore, Tamil Nadu, India

Aiza Hasnain
University of the Punjab
Lahore Punjab, Pakistan

Shaziya Hassan
University of Agricultural Sciences and Technology of Kashmir
Shalimar, Srinagar, Jammu and Kashmir, India

I. Hernández-Soto
University of Hidalgo State
Tulancingo, Hidalgo, México

Ruth Hornedo-Ortega
Universidad de Sevilla
Sevilla, Spain

Salam A. Ibrahim
North Carolina Agricultural and Technical State University
Greensboro, North Carolina

Nabia Ijaz
University of Lahore
Lahore, Pakistan

Muhammad Imran
Government College University
Faisalabad, Pakistan

Shafeeqa Irfan
University of Sargodha
Sargodha, Pakistan

Yash D. Jagdale
MIT ADT University
Pune, Maharashtra, India

Rehana Javid
University of Agricultural Sciences and Technology of Kashmir
Shalimar, Srinagar, Jammu and Kashmir, India

Guihun Jiang
Jilin Medical University
Jilin, China

Rabia Kanwal
South China University of Technology
Guangzhou, China

Muhammad Asif Khan
University of Agriculture
Faisalabad, Pakistan

Muhammad Kamran Khan
Government College University
Faisalabad, Pakistan

M.A. Kuchay
University of Agricultural Sciences and Technology of Kashmir
Shalimar, Srinagar, Jammu and Kashmir, India

Xian Lin
Sericultural & Agri-Food Research Institute
Guangdong, China and
BNU-HKBU United International College
Zhuhai Guangdong, China

Shahid Mahmood
University of Sargodha
Sargodha, Pakistan

A. Raouf Malik
University of Agricultural Sciences and Technology of Kashmir
Shalimar, Srinagar, Jammu and Kashmir, India

Imtiyaz Ahmad Malik
Agriculture Production & Farmers Welfare Department
Government of UT of Jammu and Kashmir, India

Yesid Alejandro Mariño Macana
Universidad Central
Bogotá, Colombia

Gabriela Medina-Pérez
Universidad Autónoma del Estado de Hidalgo
Hidalgo, México

Nisar A. Mir
Chandigarh University
Mohali-Punjab, India

Ghulam Mueen-ud-Din
University of Sargodha
Sargodha, Pakistan

Mian Anjum Murtaza
University of Sargodha
Sargodha, Pakistan

Rafiya Mushtaq
University of Agricultural Sciences and Technology of Kashmir
Shalimar, Srinagar, Jammu and Kashmir, India

Sabeera Muzzaffar
University of Kashmir
Srinagar, Jammu and Kashmir, India

Muhammad Nadeem
University of Sargodha
Sargodha, Pakistan

Jessica Pandohee
Curtin University
Bentley, Wester Australia, Australia

S. Parameshwari
Periyar University
Salem, Tamil Nadu, India

S. Perez-Ríos
University of Hidalgo State
Tulancingo, Hidalgo, México

Muhammad Modassar Ali Nawaz Ranjha
University of Sargodha
Sargodha, Pakistan

Syeda Shehar Bano Rizvi
University of the Punjab
Lahore Punjab, Pakistan

Nimra Saeed
Government College University
Faisalabad, Pakistan

Amna Sajid
University of the Punjab
Lahore Punjab, Pakistan

Parthasarathy Seethapathy
Amrita School of Agricultural Sciences
Coimbatore, Tamil Nadu, India

Muhammad Shafiq
University of the Punjab
Lahore Punjab, Pakistan

Bakhtawar Shafique
University of Sargodha
Sargodha, Pakistan

Jadala Shankaraswamy
Sri Konda Laxman Telangana State Horticultural University
Wanaparthy, Telangana, India

Mohd Aaqib Sheikh
Sant Longowal Institute of Engineering & Technology
Longowal, Punjab, India

Munazah Sidiq
University of Kashmir
Srinagar, Jammu and Kashmir, India

Sithara Suresh
Saintgits College of Engineering
Pathamuttom, Kottayam, Kerala, India

Muhammad Rizwan Tariq
University of the Punjab
Lahore Punjab, Pakistan

Ana M. Troncoso
Universidad de Sevilla
Sevilla, Spain

Aida Vanessa Wilches
Universidad Central
Bogotá, Colombia

Baojun Xu
United International College Zhuhai
Guangdong, China

Syeda Mahvish Zahra
University of Sargodha
Sargodha, Pakistan
and
Allama Iqbal Open University
Islamabad, Pakistan

Saadia Zainab
Henan University of Technology
Zhengzhou, China

Aliza Zulifqar
Government College University
Faisalabad, Pakistan

1 History, Distribution, Production and Taxonomic Classification of Plum

Laura Natali Afanador-Barajas,[1] *Aida Vanessa Wilches,*[1] *Yesid Alejandro Mariño Macana*[1] *and Gabriela Medina-Pérez*[2]

[1] Programa de Biología, Facultad de Ingeniería y Ciencias Básicas, Universidad Central, Bogotá, Colombia

[2] Instituto de Ciencias Agropecuarias, Universidad Autónoma del Estado de Hidalgo. Av. Universidad s/n Tulancingo, Hidalgo, México

CONTENTS

DOI: 10.1201/9781003205449-1

1.1 INTRODUCTION OF PLUM

Plums rank third after apples and pears as far as their job in natural product creation in the cooler and calmer portions of the world (Zohary et al. 2012). Plums are a highly commercialized stone fruit and are consumed fresh, in jams or dried, known as prunes. Despite its agricultural importance and long history of cultivation, there are still many mysteries concerning the species' origins and connections. One of most cultivated is the European plum (*Prunus domestica*), a hexaploid (2 n = 6x = 48 chromosomes) fruit tree that is grown all over the world (Zhebentyayeva et al. 2019; Zohary et al. 2012).

Plums and prunes are among the fruits in the Rosaceae family, which also includes cherries, apples, nectarines, peaches, pears, and various berry harvests. The genus *Prunus*, which includes all real plums, is a prominent part of the Rosaceae family. Approximately 200 *Prunus* species are thought to exist. Principally, they are located in northern temperate areas. There are a variety of plums, the most significant of which are European plums (*Prunus domestica* L.) and Japanese plums (*Prunus salicina* Lindl). There are also American plums, damsons, Mirabelle plums, and more varieties. Recently, a few 'interspecific hybrids' have been developed by crossing plums with apricots or sweet cherries (Jayasankar et al. 2015). Figure 1.1 shows some different varieties of most cultivated plums from Europe.

Luna-Vázquez et al. (2017) compile studies of principal phytochemicals in stone fruits. Plums are low in calories, with simple sugars, vitamins, minerals, proteins and lipids. Some phytochemicals are recognized as polyphenols, carotenoids, triterpenes and unstable mixtures. Phenolic compounds are incompletely credited to their cancer prevention agent: polyphenols. Different health properties have been revealed with the consumption of plums, such as increased antioxidants, antiallergenic properties, better cognitive function and lower cardiovascular risk (Igwe and Charlton 2016). Particularly, prunes or dried plums have an exclusive nutritional bioactive profile and are suggested to be beneficial for bone health (Wallace 2017).

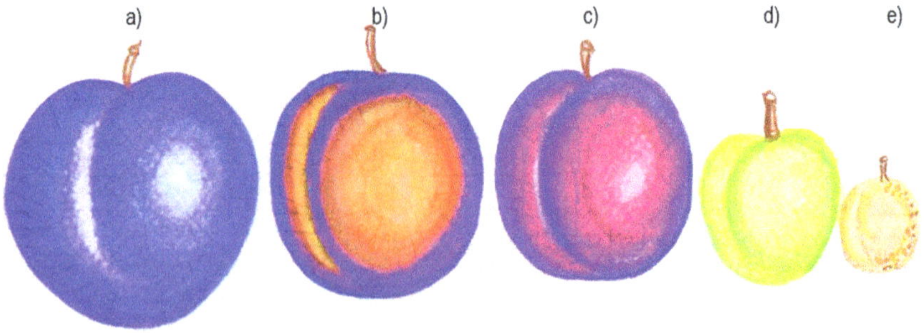

FIGURE 1.1 Types of plums *Prunus domestica*: a) prune monsieur native, b) damson, c) prune royale tours, d) greengage and e) Mirabelle plum. Illustrated by Laura Natali Afanador-Barajas.

The European plum likely began in a district near the Caspian and Caucasus Seas and is no less than 2,000 years old. The species of plum localized in the Old World, beginning in Europe or Asia, is the damson plum (*Prunus insititia*); antiquated works interface early development of those plums with the district around Damascus. The Japanese plum was first tamed in China millennia prior, yet it was broadly evolved in Japan; from that point, it was acquainted with the remainder of the world. Japanese plums have a more extended time span of usability than most European assortments and are in this way the most well-known plum sold economically. Plum production is developed all through the world, and numerous assortments are adjusted to a scope of soils and climatic conditions (Milošević and Milošević 2018; Okie and Hancock 2008; Faust and Surányi 1999; Siddiq 2006).

1.1.2 History of Plums

1.1.2.1 Origin of Plums

Different species of plums evolved in Europe, Asia, and North America and were domesticated independently. Historical evidence proposes *Prunus domestica* originated in Europe as *P. cerasifera* and *Prunus spinosa* originated in western and central Asia. *P. salicina* originated in China, and the *Prunocerasus* species, such as *P. americana* (Marshall plum), originated in North America (Topp et al. 2012; Milošević and Milošević 2018). The highest diversity of plums among different *Prunus* species and have been tamed around three continents (Faust and Surányi 1999). Figure 1.2 shows the places of origin, the expansion and the dispersion of plums around the world.

Plums are distributed throughout Europe, North America and Asia (Okie and Hancock 2008). The origin of these fruit trees of the genus *Prunus* is located in the central (the Caucasus area) and eastern (China) regions of Asia (Figure 1.2). Its

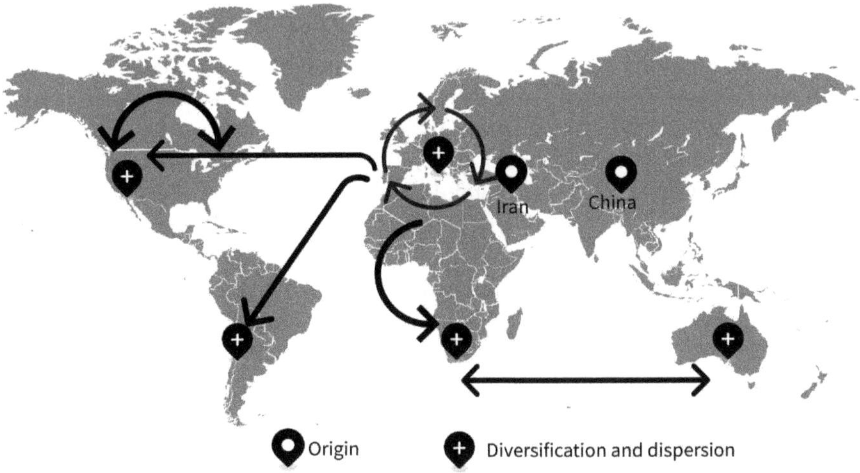

FIGURE 1.2 Places of origin, expansion and dispersion of plums around the world.

expansion toward the Near East and the Mediterranean went in parallel with trade routes and historical events that put the Persian Empire in contact with Europe in the 4th century BCE. This first arrival in Europe would occur through Armenia or Iran to Greece and Italy. The species also arrived through the southern Mediterranean to Spain through the Arabs in the second half of the 7th century CE. Finally, the expansion of these fruit trees to North America, Mexico and South Africa would be produced from Europe (mainly from Spain) in the 16th and 17th centuries. It was not until the 18th century that they arrived in Australia (Martínez-Gómez 2017).

P. domestica L. was utilized by humans at least 6,000 years ago, according to stone remains, and it was commonly grown in Roman times. Because of the lengthy history of domestication, there is a wide range of morphology (Urrestarazu et al. 2018; Topp et al. 2012). *Prunus salicina* was initiated in the Yangtze River and spread across eastern China. The cultivar known as 'Zhui Li' has a history of more than two eras ago. The species labeled as *P. consociiflora*, *P. gymnodonta*, *P. ussuriensis* and *P. thibetica* are all *P. salicina* plums with minor differences and are not currently considered differentiated species (Faust and Surányi 1999).

Watkins (1976) investigates the split of the *Prunocerasus* and *Euprunus* species. He claims that if they had separated earlier, the parallels between them would not be as strong. Most diploid species have a high degree of interspecific compatibility, and they have been hybridized during the last two centuries to increase adaptability, fruit quality, and yield. *Prunus* research has largely focused on crop species and their relatives. The geographic origin of *Prunus* has unavoidably been assumed to be Eurasia, given that the genus's core of crop diversity is in that region. A complex interaction of geologic tectonic processes and climatic oscillations from the early Eocene period determined *Prunus*'s worldwide spread from its place of origin, which either favored or impeded migration into both the Old and New Worlds at various episodic periods of geologic time. Furthermore, we have given evidence to support the notion that the primary evolved from a hybrid (Chin et al. 2014).

1.1.2.2 Polyploidy of Plums

Polyploidy is a common occurrence that plays an important role related to the expansion of new species or forms. Polyploidy is also significant from a practical standpoint because plants with this trait are frequently highly robust and may be more resistant to cold and parasitic (OECD 2002). In *Prunus*, the indispensable number of chromosomes in vegetative cells is eight. Polyploidy, because of interspecific hybridization, occurred during the phylogeny of the variety and is liable for self-sterility and intersterility (OECD 2006).

The *Prunus* species of European plums are hexaploid, despite the fact that some of them are genuine diploids. Specifically, *Prunus domestica* is a polymorphic allopolyploid (2n =6x= 48) (Faust and Surányi 1999). Two additional ancient wild plums, the diploid cherry plum *Prunus cerasifera* and the tetraploid blackthorn *Prunus spinosa* L., gave rise to the tetraploid species (Jayasankar et al. 2015). The European plum may be an amphidiploid derived from *Prunus cerasifera* (2n=16) and *Prunus spinosa* (2n=32) (Hartmann and Neumüller 2009).

Table 1.1 shows many species of plum, the principal ones being *P. domestica*, commonly called the European plum, 2n=48, hexaploid, from Europe; *Prunus*

TABLE 1.1
Types of Plums and Different Origins and Genetic Compositions

Plums	Native	Chromosome number
P. domestica (European plum)	Europe	2n=48 (hexaploid)
P. domestica ssp. P. insititia (damson plum)	Europe	2n=48 (hexaploid)
P. cerasifera (cherry plum)	Europe	2n=16 (diploid)
P. spinosa (sloe)		2n=32 (tetraploid)
P. simonii	Asia	2n=16 (diploid)
P. salicina (Japanese plum)	China	2n=16 (diploid)
P. nigra		2n=16 (diploid)
P. americana (American plum)	North America	2n=16 (diploid)
P. hortulana	North America	2n=16 (diploid)
P. angustifolia	North America	2n=16 (diploid)
P. mexicana	North America	2n=16 (diploid)
P. munsoniaca	North America	2n=16 (diploid)

domestica ssp. P. insititia, or damson plum, 2n=48, hexaploid, from Europe; *Prunus salicina* with a 2n=16, diploid, from China; and *Prunus americana*, known as the American plum, 2n=16, diploid, from North America. The present plum industry is based on *Prunus domestica* and *Prunus salicina*. Neither of these species takes wild ancestors, and both entered into human use highly developed. The garden plum and the Japanese plum emerged as important fruit crops around 300 BCE (Zohary et al. 2012).

1.1.2.3 Archaeological Evidence

Stones of all European species of plum (*P. domestica*, *P. cerasifera* and *P. spinosa*) have been uncovered dating back to 4000–6000 BCE in Ukraine and Germany. These archeological discoveries of European plums matches more specifically in Neolithic times (Faust and Surányi 1999). *Prunus domestica* appears to have begun in southern Europe or western Asia between the Caucasus Mountains and the Caspian Sea, covering the beginning of *P. cerasifera* and moving into western Europe. The most punctual archeological remaining parts of *P. domestica* in Europe are inferable from the Roman period (Sarigu et al. 2017; Browicz and Zohary 1996; Okie and Ramming 1999; Zohary et al. 2012). Plum stones had appeared in different locations in Europe from the Neolithic and Bronze Ages, such as Italy, Germany, Austria, and Switzerland. The stones are variable in shape, but they decrease within the morphological range compared to actual plums, such as *cerasifera* and *insititia* (Zohary et al. 2012).

Based on archeological findings, confined stones of *Prunus spinosa* and *Prunus insititia*. were found in resident disinterring near to lake at Switzerland. From the Caspian Sea and the Caucasus Mountains, the plum was spread to antiquated Egypt, Syria, Mesopotamia and Crete around 6000 BCE (Faust and Surányi 1999). A lot later, because of wars, relocation and exchange developments, plums were known in Rome and Greece. The previous account of the history of plums comes from the Greek verse writer Archilochus (680–645 BCE) and the versifying artist Hipponax (541–487 BCE), whereas the old Greek thinker Theophrastus (370–287 BCE) depicted three plum cultivars. The Romans brought homegrown plums from Syria after they vanquished the region in 65 BCE (Zohari 1986).

1.2 DISTRIBUTION OF PLUMS

1.2.1 Generalities about Plum Distribution

Diversity in plants is driven by many factors related to climate, geography, soil and nutrient content (Lompo et al. 2021). Recently, improvement in predicting species distribution attempts to contribute to understanding the ecological and evolutive patterns of diversity that will be an important tool in making proper decisions in reforestation plans, conservation and predicting the suitability of a habitat for many species (Elith et al. 2006). Undoubtedly, climate change will cause fast alterations in these patterns, causing problems for people who cultivate these important fruits for their subsistence. Particularly, plums are native in temperate regions (Okie and Ramming 1999), and many of the native plums have been improved with better flavor, shape, color and nutritional content.

Ten countries lead plum production, China being the most important. The other countries are Chile, Spain, India, Iran, Italy, Romania, Serbia, Turkey and the United States (Milošević and Milošević 2018). According to FAOstat data, plums are widely distributed around the world. but it is important to distinguish between botanically recognized species and varieties, although this classification may vary in some articles. For Faust and Surányi (1999), *Prunus spinose* and *Prunus cocomilia*, *Prunus spinose* L., were the two former species. Both were well distributed and originated in Europe, North Africa and Northern Turkey; specifically, *P. cocomilia* is described in Italy. *Prunus spinossa macrocarpa* can also be considered an early species, but its exact origin and history are very controversial (Milošević and Milošević 2018), even today.

1.2.2 Principal Groups

1.2.2.1 Cherry Plum (*Prunus cerasifera*)

The well-distributed *Prunus cerasifera* (cherry plum) originated in west Asia, is native to southeast Europe, and is well known as a common plum; however, the exact origin of this species is not clear. It is considered the diploid European plum's progenitor and is cross-fertile with American species and Asiatic species, which constitutes it as a basic species because it can reproduce with many other species.

The cherry plum is also known as the Myrobalan plum (derived from the Greek words *Myron*, which means juice of plant, and *balanos*, which means nut). In Europe, it is possible to find it in the United Kingdom, and it is naturalized in the northeast and far west United States as the Stanley prune. Some common varieties are: 'Thundercloud', 'Krauter Vesuvius', and 'Newport'. Some authors consider it a cultivated form of *P. cerasifera* macrocarpa and a wild form named *P. divaricate*, more common in the Caucasian region but native from Macedonia to northern Persia (Faust and Surányi 1999).

Some species derived from *P. cerasifera* are spp ursine (Turkey and Syria); and varieties and forms of *P. cerasifera* based on local adaptations have been described as var.iranica in Iran, var nairica in Armenia and caspica in Caucasia. There are many hybrids of cherry plums, including the Mariana plum, the Methley plum, and *P. ferganica*, and ornamental forms, such as *planteriensis* Hort, *pendula* Hort, *acutifolia* Hort and *pisardiii,* distributed in France.

Garden plums, derived from *cerasifera*, may include common *P. domestica* and *P. insititia*. The first one well cultivated may be a cross between *cerasifera* and *spinosa*; found in Turkey, northwestern Xinjiang (China), Serbia, Hungary and Bulgaria, with possible origin in Iran (Faust and Surányi 1999).

1.2.2.2 *Prunus domestica*—The Most Important Plum Industry

Prunus domestica is one of the two most important bases of the plum industry and commerce; the other is *Prunus salicina* (Faust and Surányi 1999). This species was described by Linnaeus in 1753, and his description was published in *Species Plantarum* 1:475 (Méndez 2015). Actually, *Prunus salicina* are the most important species in the plum industry and are widely cultivated (Faust and Surányi 1999). This variety is known as the Italian or European plum, but the first cultivars were in Syria and probably distributed in Rome and then to the rest of Europe, especially in the west countries (Okie and Ramming 1999).

During the Crusades, the Romans introduced European plums to the western part of Europe (Okie and Ramming 1999). It is still not clear for breeders where they can find its wild form, but is possibly near the Caspian Sea (Milošević and Milošević 2018). It is the most economically important fruit tree, particularly in Europe and southwest Asia, where it is vastly cultivated and is considered a younger species resulting from the natural cross between *Prunus cerasifera* and *Prunus spinose*, but it is necessary to use other types of analysis to determine this idea.

The trees are well adapted to cooler regions and may be eaten in a variety of ways. In the United States, California is the main producer, but it can also be found in Idaho, Washington, Oregon, Michigan and New York (Okie and Ramming 1999). A subspecies, *P. insititia*, is cultivated in Connecticut, Maine, Massachusetts, New Hampshire and Vermont.

1.2.2.3 Yellow Egg Plums—*Prunus institia*

This group comprises *P. insititia* L., with all its derived varieties. It originated in Turkey and is derived from *Prunus domestica* (*Prunus domestica* subsp institia) and has divisions as *P. pomarium*, *P. insititia glaberrima*, alpine-orientalis y var leopodiensis, *Prunus insititia syriaca* and var Juliana (Faust and Surányi 1999).

1.2.3 Worldwide Distribution

The two major types of plum are the Asiatic and European species. The first ones include two species, *Prunus salicina* (formerly *P. triflora* Roxb.) and *Prunus simonii*, also known as the Simon or apricot plum found in China, Japan, and Central Asia; *Prunus simonii* was used as a parental species in Californian cultivars. *P. salicina*, the Japanese plum, has its origin in the Yangtze River in China, according to Yoshida (1987), and then spread to the east lands. Small, sun-loving deciduous trees found in

TABLE 1.2
Major Distribution of Principal Species Cultivated over the World

Country	World plum distribution	Source
Argentina	*Prunus domestica* (var Agen, President and Reineclaudes) *Prunus salicina* (Black Amber, Santa Rosa and Red Beaut)	(Colamarino 2010)
Bulgaria	*Prunus domestica*	(Faust and Surányi 1999)
Brazil	*Prunus salicina* (Carmesin, Rosa Paulista, Gran Cuore, Gema de Ouro, Golden Talisma, Rosa Mineira, Januaria and Kelsey-31)	(Milošević and Milošević 2018)
Chile	*Prunus domestica* (var DÁgen, President and Imperial Epineuse)	(Colamarino 2010)
China	*Prunus domestica* *Prunus salicina*	(Faust and Surányi 1999)
Colombia	*Prunus salicina* Lindl. Horvin	(Orjuela, Camacho, and Parra-Coronado 2016)
Costa Rica	*Prunus domestica*	(Calvo 2009)
France	*Prunus cerasifera*	(Faust and Surányi 1999)
United Kingdom	*Prunus cerasifera* (Myrobalan)	(Faust and Surányi 1999)
Hungary	*Prunus domestica*	(Faust and Surányi 1999)
Iran	*Prunus cerasifera*	(Faust and Surányi 1999)
Italy	*Prunus domestica*	(Okie and Ramming 1999)
Japan	*Prunus salicina*	(Yoshida 1987)
South Korea	*Prunus salicina*	(Yoshida 1987)
Mexico	*Prunus domestica* *Prunus sativa*	(Méndez 2015)
Syria	*Prunus cerasifera*	(Milošević and Milošević 2018)
Uruguay	*Prunus domestica* (Stanley and Reineclaudes) *Prunus salicina* (var Santa Rosa, Obil´naja, Leticia and Rosa Nativa)	(Mercado 2016)
United States	*Prunus angustifolia* *Prunus nigra* *Prunus hortulana* Bailey *Prunus munsoniana*	(Faust and Surányi 1999)
Vietnam	*Prunus salicina*	

Japan, Korea, Vietnam and Australia produce Japanese plums. However, these round and smooth or wrinkled fruits are actually native to China. In the United States, especially in California, the plum cultivars now include all the fresh-market plums developed by intercrossing various diploid species with the original species. Japanese plums are well distributed in Australia, Vietnam, Japan and Korea. In California, this species is usually found at supermarkets. A wide variety of fruits resulted from several genetic crosses between the original species and the diploid species (Okie and Ramming 1999).

1.2.3.1 Species and Varieties in the United States

The original five American species of plums are *Prunus americana*, or wild plum, *Prunus angustifolia*, *Prunus nigra*, *Prunus hortulana* Bailey and Prunus *munsoniana. P. americana* is well distributed in the United States, especially in the northern and eastern regions; it is probably native to Missouri and is well cultivated in Utah and at the Gulf of Mexico, as well as other states, such as Georgia and Massachusetts. *Prunus angustifolia* is well distributed alongside the Missouri River. It comes from a small deciduous tree and is also called a Chickasaw plum. It can be found in Delaware, Florida and Texas because it grows well in sandy soils. *Prunus nigra* is also known as the Canada plum, but its range includes New England, Michigan, New York, northern Ohio and Wisconsin. *Prunus hortulana* Bailey has been cultivated in many regions, including central Kentucky, Tennessee, Iowa and Oklahoma. This variety is a hybrid between *P. americana* and *P. angustifolia*. Finally, *Prunus munsoniana*, or the wild goose plum, is cultivated in Kentucky, Tennessee, Mississippi, Texas, Minnesota and Kansas (Faust and Surányi 1999). Numerous American plum species resulted from crosses of commercial varieties.

Prunus avium's (wild cherry) origin is in Europe, northern Africa and southwestern Asia, but it has been well distributed in the United States since colonial times. It's not as sweet as other species that have been derived from this type of plum. Their seeds are dispersed by birds and small mammals, such as squirrels. Varieties include Lapin Starkrimson and Stella, 'Thomas' Stark gold.

Prunus armeniaca (dwarf apricot) has two types. The 'Homedale' Stark Sweetheart is native to northern China, and its name is derived from Armenia (western Asia). It was introduced by Stark Bro's of Louisiana, Missouri, but it's not as well developed as other types of plums because of its low tolerance to freezing temperatures and because it's extremely susceptible to insects and other pests. The Wilson is a small-dwarf apricot native to northern China and is considered a cultivar that adapts well in small spaces.

Prunus carolinianus (cherry laurel) is widely distributed to North Carolina and Florida at the southeast of the United States and to Louisiana, Texas and Arkansas in the central part of the country. It is characterized as being well adapted to drought places.

Prunus cistena (purple-leaf sand cherry) resulted from controlled reproduction between *P. pumila* and *P. cerasifera*, resulting in a hybrid. It has red leaves is a small shrub, is tolerant to sun exposure and is more abundant in Canada but is also found in the United States. *Prunus glandulosa* 'Sinensis', or commonly called the dwarf flowering almond, has adapted to urban areas but requires well drained soils.

1.2.3.2 Species and Varieties in Latin America

Chile is one of the most important plum cultivators in Latin America; "Sweet Pekeetah" was the first variety produced at Universidad de Chile for genetic improvement and fruit quality (Bravo 2010). This plum's culture comprises 32° to 36° south latitude, at the central valley from the V to VII regions that comprise Maipo Valley, Curicó and Maule Valley. The varieties used to pulp dehydration are D'Agen, President and Imperial Épineuse, all derived from *Prunus domestica*; D´Agen concentrates 95% of the planted area, and due to its high sugar level it is particularly suitable for dehydration, allowing for a high-quality product.

Argentina is the leading producer of plums in the Southern Hemisphere and is among one of the three largest exporters of the same group. Varieties of both *Prunus domestica* and *Prunus salicina* can be found in Mendoza and Rio Negro. Japanese plum varieties (*P. salicina*) include Black Amber, Red Beauty and Santa Rosa. The European plum varieties (*P. domestica*) Agen, President and Reineclaudes are the most cultivated (Colamarino 2010). Uruguay produces fresh fruit based in the two principal species (*Prunus domestica* and *Prunus salicina* Lind.) and some varieties include Stanley and Reineclaudes. Uruguay also produces Santa Rosa, Obil'naja, Methley, Leticia and Rosa Nativa varieties with high juicy content and Golden Japan and Crystal, with particularly yellow and white pulp (Mercado 2016). Colombia reports *Prunus salicina* Lindl. Horvin in 16 small towns at Nuevo Colón, principally Aposentos, Potreros 5° 21' 11" North, 73 ° 27' 24" West in the Boyaca region, where 63% of the country's plums are produced (Orjuela et al. 2016). Brazil cultivates the improved Japanese plum varieties Carmesin, Rosa Paulista, Gran Cuore, Gema de Ouro, Golden Talisma, Rosa Mineira, Januaria and Kelsey-31 (Milošević and Milošević 2018).

Mexico cultivates many subspecies derived from *Prunus domestica* in Puebla, Veracruz, Chiapas, Jalisco, Nayarit and Sinaloa, among others, and also *Prunus sativa* (Méndez 2015). Costa Rica has plum cultivars in high zones at Los Santos, in La Pastora in San Marcos de Tarrazý and Copey de Dota. In Copey de Dota, some growers are replacing their apple plantations with plum, as the production is less expensive. However, its production and commercialization is recent, in small quantities within the national economy; in addition, it is a product little rooted in the national economy (Calvo 2009).

The European-type plums (*Prunus domestica*) are elongated, have a dark skin color and the stone is easily separable from the pulp. They are the most appropriate for industrial processing. Within this group are the Stanley and Reineclaudes varieties, both with skin tones ranging from purple to blue. Japanese plums (*Prunus salicina* Lind.) are medium to large spherical fruits, with a sweet aroma that is accentuated with ripening, as well as abundant juice, which is why they are preferred for fresh consumption (Sottile et al. 2010).

1.3 PLUM PRODUCTION

1.3.1 Background

The status of the plum production, market and trade around the world is reviewed. China is the most important producer of plum and sloes, with seven million tons and with revenue amounted to USD 10 billion (Wood 2019), followed by Romania, Serbia, Chile and the United States (FAOSTAT 2021).

1.3.2 International Production

Plums usually need a subtropical and temperate climate, such as those in Europe, Asia and America. Of the nearly 40 existing species, the most important species commercially around the world are *Prunus domestica* and *Prunus salicina* (Okie and Ramming 1999; Topp et al. 2012).

World plum production increased from 10.8 million tons in 2009 to 12.6 million tons in 2019, as per the estimated data from the Food and Agriculture Organization of the United Nations (FAOSTAT 2021). The world's major plum producers are China (11%), Romania (1%) and Serbia (1%) (Figure 1.3). The United States contributes 0.5% (FAOSTAT 2021). Most plums are produced in Asia (58%), followed by Europe (29%), America (10%), Africa (2%) and Oceania (1%) (Figure 1.3).

China has established itself as one of the main plum producers and has been increasing its harvest due to national economic reform (Liu et al. 2007). In the 1980s the production reached 1 million tons, during the 1990s it reached approximately 4 million tons and finally in the last decade it reached approximately 7 tons (FAOSTAT 2021). GuangDong Food and Drug Vocational College is one of the companies that preserved plum production lines in GuangDong Province, an area in China that supplies more than 65% of preserved plums in the global market (Wang et al. 2017).

In cooler areas European plums are produced and usually are more acclimatized than Japanese plums that are produced in hotter areas (Topp et al. 2012; Wangchu et al. 2021). One of the limitations Japanese plums present are frosts in the winter

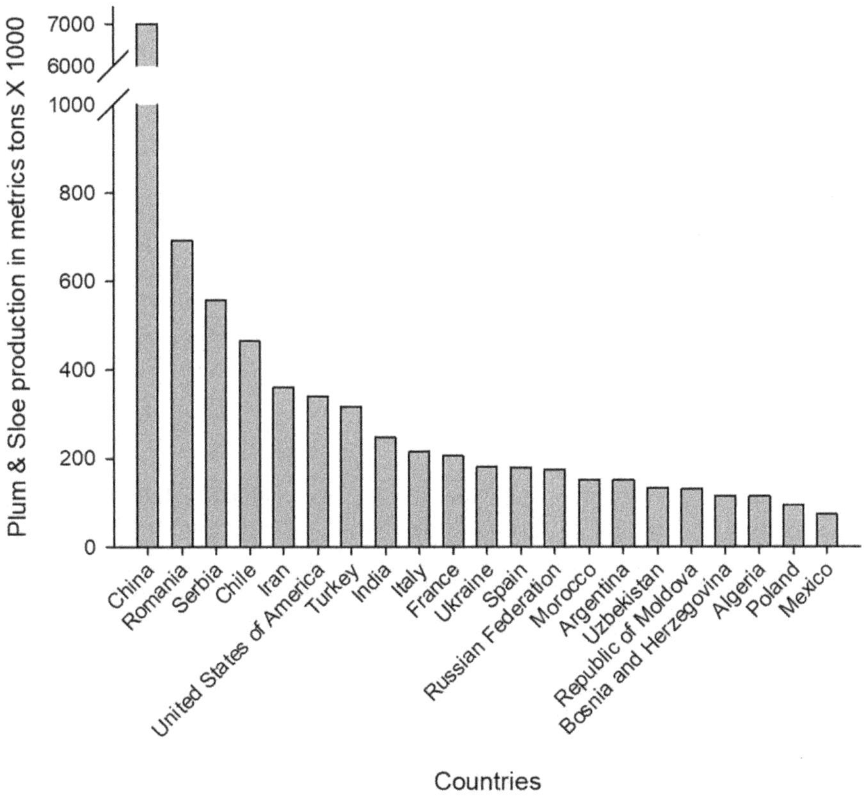

FIGURE 1.3 Principal countries associated with plum and sloe world production.

and early spring, and some physical and chemical characteristics of the soil (Okie and Hancock 2008). The European market sells more European plums, about 90% of plum production, while Asian markets sell more Japanese plums, about 82%. They are sold as separate crops and marketed as distinct commodities (Topp et al. 2012).

Europe plum production in 2019 was 2,895,801 tons (FAOSTAT 2021). The principal countries in Europe with the highest production of *Prunus domestica* are Romania, Bulgaria, Hungary and Germany (Dalla and Baric 2012). However, since 2010 the plum yielding in countries such as Bulgaria and Romania has decreased due to early fruit drop from sharka disease, a viral disease caused by the plum pox virus (PPV) (Wangchu et al. 2021). In 1997 Bulgaria produced 112,589 tons of plum fruit, which decreased by 50% in 2019. In Romania from 2018 to 2019 the production of plums decreased by 17%. During the 1960s Germany produced approximately 1 million tons and in 2019 only 52,140 tons (FAOSTAT 2021), making it sixth in plum production within the European Union (Dalla and Baric 2012). Even so, cultivars often self-fertilize with better resistance to PPV and with high yielding (Mónika Molnár et al. 2016). Contrary to these countries and despite the fruit production in Serbia showing signs of reduction, plum production has increased from 459,712 tons in the period 1981–2011 (Lukac Bulatovic et al. 2013) to 558,930 in 2019 (FAOSTAT 2021).

Plum production in the Americas in 2019 was 1,075,914 tons. South America contributed 66% of production, and North America contributed 34%. Chile, with 465,000 tons, and Argentina, with 150,000 tons, are two of the most important countries in Latin America when it comes to plum production. The production is located in cold areas as the states of Rio Grande do Sul that produces about 50% of the national harvest and the upper mountains of states such as Minas Gerais, Santa Catarina and São Paulo (Eidam et al. 2012).

In the United States close to 53% of the plum yield are European plums, 38% are Japanese plums and 9% are a mixture of *Prunus americana* and others species (Surányi and Erdős 2004). California during the 2000s was the principal US producer, with close to 300,000 tons in annual yield (Jayasankar et al. 2015). However, since 2011 production has decreased to close to 67%. The last register for 2019 was approximately 100,000 tons (USDA 2006).

1.3.3 Marketing and Trade

The world market revenue for plums and sloes came to USD 15.7 billion in 2018, a 12% increase from 2017. In the last decade the market value of plums has increased by about 5%, with some years seeing a decreases in value. By the end of the past decade market value totaled USD 16.3 billion. China, with USD 10 billion, is the principal marketer, followed by the United States, with USD 757 million (Wood 2019).

California is the major marketer of plums used for drying. The value of plum production in California State in 2020 was USD 122,233,000 (Ross et al. 2020). The United States is the most important plum seller internationally, with 30% of the worldwide average. Canada and Taiwan were the main importers of US plums in 2020, followed by Mexico and Hong Kong (Boriss et al. 2006). In 2019 the leading importers were Canada (37%), Mexico (22%), Hong Kong (20%) and, finally, Taiwan (11%) (FAOSTAT 2021).

Following the United States, Chile increased its sales from USD 35 million to USD 100 million between 2001 and 2008 in the exportation of plums (Bravo 2010). The main Chilean market for plums is the United States. For the years 2016 and 2017, the United States imported about USD 54 million in fresh plums and USD 40 million in dried plums (USDA 2017). In 2019 Chile exported 76,305 tons of plum and sloes to China and 23,860 to the United States, with transactions of USD 121 million and USD 35 million, respectively. The United States imported USD 52 million in plum and sloes from Chile (FAOSTAT 2021).

In Germany about 50,000 tons of European plums are sold by producer markets annually. In total about 300,000 tons of plums are harvested, including fruits produced for direct marketing, liquor production and for homegrown trees. In Germany half of the plum production is used in bakeries, 30% for fresh consumption, and the rest for alcohol production as brandy (Wangchu et al. 2021).

1.4 TAXONOMY

1.4.1 Botanical Characteristics of Plums

The study of taxonomy involves studying an organism's morphological characters in detail and classifying; plums are part of the genus *Prunus*. The botany defines plums, especially *P. domestica*, as trees characterized to be deciduous, small and branched (Figure 1.4). The plums are approximately 4–15 m in height, the branches are rosy or brown and glabrous, with a few spines that change from reddish brown to grayish green. *P. domestica*'s leaves have a deep green color and are alternated; the form can be obovate or elliptic, the margins are serrulate, and it has parallel veins. The flowers are solitary or in fascicles of three, with spreading white or pink petals and approximately 15–30 stamens, diverse with filiform strings. The flower stigma is

FIGURE 1.4 Illustration of *Prunus domestica*, characteristics of leaves, fruit, seed and flower. Illustrated by Laura Natali Afanador-Barajas.

usually truncated. The fruit mostly has a fleshy exterior and is edible. The mesocarp is fleshy, the endocarp is generally ellipsoid and horizontally compacted, containing a hard seed known as a stone with a bony pit. Drupes can be purple, purpleblack, red, green, yellow or golden yellow, are usually globose to oblong and are rarely subglobose (see Figure 1.4) (Lim 2012; Waugh 1901).

1.4.2 Principal Taxonomy of *Prunus* Classification

Prunus is the principal genus in most significant species of plum (Table 1.3), in the family Rosaceae and subfamily Amygdaloideae. The European plum is *Prunus domestica*, which is hexaploidy. Japanese plum (*Prunus salicina*) and sloe (*Prunus spinosa*) have diploidy, and *Prunus simonii* shows tetraploidy (Hussain et al. 2021). Waugh (1901) pronounced the genus *Prunus* as trees. Flowers have spreading white or pink petals, with approximately 15–30 stamens, are diverse with filiform strings. The flower stigma is usually truncated. The fruit is mostly with carnosus with a fleshy exterior and edible. Also, fruit contains hard seeds known as stones with a bony pit. An example of these characteristics in *Prunus domestica* is shown in Figure 1.4.

The NCBI database (Schoch et al. 2020) includes two subspecies for *Prunus domestica* (*P. domestica* subsp. insititia and *P. domestica* subsp. syriaca), and only one variety (*P. domestica* var. Juliana). For *Prunus salicina* only one variety is included in the database (*Prunus salicina var. cordata*). Similarly, *P. japonica* has the variety *P. japonica var. nakaii*. Different varieties of *Prunus cerasifera* (cherry plum) include var. *cerasifera*, f. atropurpurea, var. divaricate and var. Pissardii. There are also different hybrids with other species of *Prunus* genus, such as *Prunus*

TABLE 1.3
Taxonomic Classification in the Different Species of Plums. According to the Taxonomy of the NCBI Database (Schoch et al. 2020)

Classification	European plum	Asian plum	American plum
Kingdom	Viridioplantae	Viridioplantae	Viridioplantae
Phylum	Streptophyta	Streptophyta	Streptophyta
Class	Magnoliopsida	Magnoliopsida	Magnoliopsida
Superorder	Rosanae	Rosanae	Rosanae
Order	Rosales	Rosales	Rosales
Family	Rosaceae	Rosaceae	Rosaceae
Genus	*Prunus*	*Prunus*	*Prunus*
Species	*Prunus domestica* (damsons and domestica plums) *Prunus spinosa* *Prunus cerasifera* (Myrobalan or cherry plum)	*Prunus salicina* *Prunus simonii* (Simon plums) *Prunus triflora* (Japanese plums)	*Prunus americana* *Prunus nigra* *Prunus angustifolia* *Prunus hortulana* *Prunus munsoniana* *Prunus subcordata* *Prunus maritima* *Prunus pumila*

cerasifera x *Prunus dulcis*; *Prunus cerasifera* x *Prunus munsoniana*; *Prunus cerasifera* x *Prunus persica*; *Prunus cerasifera* x (*Prunus persica* x *Prunus dulcis*); *Prunus cerasifera* x *Prunus salicina*; and (*Prunus cerasifera* x *Prunus salicina*) x (*Prunus cerasifera* x *Prunus persica*).

Waugh (1901) found some contradictions in the taxonomy of some members of the *Prunus* genus. Consequently, Bortiri et al. (2001) made a summary of the principal inconsistencies. First, the classification of *Prunus* into four different genera (*Prunus*, *Padus*, *Cerasus* and *Amygdalus*), and later changed to two genera (*Prunus* and *Amygdalus*). Second, a division of five genera formed by *Prunus*, *Armeniaca*, *Persica*, *Amygdalus* and *Cerasus* (with *Padus* and *Laurocerasus*). Third, *Prunus* as a single genus divided into seven units (*Amygdalopsis*, *Amygdalus*, *Armeniaca*, *Cerasus*, *Ceraseidos*, *Laurocerasus* and *Prunus*). Fourth, *Prunus* with the previous seven parts as subgenera; and fifth, *Prunus* were classified into five subgenera as *Amygdalus*, *Cerasus*, *Laurocerasus*, *Prunophora* and *Padus* and *Prunus* distributed in *Euprunus*, *Prunocerasus* and *Armeniaca*. Finally, sixth, *Prunus* was separated into three genera conformed by *Prunus*, *Laurocerasus* and *Padus* (Chavez and Chaparro 2020).

1.4.3 *Prunus* Phylogenetic Classification

Plant classification has been difficult due to intergenic hybridization, which has changed the configuration of taxa. As a result, the definition of species is broadened to include apomixis and hybridization. Mowrey and Werner (1990) reported a phylogeny and systematics in *Prunus* using isozymes to focus on the phylogenetic connections in *Prunus*. *Prunocerasus* was observed to be polyphyletic, with a clade formed by *Prunus americana*, *Prunus angustifolia*, *Prunus hortulana*, *Prunus munsoniana* and *Prunus subcordata*. Other clade were formed by *Prunus maritima* and *Prunus umbellata*. Also, Bortiri et al. (2006) demonstrated the utility of morphological characters in phylogenetic analysis of *Prunus*. While Bortiri et al. (2001) used a combination of ITS sequences, cpDNA, and morphological features to support several nodes that had previously been discovered in *Prunus*. Large groups were supported by several synapomorphies, which provided extra resolution for several clades.

Hodel et al. (2021) studied nuclear and chloroplast phylogenies of *Prunus* species using nuclear markers of DNA. They found that various species of *Prunus* had histories consistent with hybridization and allopolyploidy. Also, they observed that the *Prunus* group is monophyletic, but there is some conflict in the gene tree with a substantial disharmony at several nodes, including the node of the racemose group. They suggest that a different form of tree topologies that conflicted with the species tree were consistent with a paraphyletic racemose group. In the case of European plums, they found a monophyletic clade between *Prunus salicina* and *Prunus domestica*.

Chin et al. (2014) found the importance of studying phylogenetic analysis in *Prunus* to understand the geographic beginning and ancestral genes of cladogenesis using the Bayes-DIVA method. Their outcomes showed that contemporary genus appeared almost ~61 Myr in eastern Asia and enhancement of all significant heredities might have been set off by a worldwide temperature boost of the early Eocene epoch.

We constructed a phylogeny using the 18S marker with some of the most representative species of plum. Figure 1.5 shows three principal clades of species

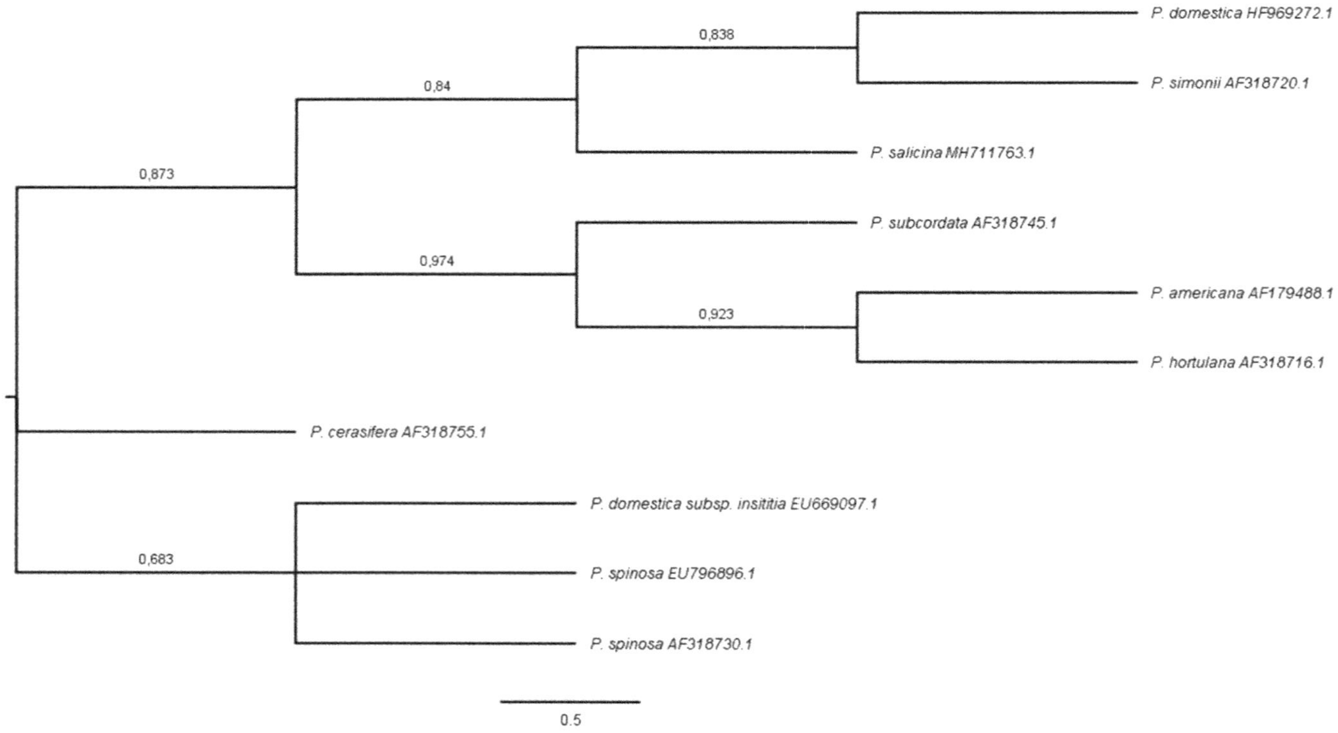

FIGURE 1.5 Maximum likelihood phylogeny of gene 18S rRNA with some of most representative species of plums.

plums more related with the geographical distribution. The first clade is formed by Asian plums (*P. salicina* and *P. simonii*), the second clade are formed by American plums (*P. subcordata*, *P. americana* and *P. hortulana*) and third clade is formed by European plums (*P. domestica*, *P. spinosa* and *P. cerasifera*).

1.5 CONCLUSION

Plums are an edible plant with diverse origins, dissemination and geographical distribution (mainly in Europe, the Americas and Asia) that constitutes important resources of the diversity of fruits. These species of *Prunus* have great potential to increase the world production and commercialization of other types of plums in the future. The classification of plums using taxonomy and phylogenetic analysis offers a geat form to appreciate the species of plant relationships. In summary, the study of the history, distribution and production of plums can help maintain the germplasm of plum species.

1.6 REFERENCES

Boriss, H., Specialist, J., Brunke, H., Specialist, A., and Kreith, M. 2006. *Commodity Profile: Plums, Fresh Market*. Agricultural Marketing Resource Center.

Bortiri, E., Heuvel, B., and Potter, D. 2006. Phylogenetic Analysis of Morphology in Prunus Reveals Extensive Homoplasy. *Plant Systematics and Evolution* 259 (1): 53–71.

Bortiri, E., Sang-Hun, O., Jiang, J., Baggett, S., Granger, A., Weeks, C., Buckingham, M.P., and Parfitt, D.E. 2001. Phylogeny and Systematics of Prunus (Rosaceae) as Determined by Sequence Analysis of ITS and the Chloroplast TrnL-TrnF Spacer DNA on JSTOR. *Systematic Botany* 26 (4): 797–807.

Bravo, J. 2010. Chile y El Mercado Mundial de La Fruta Industrializada. In *Oficina de Estudios y Políticas Agrarias—Odepa*, 1–14. www.odepa.gob.cl/odepaweb/publicaciones/doc/2311.pdf.

Browicz, K., and Zohary, D. 1996. The Genus Amygdalus L. (Rosaceae): Species Relationships, Distribution and Evolution Under Domestication. *Genetic Resources and Crop Evolution* 43 (1): 229–247.

Calvo, I. 2009. *El Cultivo Del Ciruelo (Prunus Domestica)*. Boletin Técnico No. 9. Proyecto Microcuenca Plantón—Pacayas 9: 8. www.scribd.com/document/139519568/CIRUELO-Cultivo.

Chavez, D.J., and Chaparro, J.X. 2020. *The North American Plums (Prunus Spp.): A Review of the Taxonomic and Phylogenetic Relationships, Prunus*. IntechOpen.

Chin, S.W., Shaw, J., Haberle, R., Wen, J., and Potter, D. 2014. Diversification of Almonds, Peaches, Plums and Cherries—Molecular Systematics and Biogeographic History of Prunus (Rosaceae). *Molecular Phylogenetics and Evolution* 76 (1): 34–48.

Colamarino, I. 2010. *Ciruelas Frescas: Informe Sectorial de Las Cadenas de Origen Agrícola y Forestal*, October. www.alimentosargentinos.gob.ar.

Dalla, J., and Baric, S. 2012. Tree Fruit Growing—Research and Production in Germany: A Statistical and Bibliometric Analysis of the Period 1950–2010. *Erwerbs-Obstbau* 54 (1): 11–30.

Eidam, T., Pavanello, A., and Ayub, R., antonio. 2012. The Plum Culture. *Revista Brasileira de Fruticultura* 34 (1): 1–2.

Elith, J. et al. 2006. Novel Methods Improve Prediction of Species' Distributions from Occurrence Data. *Ecography* 29 (2): 129–151.

FAOSTAT. 2021. *Food and Agriculture Organization of the United National*. www.fao.org/faostat/en/#search/plum.

Faust, M., and Surányi, D. 1999. Origin and Dissemination of Plums. In *Horticultural Reviews*, edited by Jules Janick, 179–231. John Wiley & Sons, Ltd.

Hartmann, W., and Neumüller, M. 2009. Plum Breeding. *Breeding Plantation Tree Crops: Temperate Species* 161–231.

Hodel, R.G.J., Zimmer, E., and Wen, J. 2021. A Phylogenomic Approach Resolves the Backbone of Prunus (Rosaceae) and Identifies Signals of Hybridization and Allopolyploidy. *Molecular Phylogenetics and Evolution* 160: 1055–7903.

Hussain, S.Z., Naseer, B., Qadri, T., Fatima, T., and Bhat, T.A. 2021. Plum (Prunus Domestica): Morphology, Taxonomy, Composition and Health Benefits. In *Fruits Grown in Highland Regions of the Himalayas*, 169–179. Springer International Publishing.

Igwe, E.O., and Charlton, K.E. 2016. A Systematic Review on the Health Effects of Plums (Prunus Domestica and Prunus Salicina). *Phytotherapy Research* 30 (5): 701–731.

Jayasankar, S., Dowling, C., and Selvaraj, D.K. 2015. Plums and Related Fruits. In *Encyclopedia of Food and Health*, 401–405. Elsevier Inc.

Lim, T.K. 2012. Edible Medicinal and Non-Medicinal Plants. *Fruits* 4.

Liu, W., Liu, D., Zhang, A., Feng, C., Yang, J., Yoon, J., and Li, S. 2007. Genetic Diversity and Phylogenetic Relationships Among Plum Germplasm Resources in China Assessed with Inter-Simple Sequence Repeat Markers. *Journal of the American Society for Horticultural Science* 132 (5): 619–628.

Lompo, O., Dimobe, K., Mbayngone, E., Savadogo, S., Sambaré, O., Thiombiano, A., and Ouédraogo, A. 2021. Climate Influence on the Distribution of the Yellow Plum (Ximenia Americana L.) in Burkina Faso. *Trees, Forests and People* 4: 100072.

Lukac Bulatovic, M., Rajic, Z., and Dokovic, J. 2013. Development of Fruit Production and Processing in the Republic of Serbia. *Economics and Agriculture* 60 (1): 141–151. www.ea.bg.ac.rs/index.php/EA/article/view/515.

Luna-Vázquez, F.J., Ibarra-Alvarado, C., Rojas-Molina, A., Rojas-Molina, J.I., and Bah, M. 2017. Prunus. *Fruit and Vegetable Phytochemicals: Chemistry and Human Health* 2: 1215–1226, September.

Martínez-Gómez, P. 2017. Predicción Científica y Prescripción En Mejora Genética Vegetal En Cuanto Ciencia Aplicada de Diseño: El Caso de La Mejora de Frutales Del Género Prunus. *Acta Agronomica* 66 (1): 115–127.

Méndez, J.C. 2015. *Evaluación de Un Complejo Hormonal y Micronutrientes En El Cultivo de Ciruelo (Prunus Domestica) y Sus Efectos En La Calidad Del Fruto*. Uaaan, Buenavista Saltillo Coah. http://repositorio.uaaan.mx:8080/xmlui/handle/123456789/6594.

Mercado, M. 2016. *Ciruela*. Montevideo. www.mercadomodelo.net/c/document_library/get_file?uuid=3f8244ff-a168-4e04-b173-cd9c2ac75056&groupId=42766.

Milošević, T., and Milošević, N. 2018. Plum (*Prunus* spp.) Breeding. In *Advances in Plant Breeding Strategies: Fruits*, edited by J. Al-Khayri, S. Jain, and D. Johnson, 165–215. Springer.

Mónika Molnár, Á., Ladányi, M., and Kovács, S. 2016. Evaluation of the Production Traits and Fruit Quality of German Plum Cultivars. *Acta Universitatis Agriculturae Silviculturae Mendelianae Brunensis* 64 (1): 109–114.

Mowrey, B.D., and Werner, D.J. 1990. Phylogenetic Relationships Among Species of Prunus as Inferred by Isozyme Markers. *Theoretical and Applied Genetics* 80 (1): 129–133.

OECD. 2002. *Consensus Document on the Biology of Prunus Sp. (Stone Fruits)*. www.oecd.org/env/ehs/biotrack/46815738.pdf.

OECD. 2006. Section 6 — Stone Fruits (Prunus Spp.). In *Safety Assessment of Transgenic Organisms*, vol. 2, 175–203. OECD Publishing.

Okie, W.R., and Hancock, J.F. 2008. Plums. *Temperate Fruit Crop Breeding: Germplasm to Genomics* 1: 337–358.

Okie, W.R., and Ramming, D.W. 1999. Plum Breeding Worldwide. *HortTechnology* 9 (2): 163–176.

Orjuela, M., Camacho, J., and Parra-coronado, A. 2016. Evaluación de Crecimiento En Cosecha de Ciruela (Prunus Domestica L.), Variedad Horvin. En *El Municipio de Nuevo Colón Boyacá*, 601–605. XII Congreso Latinoamericano y Del Caribe de Ingeniería Agrícola (CLIA). www.researchgate.net/publication/309352790.

Pio, R., Farias, D. da H., Bianchini, F.G., Peche, P.M., and Bisi, R.B. 2018. Selection of Plum Cultivars for Subtropical Regions. *Ciência Rural* 48 (11).

Ross, J., Davis, V., Foster, C., and Ray, T. 2020. *National Agricultural Statistics Service*. www.nass.usda.gov/.

Sarigu, M., Grillo, O., Bianco, M. L., Ucchesu, M., d'Hallewin, G., Loi, M.C., Venora, G., and Bacchetta, G. 2017. Phenotypic Identification of Plum Varieties (Prunus Domestica L.) by Endocarps Morpho-Colorimetric and Textural Descriptors. *Computers and Electronics in Agriculture* 136: 25–30, April.

Schoch, C.L., Ciufo, S., Domrachev, M., Hotton, C.L., Kannan, S., Khovanskaya, R., Leipe, D., McVeigh, R., O'Neill, K., Robbertse, B., Sharma, S., Soussov, V., Sullivan, J.P., Sun, L., Turner, S., and Karsch-Mizrachi, I. 2020. *NCBI Taxonomy: A Comprehensive Update on Curation*. Resources and Tools Database.

Siddiq, M. 2006. Plums and Prunes. In *Handbook of Fruits and Fruit Processing*, edited by Y.H. Hui, 1st ed., 553–564. John Wiley & Sons, Ltd.

Sottile, F., Peano, C., Mezzetti, B., Capocasa, F., Bellini, E., Nencetti, V., Palara, U., Pirazzini, P., Mennone, C., and Catalano, L. 2010. Plum Production in Italy: State of the Art and Perspectives. *Acta Horticulturae* 874 (2): 25–34.

Surányi, D., and Erdős, Z. 2004. Comparative Study of Plum Cultivars Belonging to Different Taxons During 1980–1996. *International Journal of Horticultural Science* 10 (4): 13–19.

Topp, B.L., Russell, D.M., Neumüller, M., Dalbó, M.A., and Liu, W. 2012. Plum. In *Fruit Breeding*, edited by Marisa Luisa Badenes and David H. Byrne, 571–621. Springer.

Urrestarazu, J., Errea, P., Miranda, C., Santesteban, L.G., and Pina, A. 2018. Genetic Diversity of Spanish Prunus Domestica L. Germplasm Reveals a Complex Genetic Structure Underlying. *Plos One* 13 (4): 1–14.

USDA. 2006. *List of Tables*. USDA's National Agricultural Statistics Service, California Field Office Publication. www.nass.usda.gov/Statistics_by_State/California/Publications/Annual_Statistical_Reviews/2006/2006cas-all.pdf.

USDA. 2017. *Strategic Plan 2013–2018*. www.ers.usda.gov/media/9361/strategic-plan-2013-18.pdf.

Wallace, T.C. 2017. Dried Plums, Prunes and Bone Health: A Comprehensive Review. *Nutrients* 9 (4): 1–401.

Wang, S., Zhou, A., Yang, X., Yang, J., Huang, K., and Chen, H. 2017. Study on the Heat Pump Technology of Shuanghua-Plum Preserved. *Science and Technology of Food Industry* 38 (12): 227–232.

Wangchu, L., Angami, T., and Mandal, D. 2021. Plum. In *Temperate Fruits*, edited by D. Mandal, U. Wermund, L. Phavaphutanon, and R. Cronje, 1st ed., 297–331. Apple Academic Press.

Watkins, R. 1976. Cherry, Plum, Peach, Apricot and Almond: Prunus Spp. (Rosaceae). In *Evolution Crop of Plants*, edited by N.W. Simmonds, 242–247. Longman.

Waugh, F. 1901. *Plums and Plum Culture: A Monograph of the Plums Cultivated and Indigenous in North America, with a Complete Account of Their Propagation, Cultivation and Utilization*. Orange Judd Co.

Wood, L. 2019. *World Plums and Sloes Market Analysis, Forecast, Size, Trends and Insights*. www.businesswire.com/news/home/20190904005651/en/World-PlumsZZZ-and-Sloes-Market-Analysis-Forecast-Size-Trends-and-Insights-Report-2019.

Yoshida, M. 1987. Plums. In *The Origen of Fruits*, vol. 42, 49–53. Fruit of Japan.

Zhebentyayeva, T., Shankar, V., Scorza, R., Callahan, A., Ravelonandro, M., Castro, S., DeJong, T., Saski, C.A., and Dardick, C. 2019. Genetic Characterization of Worldwide Prunus Domestica (Plum) Germplasm Using Sequence-Based Genotyping. *Horticulture Research* 6 (1): 1–13.

Zohari, D. 1986. The Origin and Early Spread of Agriculture in the Old World. In *Developments in Agricultural and Managed Forest Ecology*, vol. 16, 3–20. Elsevier.

Zohary, D., Hopf, M., and Weiss, E. 2012. *Domestication of Plants in the Old World: The Origin and Spread of Domesticated Plants in South-West Asia, Europe, and the Mediterranean Basin*, 4th ed. Oxford University Press.

2 Orchard Planning, Establishment, and Soil Management of Plum

I. Hernández-Soto[1], A.J. Cenobio-Galindo[1], G. De Vega-Luttmann[1], S. Perez-Ríos[1], M.J. Franco-Fernández[1], O. Fernández-Fernández[2], Gabriela Medina-Pérez[1], Yash D. Jagdale[3], and Mohammad Javed Ansari[4]

[1] Institute of Agricultural Sciences, Autonomous University of Hidalgo State. Av. Universidad s/n Tulancingo, Hidalgo, México

[2] Chapingo Autonomous University Km. 38.5 Mexico-Texcoco. Chapingo, State of Mexico, 56230

[3] MIT School of Food Technology, MIT ADT University, Pune, Maharashtra-412201, India

[4] Department of Botany, Hindu College Moradabad (Mahatma Jyotiba Phule Rohilkhand University, Bareilly, UP), 244001, India

CONTENTS

2.1 INTRODUCTION

Plum is a fruit that contains a woody seed. It is classified in the *Prunus* genus of the Rosaceae family (Okie and Hancock 2008). Plums are fruit trees of temperate zones with round or oval fruit which can be different sizes and color and present a

DOI: 10.1201/9781003205449-2

juicy, soft, and primarily sticky pulp. The fruits are mainly used for fresh or canned consumption (Neumüller 2011). However, plums grow in other areas, and some of the most representative varieties are German prunes, Stanley prunes, Prune d'Agen, and Italian prunes. Currently, work is being done to improve the quality of plums to ensure that all varieties adapt to different climates and produce attractive, high-quality fruits for profitable marketing (Okie 1995).

Several factors influence the yield of a plum fruit orchard. The most important is the sensitivity of the flowers to climatic conditions at the time of flowering (Milosevic and Milosevic 2011). In addition, plums are often self-fertile; although some varieties require bees to cross-pollinate the flowers, it is also a widespread practice to increase profit in orchards (Okie and Hancock 2008). As a result, reports suggest a steady increase in the world production of plums, from 8.1 million tons in 1999 to more than 9.9 million tons in 2003 to 10.3 million tons in 2008 (Milosevic and Milosevic 2011).

However, the plum-germination process presents a problem: the pit's thickness. Therefore, for these species, the vegetative propagation and stratification of the seeds are necessary, at 4–5 °C for three to four months (Tehrani 1972). When the seedlings reach a height of about 50 cm, it is recommended to plant them directly in the field; otherwise, one year of cultivation in the nursery is suggested.

2.2 ORCHARD LAYOUT

In preparation for planting, it is convenient to dig a hole about 50 cm deep. Next, plantation is carried out, the most favorable time being during the winter stop, after the coldest period. When the ground has been cleared, the openings of the holes are of sufficient dimensions to place the plant properly. If the digging work has not been carried out, the holes are usually given dimensions of 70 cm deep and about 40 cm long and wide (Tóth and Surányi 1980). Generally, the spacing of the plants are 6–7 m at the proper frame. The farmer currently plants in streets, about 5 m between plants, with 7 m lanes between lines. Blažek and Pištěková (2009) tested 16 novel plum cultivars in a high-density experimental orchard with a spacing of 5.5 m, on St. Julian A rootstock. Tophit, Jojo, Elena, and President varieties produced the most significant yields and fruited the earliest, whereas Ruth Gerstetter, Anna Späth, and Topgigant Plus produced the least. The cultivars Katinka, Jojo, Topper, and Empress, had the maximum yield efficiency, whereas Katinka, Jojo, Topper, and Empress had the lowest.

2.2.1 Orchard Planting

Plum trees are planted in the winter months, between December and January, when they are dormant (Norton 2007). The orchard should be carefully cleared of weeds before planting, and plowing the plantation area is recommended. Terraces should be maintained inwards in steep areas to aid soil conservation. The orchard area must be adequately mapped out two months before planting. An orchard should be laid out using a contour or terrace method in sloppy soil. A square system is recommended in flat land.

2.2.2 Associated Crops

The biological control of aphids can be achieved by planting oats as an inter-row crop in plum orchards. This technique relies on introducing a plant species that hosts innocuous aphids for the crop in question and shares natural enemies with the pests in question (Andorno et al. 2004). Because it is regularly plagued by the aphid *Rhopalosiphum padi*, oats (*Avena sativa*), an annual plant belonging to the Poaceae family, have been examined and identified as a suitable species utilized as an alternate host for natural enemies of aphids. *R. padi* serves as an alternate host for parasitoid natural enemies of pest aphids; hence, this is a helpful signal. *Aphidius colemani*, *Aphidius ervi*, and *Lysiphlebus testaceipes* are the natural parasitoid enemies of *R. padi* (Andorno et al. 2015). Different grasses, such as oats (*Avena sativa*), infested by specialized aphids of this type of plant, such as *Rhopalosiphum paid*, are cultivated in commercial insectaries and introduced in high populations in crops (Gómez-Marco et al. 2012).

Researchers from Veracruz state in Mexico studied the agroforestry systems: the corn-avocado-plum and corn-plum systems form a well-established ecosystem with ecological interactions to sustain the genetic biodiversity of corn and fruit trees (avocado and plum). Furthermore, these systems are resilient because they continue to last over time, with an alteration seen in the decrease in production (Luna Tetelano 2012).

In the Levante region of Spain, it is common to find plum plantations associated with orange trees in the citrus zone, which, with the same cultivation care given to the latter, provide good yields. Likewise, in this area it is very common to find many trees on the margins and extensions of the farms (Gomez 2015).

2.3 LAND PREPARATION

Land-preparation activities are centered on eradicating existing difficulties, such as the working sole, for which subsoilers are used to allow soil aeration without inverting the soil layers. Before planting, organic amendments (at a rate of 20–25 tons per hectare [ha]), pH correction, and the administration of specific nutrients (mainly phosphorus, potassium, and calcium) should all be done to prevent hindering the crop's natural development in its early years. The adaptability to soil conditions, resistance or tolerance to specific diseases, and differing pathophysiology depend on the rootstock used. In the case of the variations, their extensive evolutionary and adaption process has enabled them to develop defensive mechanisms (tougher skins, segregation of poisonous chemicals, etc.) against predatory creatures, such as insects, fungi, and bacteria. They may also include more nutritious ingredients (minerals, vitamins, proteins, etc.) and are increasingly valued in the market, particularly in the catering industry, for their organoleptic qualities.

On the other hand, traditional variants have several drawbacks, including the following: They have decreased production when inputs are used intensively, and their fruits, in some situations, do not withstand long-distance travel well or are diverse. These historic cultivars are difficult to come by since they are rarely produced for commercial purposes. New fruit types have features that are opposed to those that have been around for a long time. They have expanded rapidly, particularly in production circumstances (rich and deep soils, sufficient water, etc.) and more intense technological usage. However, it must be remembered that to achieve significant output, these plants must be kept in ideal circumstances, which means using a lot of fertilizer and maintaining strict phytosanitary and herb management. Other variables to consider in organic fruit growing might include the search for late or early types that allow for the avoidance of pests and illnesses that are particularly destructive in the region. For example, early types of fruit trees can protect inland portions of the peninsula from Mediterranean fly damage.

2.4 AGROCLIMATOLOGY

Winter chilling (Ruiz et al. 2018; Fadón et al. 2020) is required for plum plants to reach a state of dormancy (Ruiz et al. 2018). For 850–1,000 hours, the ambient temperature should be between 2.5 and 12.5°C. Plum trees can thrive in various soil conditions. However, deep, well-drained soils with a depth of at least 600 mm, ranging from sandy loam to sandy clay loam, are suggested. Unlike most stone fruits, plums are more tolerant of thick or damp soils. The pH of the soils should be between 5.5 and 6.5 (El-Shall et al. 2010). Establish orchards at heights below 2,400 m above sea level, with average temperatures between 6 and 28°C. The soil should be deep, regular to good drainage, with a pH of 4.5–6, and slopes from 4% to 15% (Nuttonson 1947). The topography could be irregular, from flat terrain to very steep slopes. Plum trees can be planted up to 700 m above sea level in temperate zones, but in the case of tropical latitudes, the plum tree is produced in areas ranging from 1,500 to 2,300 m above sea level. In addition, it requires low temperatures to establish the inactivity phase and fill its requirement of cold hours, which in our conditions is obtained at higher altitudes above sea level. The optimum temperature for its development is between 12 and 22°C; however, depending on the variety, it usually resists low temperatures quite well. In the case of tropical areas, plum requires well-distributed rainfall of more than 1,400 mm per year (Montgomery and Remondo 1964; Häberlein 1990; LAZĂR et al. 2011). The literature mentions a requirement of not less than 700 mm per year. There is no information reported regarding relative humidity. The location of the planting land is of utmost importance concerning the presence of strong winds since it can cause the breaking of delicate branches and cause the fall of flowers and fruits. The plum tree requires loose, deep, and well-drained soils that are rich in organic matter. Furthermore, the literature indicates that it can tolerate humid and shallow soils, given its superficial root system.

2.5 AGRONOMIC MANAGEMENT OF PLUM CULTIVATION

The plum needs flower induction as a stimulus for the terminal buds to achieve flower differentiation; then, hydrogenated cyanimide is applied at a dose of 10–20 cc/liter immediately after pruning, mainly at the branches and trunk (places where these buds are located). The product simultaneously causes a cooling of the buds through thermal shock and leaf fall, stimulating the budding of flower buds. For sowing, vigorous, healthy, and erect grafted plants (generally on a peach pattern) that showed the best root development were chosen. The hole should be 30 x 30 cm; then, organic compost can be placed at the bottom to improve the physical structure and 50 g of complete formula (10–10–30 or 15–15–15). Then, putting a layer of soil over the compost, the plum plants are placed. The sowing distance can range from 3.5 m x 3.5 m to 5 m x 5 m in a square for a population of 400–800 trees/ha, although sowing in crow's feet can also be used (Batlle et al. 2018).

2.6 FERTILIZATION

Fertilization in fruit trees is one factor that influences the quantity and quality of production. The handling of fertilizers in amount, application times, and types of fertilizers may vary. In addition, fundamental criteria for this fruit tree must be considered: (1) Almost 50% of the nutritional elements used by the plum tree are destined to the formation of the fruit, and the rest to wood, leaves, flowers, and roots; (2) the root absorption of the nutritional elements is not regular throughout the vegetative period of the plum tree; nutrient absorption begins in April, reaching its peak in July and gradually decreasing until the end of October; and (3) the contributions of the nutrients must be adapted to the extractions made by the tree to form the fruits, wood, and leaves, mainly in the soil (Fudge and Fehmerling 1940). Most of the nutrients used by the tree for flowering, fruit set, and the formation of the first leaves, come from the accumulated reserves of the previous year. Therefore, the contribution of nutrients after harvesting is essential for the plant to collect the necessary resources for the following year's harvest (Villarrubia Horta and Mataix Gato 1998).

The availability of organic and inorganic fertilizers, as well as the soil, influences plum tree fertilization. Phosphorus, potassium, and boron insufficiencies are common nutritional deficiencies and can be remedied by adding phosphorus or potassium fertilizers to the soil or spraying the trees with boric acid. The public's growing awareness about the environmental effects of overfertilization has reignited interest in determining the appropriate fertilization recommended in the field to preserve fruit output and quality while reducing environmental impact.

Potassium and nitrogen fertilization could influence the quality of fresh plum fruit (*Prunus salicina*) and the behavior of plums during cold storage. The experiment was carried out on a plot design divided into a factorial scheme in a five-year-old plum orchard, with potassium fertilizer applied at 55 and 110 kg/ha/year of K_2O, as KCl, and nitrogen fertilizer used at 40, 80, 120, 160, and 200 kg/ha/year of nitrogen, as urea. One hundred plum fruits were taken from each plot on the same day, when 25–50% of the peel was reddish yellow. At harvest and after 17, 27, and 37 days of storage at 00.5 °C, the pulp hardness, total soluble solids, and titratable acidity were assessed. The use of nitrogen affected the quality of fresh fruit, with the best results obtained when 40 kg/ha/year of nitrogen was applied. Fruit quality was maintained during storage (nitrogen and potassium doses of 40 and 110 kg/ha/year, respectively) for 27 days.

Wójcik (1997) investigated the influence of boron fertilization on the growth, yield, and quality of plum fruits. The experiment was conducted in a nine-year-old orchard that received 2 kg B kg/ha/year of boron fertilizer and boron sprinkles in the spring and autumn. According to the findings, all boron treatments increased the number of one-year shoots while decreasing their average length. Boron fertilization did not influence fruit set percentages at two, four, or five weeks after blooming. Furthermore, boron fertilization did not affect the yield or average weight of the fruit. Through two separate field studies in an experimental orchard, boron foliar sprays administered in the spring or autumn resulted in a considerable increase in soluble solids content in the fruit at harvest time. Vangdal et al. (2007) evaluated the effects of potassium and magnesium foliar fertilization on plum fruit quality. In experiment one, the results of increased magnesium, potassium, and a combination of magnesium and potassium on control trees that received a traditional foliar fertilization program were evaluated. The results of the treated trees were compared versus the results of control with standard foliar fertilization program results trees (that only received soil fertilization). Fruit yield and size were used as response variables and the key determinants affecting fruit quality. The primary nutrients in each example were determined by analyzing the leaves and fruits. The effects of foliar magnesium and potassium spraying on the quality of the fruit were not found in this study. Because the treatments did not influence the content of primary nutrients in the leaves and fruits, significant differences in fruit quality were expected compared to the control. Because the fruit was greener and had less blush color, was stiffer, and had a lower level of soluble solids and a greater concentration of titratable acidity, the foliar fertilization program tended to delay fruit ripening. Foliar fertilization resulted in increased nitrogen levels in the leaves and lower levels of the other major nutrients. The foliar fertilization program was able to delay ripening because the fruit was greener, had less blush color, was stiffer, had a smaller amount of soluble solids, and greater content of titratable acidity. Foliar fertilization increased nitrogen levels in the leaves while decreasing levels of the other vital nutrients. A study was evaluated for the effect of nitrogen fertilization on the biology of flowers and plum fruit setting; fertilization was carried out with various amounts of nitrogen (50, 80, or 120 kg N/ha) in early March, and in June there was a post-fertilization (+0, +20, or +30 kg N/ha, respectively). The results showed that applying a high nitrogen level in fertilization resulted in faster growth of pollen tubes after self-pollination and cross-pollination. This was observed after self-pollination and a nitrogen level containing 120 + 30 kg N/ha. The longevity of the eggs was reduced by increasing the nitrogen level. The embryo sac and haustorium development was accelerated using a nitrogen level containing 80 + 20 kg N/ha. Over time, many alternative production systems have been developed, including organic agriculture, characterized by the absence of synthetic fertilizers and pesticides and the frequent use of sources of organic matter to maintain soil fertility (Ruiz et al. 2007). Vitanova et al. (2014) confirmed that organic fertilization is crucial for producing organic plum products. Manure fertilization and green fertilization with winter forage peas and pea and rye mixture increase organic matter content and positively affect the supply of plum plants with essential macronutrients by increasing yield. This treatment can be applied successfully in plum plantations and non-watering conditions.

Al-Dulaimi and Al-Rawi (2020) investigated the influence of biofertilizers and compost on the growth of one-year-old plum trees. In their experiment, the following factors were considered; (1) the addition of biofertilizers, (2) without biofertilizers, (3) *Azospirillum brasilemse* to the soil, (4) *Bacillus megatherium* to the soil, and

(5) the addition of *Azospirillum Brasilmse* + *Bacillus megatherium* to the ground. The second factor is (1) the addition of compost, (2) without adding, (3) adding compost prepared from palm leaves to the soil, and (4) adding compost prepared from wheat residues to the soil. The results showed that the addition of biofertilizers, especially *Azospirillum brasilemse* + *Bacillus megatherium* to the soil, significantly improves vegetative growth traits and increases shoot length diameter. Furthermore, the addition of compost prepared from palm leaves and prepared from wheat residues to the soil increased most vegetative characteristics (branch length, height of the transplants, foliar chlorophyll content, and dry weight of the leaves) (Al-Dulaimi and Al-Rawi 2020). The main reason for choosing organically grown fruits and concern for the environment is to improve the quality of the fruits. Several articles show that organic fruits have higher micronutrients, phenolic compounds, vitamins, etc. (Young et al. 2005).

Paraschiv et al. (2020) studied the yield and fruit quality of three plum cultivars (Centenar, Tita, and Stanley) grown in an organic system, as well as the effects of several fertilizers (Biohumus, Macys BC 28, and Cifamin BK). Finally, Cuevas et al. (2015) investigated the impact of organic and conventional management on plum cultivar biofunctional qualities. The research was conducted in two experimental orchards 2012–2013. Except for total carotenoids, the culture system influenced all the parameters evaluated in this study: total polyphenols, total anthocyanins, total carotenoids, and antioxidant capacity (malic, citric, tartaric, succinic, shikimic, ascorbic, and fumaric) (FRAP, ABTS).

The generation of phytochemical substances in plums was affected by organic management. When cultivated organically, plums exhibited significantly higher polyphenol and anthocyanin contents, as well as increased antioxidant capacity (Chocano et al. 2016). Over six years, they investigated the performance of five different organic management practices on an organic plum farm. All the experimental plots had crushed pruning residues incorporated into the soil every year.

As previously stated, just one plot was given this treatment (crop biomass treatment). The remaining plots were shown one of three treatments: (1) biofertilizer (*Azospirillum brasilense* N-fixing bacteria and *Pantoea dispersa*; bacteria with the ability to solubilize phosphates and stimulate plant growth), (2) an annual compost treatment, (3) a biennial addition of green manure (every year, the following ecosystem sustainability indicators relating to soil microbiological properties and postharvest management are calculated), and (4) the planting and subsequent inclusion of a 60% *Avena sativa* and 40% *Vicia sativa* mixture (treatment with green manure). In addition to plum production, sustainability ecosystem indicators were examined each year: organic carbon, water-soluble C, humic substances and humic acid C, microbial biomass C, respiration, ATP, dehydrogenase activity, and hydrolase enzyme activities (alkaline phosphatase, urease, and beta-glucosidase). The findings revealed that, except yield and dehydrogenase activity, the interaction of treatment and time significantly influenced all these parameters. Composts outperformed green manure cover crops and biofertilizer treatments in terms of plum output. They demonstrated more significant improvements in soil microbiological properties and bigger increases in soil pool. For the six-year trial, compost-treated soils produced greater average plum yields and had higher organic carbon levels,

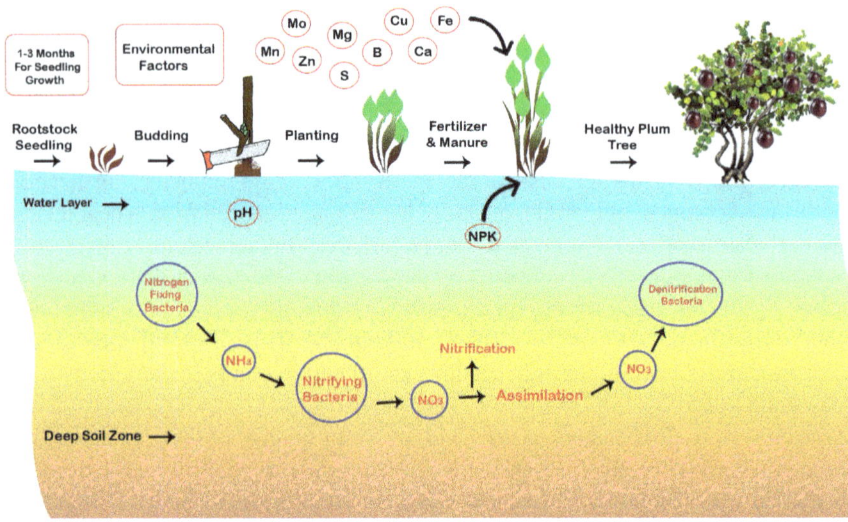

FIGURE 2.1 Schematic diagram of the life cycle of a plant from seedling to fruit bearing tree.

humic substance, and humic acid than the other treatments. Microbial population size and activity were also more prominent and more active in compost-treated soils, as evidenced by higher microbial biomass C, soil respiration, ATP (adenosine triphosphate) content, and dehydrogenase and hydrolase activity. When compost was applied every two years, production yields were higher than when compost was used annually. Figure 2.1 shows the schematic diagram of the life cycle of a plant from seedling to fruit-bearing tree.

2.7 IRRIGATION

Fruit trees need to absorb water from the soil to grow and develop. When the moisture content is low, absorption is difficult, so it is necessary to water to replace it and make it available for the plants. There are different irrigation methods. There is no one better than another, but each is better adjusted to each particular situation, although they present differences in water application efficiency. Unfortunately, most plum orchards do not have irrigation available. A practical and low-cost measure to meet the water needs of seasonal crops is to store rain during the growing season and the dormant season by establishing holes in the soil surface around each tree. This approach can be combined with planting cover crops outside wells to keep soil and water in place (Liu 2004).

In places where the climate is characterized by high variability, the water balance is insufficient for plants' proper growth and development. Therefore, drip irrigation is used, improving fruit tree quality and yield (Dubenok et al. 2020). Treder (1998) evaluated the effectiveness of drip irrigation in plum cultivation; drip irrigation significantly increased tree growth, yield, and fruit quality (Sánchez Valverde 2013). They also assessed the influence that the use of plastic mulch and plateau cultivation has on water saving, development, and plum production; in this experiment, different treatments were considered: a control, three irrigation doses, as well as the use of plastic mulch, plateaus, and combinations of both cultivation techniques with each other. In addition, the following parameters were considered: diameter of the trunk in pattern and graft, shoot growth, production, productivity, and the morphological characterization of the fruits obtained. The results obtained indicate that both the black plastic mulch cultivation technique and the plateau cultivation technique significantly increased the vegetative development of the trees. Furthermore, it was noticed that 50% of the water could be saved with these cultivation techniques. Moreover, the plateau cultivation technique favors higher production, while plateaus with plastic cover increase early production.

Hussien et al. (2013) evaluated the impact of the flood irrigation system and the drip irrigation system on 23-year-old plum cultivars. The drip irrigation system generated an increase in the vegetative growth parameters (shoot length, leaf area, % dry weight of the leaf, number of leaves/shoots, and % foliation), fruit attributes (% flowering, fruit set, number of fruits/tree, and yield), fruit quality (weight, size, firmness, TSS, and acidity), root length, root dry weight, and efficiency in the use of water. At the same time, the flood system increased the percentage of dry weight of the leaf, the dry weight of the root, and the spread of little and broadleaf weeds.

Dubenok et al. (2020) investigated the effect of drip irrigation on plum seedling growth; the field experiment was based on a scheme of soil humidity supported in the range of 60–80%, 70–90%, 80–100% lower humidity capacity, and control without irrigation; it is recommended to use a drip irrigation regimen in the first year that maintains soil moisture in the range of 80–100% lower moisture capacity with a soaking depth of 30 cm.

Fruit quality has improved dramatically due to water limitations; in particular, fruit composition increases when applied during fruit growth. In this regard, the most recent findings of Intrigliolo and Castel (2005) conclude that a water deficit during fruit growth in phenological stages II and III reduces the average weight of the fruit. On the other hand, a drought after harvest does not affect flowering, fruit set, or short-term fruit yield growth. However, compared to the control plots, there was a 10% drop in yield in the last year of the trial. As a result of the cumulative impacts of water deficit on tree growth in young trees, postharvest dryness might lower output in the long run. Moñino et al. (2020) compared three irrigation systems over three years (2014, 2015, and 2016): Control (CON) designed to meet the trees' water needs throughout the season; regulated deficit irrigation (RDI) with two separate periods of water deficit, one before harvest without irrigation inputs during an intermediate period of fruit growth (pit ripening period) and the other postharvest with a 30% reduction in CON inputs; and Preharvest + Regulated Deficit Irrigation (RDP) with a single period of deficit irrigation begun before harvest. Both deficit irrigation treatments effectively controlled tree vigor, resulting in lower trunk cross-sectional area growth and trimmed wood weight than the CON treatment. The RDP treatment produced fruit equivalent in size to the CON treatment. Fruit hardness, soluble solids concentration (SSC), and skin color (C*) varied slightly or inconsistently across the three years.

The effect of irrigation regimes on the development dynamics of the plum root system was studied by Cochavi et al. (2019), who grew the plants for five months under three different irrigation regimes. The effects of the irrigation regime on several orders of roots, total root elongation, biomass growth, and numerous tips were investigated. The findings show that the responses differ depending on the level of stress. In the case of comprehensive irrigation treatment, the total biomass of the root system falls under moderate stress. On the other hand, low-order root production is similar to that seen in non-stressed settings. The production of low-order roots is also suppressed by severe stress; moderate stress influences the root system's growth rate but does not affect the relative contributions of different root types.

2.8 CONCLUSION

Establishment and management of plum orchards could be favored in deep, well-drained soils with a depth of at least 600 mm, a pH of 4.5–6, and slopes of 4–15%. Furthermore, previously subsoiled and nourished (mainly phosphorus, potassium, and calcium) compounds, such as minerals, vitamins, proteins, among others, must be included to complement nutrition, drip irrigation being an excellent alternative for supply. In this context, this chapter makes it possible to appropriate agronomic aspects that allow making appropriate decisions in the implementation of orchards, tending to optimize the behavior of the plant to obtain maximum production with the best product quality.

2.9 REFERENCES

Al-Dulaimi, A.S.T., and Al-Rawi, W.A. 2020. Effects of Biofertilizers and Compost Application on Vegtative Growth of Plum Transplants. *Plant Archives* 20: 2215–2220.

Andorno, A.V., Botto, E.N., La Rossa, F.R., and Möhle, R. 2004. *Estudios preliminares sobre la diversidad biológica de áfidos y sus enemigos naturales asociados a cultivos orgánicos de hortalizas bajo cubierta.* Implicancias para su empleo en el desarrollo de estrategias de control biológico. En resúmenes del XXVII Congreso Argentino de Horticultura.

Andorno, A.V., Botto, E.N., La Rossa, F.R., and Mohle, R. 2015. *Control biológico de áfidos por métodos conservativos en cultivos hortícolas y aromáticas.* Ediciones INTA.

Batlle, I., Iglesias, I., Cantín Mardones, C.M., Badenes, M.L., Ríos, G., Ruiz, D., Dicenta, F., Egea, J., López Corrales, M., and Guerra Velo, M. 2018. *Frutales de hueso y pepita*, 79–132. Sociedad Española de Ciencias Hortícolas. ISBN 978-84-948233-8-1.

Blažek, J., and Pištěková, I. 2009. Preliminary Evaluation Results of New Plum Cultivars in a Dense Planting. *Horticultural Science* 36: 45–54.

Butac, M., and Chivu, M. 2020. Yield and Fruit Quality of Some Plum Cultivar Sin Ecological System. *Academy of Agricultural and Forestry Sciences—"Gheorghe Ionescu-Şişeşti" Horticulture Section* 1: 67–74.

Chocano, C., García, C., González, D., de Aguilar, J.M., and Hernández, T. 2016. Organic Plum Cultivation in the Mediterranean Region: The Medium-Term Effect of Five Different Organic Soil Management Practices on Crop Production and Microbiological Soil Quality. *Agriculture, Ecosystems & Environment* 221: 60–70, April. https://doi.org/10.1016/j.agee.2016.01.031.

Cochavi, A., Rachmilevitch, S., and Bel, G. 2019. The Effect of Irrigation Regimes on Plum (Prunus Cerasifera) Root System Development Dynamics. *Plant Biosystems-An International Journal Dealing with All Aspects of Plant Biology* 153 (4): 529–537.

Cuevas, F.J., Pradas, I., Ruiz-Moreno, M.R., Arroyo, F.T., Perez-Romero, L.F., Montenegro, J.C., and Moreno-Rojas, J.M. 2015. Effect of Organic and Conventional Management on Bio-Functional Quality of Thirteen Plum Cultivars (Prunus Salicina Lindl.). *PLoS One* 10 (8): e0136596.

Dubenok, N., Gemonov, A. and Lebedev, A. 2020. *The Influence of Drip Irrigation on Growth of Plum Seedlings in Central Non-Black Soil Zone of European Russia.* IOP Conference Series: Earth and Environmental Science, IOP Publishing.

El-Shall, S., El-Messeih, A., and Abd El Megeed, N. 2010. The Influence of Humic Acid Treatment on the Performance and Water Requirements of Plum Trees Planted in Calcareous Soil. *Alexandria Science Exchange Journal* 31: 38–50.

Fadón, E., Herrera, S., Guerrero, B.I., Engracia Guerra, M., and Rodrigo, J. 2020. Chilling and Heat Requirements of Temperate Stone Fruit Trees (Prunus Sp.). *Agronomy* 10 (3): 409. https://doi.org/10.3390/agronomy10030409.

Fudge, B., and Fehmerling, G. 1940. Some Effects of Soils and Fertilizers on Fruit Composition. *Proceedings Ann Meet Florida Stale Hort Society* 53: 38–46.

Gomez, M.D.D.M. 2015. *La fruticultura del siglo XXI en España*, coordinadores by H. Martín, J. José, and Y. Cuevas González Julián. Serie Agricultura n °10. Cajamar, Caja Rural.

Gómez-Marco, F., Hermoso-De-Mendoza, A., Tena, A., Jacas Miret, J.A., and Urbaneja, A. 2012. Mejora del control biológico de pulgones en cítricos mediante la gestión de cubiertas vegetales. *Vida Rural* 352: 22–29.

Häberlein, E. 1990. Training for Flat Crown in Plum Trees. *Erwerbsobstbau* 32: 39–42.

Hussien, S.M., Fathi, M.A., and Eid, T.A. 2013. Effect of Shifring to Drip Irrigation on Some Plum Cultivars Grown in Clay Loamy Soil. *Egyptian Journal of Agricultural Research* 91 (1): 217–233. https://doi.org/10.21608/ejar.2013.161553.

Intrigliolo, D.S., and Castel, J.R. 2005. Effects of Regulated Deficit Irrigation on Growth and Yield of Young Japanese Plum Trees. *The Journal of Horticultural Science and Biotechnology* 80 (2): 177–182. https://doi.org/10.1080/14620316.2005.11511913.

Lazăr, R., Lazăr, C., Răducu, D., and Marinca, C. 2011. The Influence of Climatic and Soil Conditions on the Growth and Development of the Plum Trees and on the Yield Levels on a Typical Luvosol at Caransebes-Caras Severin County. *Research Journal of Agricultural Science* 43: 416–423.

Liu, W. 2004. *VIII International Symposium on Plum and Prune Genetics, Breeding and Pomology*. Plum Production in China, 734: 89–92.

Luna Tetelano, J. 2012. *Resiliencia en cultivos asociados: Maíz (Zea mays L.), Aguacate (Persea americana Miller) y ciruela (Prunus domestica L.), en comunidades de Jalacingo, Veracruz, México*. Universidad Internacional de Andalucía.

Milosevic, T., and Milosevic, N. 2011. Seasonal Changes in Micronutrients Concentrations in Leaves of Apricot Trees Influenced by Different Interstocks. *Agrochimica* 55: 1–14.

Moñino, M.J., Blanco-Cipollone, F., Vivas, A., Bodelón, O.G., and Henar Prieto, M. 2020. Evaluation of Different Deficit Irrigation Strategies in the Late-Maturing Japanese Plum Cultivar 'Angeleno'. *Agricultural Water Management* 234: 106111, May. https://doi.org/10.1016/j.agwat.2020.106111.

Montgomery, H., and Remondo, J.P. 1964. *Ciruelas y cerezas*. Editorial Acribia.

Neumüller, M. 2011. Fundamental and Applied Aspects of Plum (Prunus Domestica) Breeding. *Fruit, Vegetable and Cereal Science and Biotechnology* 5: 139–156.

Norton, M. 2007. *Growing Prunes (Dried Plums) in California: An Overview*. Norton.

Nuttonson, M.Y. 1947. Agroclimatology and Crop Ecology of Palestine and Transjordan and Climatic Analogues in the United States. *Geographical Review* 37: 436–456.

Okie, W. 1995. *Plum Breeding and Genetics*. Symposium on State of the Art in Fruit Breeding.

Okie, W., and Hancock, J. 2008. Plums. In *Temperate Fruit Crop Breeding*. Springer.

Paraschiv, M., Hoza, D., Nicola, C., Florea, A., Butac, M., and Chivu, M. 2020. Evaluation of the Biochemical Content of Fruits on Some Plum Genotypes. *Lucrari Stiintifice-Universitatea de Stiinte Agricole a Banatului Timisoara, Medicina Veterinara* 53 (3): 48–58.
Ruiz, C., Russián, T., and Tua, D. 2007. Efecto de la fertilización orgánica en el cultivo de la cebolla. *Agronomía Tropical* 57: 7–14.
Ruiz, D., Egea, J., Salazar, J.A., and Campoy, J.A. 2018. Chilling and Heat Requirements of Japanese Plum Cultivars for Flowering. *Scientia Horticulturae* 242: 164–169, December. https://doi.org/10.1016/j.scienta.2018.07.014.
Sánchez Valverde, E. 2013. *Aplicación de nuevas técnicas de cultivo para el ciruelo.* Tesis Doctoral. Departamento de Producción Vegetal y Microbiología. Universidad Miguel Hernández de Elche.
Tehrani, G. 1972. Pollen Compatibility Studies with European and Japanese Plums. *Fruit Varieties and Horticultural Digest* 26: 63–66.
Tóth, E., and Surányi, D. 1980. Plums. [The Plum [Breeding, Production, Varieties, Plant Protection]]. *Mezogazdasagi Kiado* 1: 428.
Treder, W., Grzyb, Z.S., and Rozpara, E. 1998. Influence of Irrigation on Growth, Yield and Fruit Quality of Plum Trees cv. Valor Grafted on Two Rootstocks. *Acta Horticulturae* 478: 271–278, October.
Vangdal, E., Døving, A., and Måge, F. 2007. The Fruit Quality of Plums (Prunus domestica l.) As Related to Yield and Climatic Conditions. *Acta Horticulturae* 734: 425–429, February.
Villarrubia Horta, D., and Mataix Gato, E. 1998. *Plum Tree Cultivation: Aspects of Fertilization.* Comunitat Valenciana Agraria (Espana).
Vitanova, I., Ivanova, D., Stefanova, B., and Dimkova, S. 2014. Perspectives for Development of the Biologic Plum Production in Bulgaria. *New Knowledge Journal of Science* 3 (1).
Wójcik, P. 1997. Effect of Boron Fertilization on Growth, Yield and Fruit Quality of Plum Trees (Prunus domestica L.). *VI International Symposium on Plum and Prune Genetics, Breeding, Pomology* 478: 255–260.
Young, J.E., Zhao, X., Carey, E.E., Welti, R., Yang, S.S., and Wang, W. 2005. Phytochemical Phenolics in Organically Grown Vegetables. *Molecular Nutrition & Food Research* 49: 1136–1142.

3 Recent Advances in Varietal Improvement and Rootstock Breeding of Plum

M.A. Kuchay, A. Raouf Malik, Rehana Javid, Shaziya Hassan and Rafiya Mushtaq
Division of Fruit Science, Faculty of Horticulture, Sher-e-Kashmir University of Agricultural Sciences and Technology of Kashmir, Shalimar, Srinagar, 190025, Jammu and Kashmir, India

CONTENTS

3.1 INTRODUCTION

Plum is known among all temperate fruit tree species to evoke human attention (Faust and Suranyi 1999). It is a famous stone fruit crop worldwide. The stone is formed by lignifications of endocarp which are enclosed by the meocarp and exocarp. The four main plum-producing countries are China (6,788,107 million tons [MT]), Romania

DOI: 10.1201/9781003205449-3

(842,132 MT), Serbia (391,485 MT), and the United States (368,206 MT) (FAOSTAT 2018). In India, production of plum is about 89,000 MT in an area of 23,000 hectares (ha) (NHB 2018). It holds the fourth position in temperate fruit crops, including pome fruits (apples and pears) and stone fruit (peaches) in this place. Plum contributes 2 per cent of the world's fruit production.

3.2 ORIGIN AND BRIEF HISTORICAL BACKGROUND

Commercially grown plums can be European plums (*Prunus domestica* Lindl.) or Chinese plums (*Prunus salicina* Lindl.) and have been cultivated more than any other fruit crop except apples. The earliest known information regarding plums indicates that they originated in China. However, the European plum may have involved *P. spinosa* and *P. cerasifera* as ancestors (Topp et al. 2012). Luther Burbank specifies that the Caucasus Mountains close to the Caspian Sea are the location of origin for *P. domestica* and its ancestors. The origin of the Japanese plum is actually in China. After being introduced to Japan, it was then brought to the United States in 1870 (Wu et al. 2019). Earlier evidence shows that *P. domestica*, *Prunus cerasifera* (cherry plum) and *Prunus spinosa* (blackthorn plum) were being used for the improvement of early European societies. These species date back to 4000–6000 BCE, during the Neolithic period, in Germany and the Ukraine (Zhebentyayeva et al. 2019). The oldest plum species, damson plum and St. Julien, have been cultivated since Neolithic times. Cherry plums and sloe are also observed in this location, indicating that other hybrids might also have originated there. However, Ermenyi (1975–1977) believes that every type of plum had been common in Austria and had been produced there earlier than the Roman occupation. Some plum species were excavated at some point of the overdue late Iron Age settlements at Maiden Castle in Dorset and at the site of the Romans in Silehester. According to Werneck (1961), Romans brought seeds to Lanz that were Celtic in origin, where they improved local environmental conditions. According to archaeological evidence, England was not the original site of domestic and damson plums, but they had been grown there since before the Roman occupation (Roach 1985). The America plum, which had limited utility during the Middle Woodland period from 250 BC to AD 400 (Sant and Stafford 1985). As per Roman writers, Lucius Junius Moderatus Columella (first century) noted different species of plums, such as Cereolum, Damasci, and Onychium. Plums were exported to Japan from China around 300 BC. Matsumoto (1977) uses the term 'ume' for Japanese plum varieties first observed in AD 751, was written in the same way in the Chinese style as a collection of Japanese poetry. On the other hand, Chinese monks of high dignity brought plums to Japan as a gift to the emperor. Plums are also important in nearly every part of Europe. Some old plum cultivars had been brought to England before the seventeenth century, but novel cultivars did not exist prior to the nineteenth century. In ancient times the small American plums had been planted with the aid of New England Indians and the western Indians used them for drying (Pickering 1879).

Breeding improvements in plum cultivation initiated in America with the selection of domestic American plums when large numbers of cultivars were brought to the region at end of the nineteenth century. Mariana rootstock is an orginator of wild goose plum seedling which was introduced to America in 1884 by means of Charles Eley of Smith Point, Texas (Yoshikawa et al. 1989). Japanese plum was introduced to California is one of the leading plum growing areas, in 1870 with the help of Mr. Hough of Vacaville. The first Japanese plum was distributed extensively by John Kelsey and was released by W. P. Hammon & Co. in 1884, which named the fruit Mr. Kelsey. Breeding of plum in California was started by Luther Burbank on a small farm in 1875. He had introduced top-ranking European plums, namely, the Improved French prune, the sugar prune and the standard prune from 1898 to 1911. When Burbank started breeding Japanese plums, there were only few cultivars available. He imported different cultivars from Japan, and most of them hybridized themselves or from open pollination. He also introduced the most important cultivar, the Santa Rosa, in 1907, as a true hybrid of *P. salicina*, *P. simonii* and *P. americana*, though it predominantly resembles *P. salicina*. Santa Rosa plums have reddish flesh that may be acquired from the Satsuma plum because it was the only plum carrying that trait at that time. In 1914, Burbank introduced what would become the most famous plum varieties, the Santa Rosa, the Formosa, the Beauty and the Wickson plums. During this period, varieties like the Friar and black Amber saw a 36–70 per cent decrease in consumption in favour of the Santa Rosa.

3.3 BOTANY AND TAXONOMICAL CLASSIFICATION

European plum trees are large, typically 4–15 m high with a more upright growth pattern, and flower late in the season compared to Japanese plums. Their branches are straight and spineless, and the bark of young twigs is dull brown with short, soft hairs. It requires a certain amount of cold temperatures (0–7.2 °C) to break the bud dormancy. The leaves are quite large (3–10 × 1.8–6 cm), simple, glabrous and dark green on the upper surface, while being compact pubescent to sub-glabrous on the lower surface, ordinarily at a younger age. The flowers are white and large, with a 15–25 mm diameter, and normally fashioned in groups of two to three, appearing along with the leaves. Leaf pedicel length is approximately 5–20 mm, and petals are white and 7–12 mm in length. Flowers consists of five ovate-shaped sepals with an acute apex and imbricates. They have white or greenish-white petals, rounded to obtuse tips, and imbricate on the rim to the hypanthium. The flower consists of 10–30 stamens in two whorls of unequal filaments. The pistil consists of one carpel forming an uninoculated ovary attached above other floral parts. European plum fruits possess yellow and green backs with red and blue epicarps (Waugh 1903; Pareek and Sharma 2017). The varieties belonging to the European group can be classified into four categories:

1. Prunes: Prunes are a commercially essential group, due to their high sugar content and stone in the centre that does not cling to the flesh (freestone), after being dried without pit removal. Prunes are firm, oval shaped and darkish blue or purple in colour.
2. Reine Claude: This type of prune is especially used for fresh consumption and canning purposes. The fruits are round in shape and green, yellow, slight red or gold in colour.

3. Yellow egg: Fruits of yellow egg cultivars are utilized for canning purposes and are oval in shape and yellow in colour.
4. Lombard: This type of plum is large, oval shaped, red or pink in colour and consumed as fresh fruit.

Japanese plum trees are small with rough bark and start flowering earlier than European plums. The trees have numerous long-lived fruiting spurs, which are mostly affected by frosts (Bhutani and Joshi 1995). They have simple leaves, oblong to ovate almost free of pubescence. Flowers are white, occur in groups of three on spurs and are rarely noticed on young shoots. The size and flesh colour of the fruits varies from yellow to deep red. A waxy layer deposits on the fruit surface, which inhibits the movement of water through the surface of the fruit during harvest.

The plum belongs to the family Rosaceae, subfamily Prunoideae, subgenus Prunophora Focke. The trees are small and deciduous; leaves are alternate, often serrate or crenate. Flowers are usually white, solitary or clustered, and perigynous. Fruits are drupes, 2–8 cm in length; of variable colours, including blue/purple, red/pink or yellow/green; with smooth, flattened stone (Singha and Baugher 2008). The basic chromosome number is $x = 8$. Particularly, 600 cultivars were released from local American species by new breeders in the eighteenth century (Das et al. 2011). The crossing between *P. cerasifera* ($2n=2x=16$) and *P. spinosa* ($2n=4x=32$) can lead to the origin of new hybrid (European plum) is a hexaploid ($2n=6x=42$) (Folta and Gardiner 2009). European plums were released for cultivation due to their sugar content, flavour, large stone, ripening, late blooming habit and higher chilling hours requirement to break the dormancy compared to Japanese plums (Zhukovsky 1965).

Genetic variability in plum species and cultivars is the first requirement of any plant breeding programme and can be used to develop the plum fruit industry by improvement over local plum cultivars (Khush 2002). More than 6,000 varieties and 20 species of plum have been growing in different geographical regions. Only a few of them are commercially more important, namely, *Prunus domestica* L., *Prunus cerasifera* Ehrh., *Prunus salicina* Lindl and their hybrids (Okie and Weinberger 1996).

Presently the genetic resources available worldwide can be used for future plum breeding programmes, particularly in diploid plums (Table 3.1). Japanese plum seedlings, introduced from Japan to the United States by Luther Burbank, have been inter-crossed with Chinese plums, Euroasian plums and Native American plums (Duval 1999). Abiotic and biotic stresses, low chilling resistance, and being self-fruitful are the most important characteristics for the future breeding of cultivars (Blazek 2007). The word "prune" or "plum" is typically used to describe the ume plum (*Prunus mume*), though it genetically more closely resembles an apricot than a plum as reported by Chinese, Japanese and Korean authors, and it is occasionally referred to as the Asian apricot (Stacewicz-Sapuntzakis 2013).

TABLE 3.1
Important Plum Species, Their Origin and Varieties

Species	Common name	Origin	Subspecies	Varieties
P. alleghaniensis	Allegheny plum, sloe	North-eastern United States	davisii	
P. americana	American plum, wild goose plum, hog plum	Central and eastern United States	lanata mollis	Anderson's Early, Ember, Goff, Hazel, Kahinta, Monitor, Red Coat, Underwood, Wolf
P. andersonii	Desert peach	California, Nevada	–	–
P. angustifolia	Chicksaw plum, sandplum, sandil plum	Southern United States	varians, watsoni	Bruce, Six Weeks
P. besseyi	Sand cherry	Canada, northern United States	cuneata, depressa, pumil, susquehanae	Alace, Black Beauty, Convoy, Deep Purple, Hiawatha, Manor, Mansan, Sapa, Sioux
P. fasciculata	Desert almond, desert peach brush	South-western United States	–	–
P. fremontii	Wild apricot, desert apricot	Southern California	–	–
P. geniculata	Scrub plum	Florida	–	–
P. glandulosa	Chinese bush cherry	China, Japan	–	–
P. havadii	Havard wild almond	Texas, Mexico		
P. hortulana	wild goose plum, hortulan plum	Central United States	mineri	Miner, Wayland
P. humilis	Manchurian dwarf cherry	Northern China	–	–
P. japonica	Flowering almond, Japanese bush cherry	Eastern Asia	–	–
P. maritima	Beach plum, shore plum	Coastal north-eastern United States	gravesii	Hancock, Jersey, Patricia, Burbank, Squibnocket
P. mexican	Big tree plum, Mexican Plum	South-central United States, Mexico	–	–
P. minutiflora	Texas wild almond, small flower peach bush	Texas, Mexico	–	–
P. munsoniana	Wild goose plum	China		Late Goose, Whitaker

(Continued)

TABLE 3.1 *Continued*

Species	Common name	Origin	Subspecies	Varieties
P. nigra	black plum, Canada plum	Northern United States, Canada	–	Atken, Assiniboine, Bounty, Grenville. Norther, Pembina
P. salicina	Japanese plum	China	bokhariensis, gymnodonta	Abundance, Burbank, Kelsey, Satsuma
P. subcordata	Pacific plum, sierra plum	North-eastern United States	kelloggii, oregana	G.M. Clark, Kelly Sierra
P. texana	Texas almond cherry, Texas peach brush	Texas	–	–
P. umbellata	Flatwoods plum, hog plum, sloe	South-eastern United States	injucunda mitis, tarda	–

Source: Okie 2006

3.4 CROSSING TECHNIQUE

For breeding purposes the emasculation process is essential for controlled pollination by the removal of anthers, petals and sepals with fingernails or forceps before anther dehiscence (Layne 1983). It is practised at a late balloon stage or before the flower is fully opened and is considered most favourable for removing anthers from flowers in peaches, nectarines, plums, apricots and cherries (Bailey and Hough 1975). The anthers should be collected of each genotype at the balloon stage (Figure 3.1A). The collected anthers should be kept on paper to dry at room temperature for 24 hours until anther dehiscence occurs completely. Thereafter, pollens should be stored at –20°C until the emasculated flowers are pollinated by donor parents (Williams 1970). For the cross-pollination of emasculated flowers, both old flowers and young buds should be removed from the branches, leaving only flowers at the early stage (Figure 3.1B). Flowers from a specific cultivar are then selected for emasculation and cross-pollination on the following day with a camel's-hair brush. Individual blossoms of a single shoot or an entire branch after completion of pollination with an individual pollen variety must be carefully labelled with complete information that will persist the identity of crossed varieties until the fruit is harvested (Cullinan 1937).

3.5 BREEDING METHODS

The conventional breeding methods of open pollination and hybridization at intra- and interspecific levels are popular strategies to create significant variation in desirable genes into plum cultivars (Milosevic and Milosevic 2018), which may result in the development of new types of cultivars (Goulet et al. 2017; Lopez-Caamal and Tovar-Sanchez 2014).

FIGURE 3.1 Japanese plum flower at the balloon stage (A) and after emasculation (B).

The clonal selection of cultivar *Vinete romanesti* has been carried out on an old domestic plum orchard in Romania, resulting in clones such as *Vinete romanesti* 300, *Vinete romanesti* 303 and *Vinete romanesti* 4. Clonal selection is also executed at the Fruit Growing Research Extension Station Valcea (S.C.D.P. Valcea) on plum and autochthonous cultivars that originated from the heterogeneous local populations *Vanat romanesc* and *Tuleu gras*. *Tuleu gras* clone 5 is composed of the same fruit traits as the parental cultivar, but its branches are stronger and less sensitive to breakage. *Tuleu gras* clone 11 has large fruits (5–7 g) compared to its parental cultivars (Botu et al. 2007a). Relationships among the cultivars and the manipulation of genetic variability are crucial for choosing parental lines based on target traits for long-running breeding programmes. The selection of a plum's promising genotype (ASR/BL-2) was examined in Punjab, India., in 1995 The selection attained large fruits with attractive colours that were free from sourness near the pit. This cultivar was named Aloo Bukhara Amritsar (Bal 2004).

Sexual hybridization plays an important role in several forms that have been utilized by different crossing types to get plants with desirable traits. Since 1950, a large number of plum varieties has been introduced by simple crossing. The best example of a simple cross are the Stanley variety, a cross between the d'Agen and the Grand Duke in the United States; the Valor, a cross between the Imperial Epineuse and the Grand Duke in Canada; the Cacanska Lepotica, a cross of the Wagenheim and the Pozegaca in Serbia; the Jojo, a hybrid of the Ortenauer and the Stanley in Germany; and the Centenar, a cross between the Tuleu gras and the Early Rivers in Romania. The cross between Black Sunrise and Autumn Giant showed an increase in the size of the fruit, and the hybrid of Larry Ann and Fortune exhibited increase in the firmness of the fruit (Lugli et al. 2010). The Dofi-Sandra, derived from Black Gold and Burmosa, was released at the Horticulture Department at Florence University in 1989. This variety was characterized by high yield and good organoleptic attributes (Bellini and Nencetti 2002). The Valcean cultivar was produced by using a double-crossing method [H8–12 (Renclod Althan x Wilhelmina Spath) x H

5–23 (Renclod Althan x Early Rivers)] in Romania. For the last two decades, the pyramidal-type cross [(A x B) x C] has been more frequently utilized. For this reason, interspecific hybridizations have been properly utilized for obtaining complex genotypes. Hybridization among diploid species may also be so easy. For example, the Santa Rosa variety is a cross between *P. salicina*, *P. simonii*, and *P. americana*.

Interspecific hybrids developed between the plum and apricot have been named the Plumcot® (e.g., Red Velvet, Royal Velvet, Flavor Supreme, Flavor Queen, Rutland, Plum Parfait, Spring Satin and Yiksa) and have appeared on the international market in recent years. The crossing of (*P. domestica* x *P. armeniaca*) x *P. domestica* and (*P. salicina* x *P. armeniaca*) x *P. salicina* resulted in the Pluot® hybrid. The third interspecific hybrid is the Aprium®, a cross of *P. domestica* x *P. armeniaca* and *P. armeniaca*. The Plumcot received a 75 per cent share in the new hybrid from the plum parent, while the Pluot acquired a 25 per cent share from its apricot parent. Apriums are also called Plumcots and are more similar to apricots than plums in that the genetic participation share is 75 per cent and 25 per cent respectively (Zhivondov and Uzundzhalieva 2012).

Some varieties are frequently used as donor parents due to highly valuable fruit quality, as well as shipping and keeping quality. The first varieties used in this manner were the Burbank and the Friar, followed by the Laroda, the Methly, the Red Beaut and the Wickson, for the discovery of new cultivars, which included the Eldorado, the Gaviota, the Nubiana and the Queen Ann.

In rootstock breeding, the Ishtara was the first interspecific hybrid of *P. domestica*, *P. cerasifera* and *P. armeniaca*. The Jaspi, a new plum rootstock, was created by crossing Methley and Mariana rootstock. Dospina 235 is hybrid of *P. domestica* with *P. spinosa*, and Docera 6 (*P. domestica* x *P. cerasifera*) was developed at the University of Munich. Rootstock breeding with specific traits needs intra- and interspecific crosses of resistant genes to achieve new recombinant genotypes that can be grown in different agroclimatic conditions (Dejampour et al. 2007). Crossing between the Aloo Bukhara Peshawari plum with Earli Grande and Shan-i-Punjab peaches showed a satisfactory fertilization gain in both of these peach cultivars as a good source of pollen donor parents (Bal et al. 2017).

Natural polyploidy, which contains more than two sets of chromosomes in genus *Prunus*, results from both the doubling of somatic chromosomes and the meiotic restitution of gametes (Ramsey and Schemske 1998). The Herkules variety exhibited pentapoloidy and is a cross of the hexaploid *P. domestica* Ontario and the diploid *P. salicina*. The other example is the Formosa, which shows pentaploidy character developed through the pollination of unreduced pollen (2n) (Neumuller 2011; Glowacka et al. 2021). Experiments had been performed with embryos of the Japanese plum cultivar Kelsey to induce tetraploidy with the aid of colchicine. Such polyploids may be useful in developing new, large-fruited cultivars and as breeding parents in crosses with the European plum (*P. domestica*). In 1993, crossings between the hybrids (Ortenauer × Stanley 34) and Hanita resulted in high fruit quality and a sweet taste. Colora, a hybrid between Ortenauer × Ruth Gerstetter, shows green, yellow and red colour. More than 90 per cent of plum cultivars were developed through conventional breeding methods. Predominantly, plum species have diploid sets of chromosomes. Some are tetraploid in nature, whereas *P. domestica* L. is hexaploid (Urbanovich et al. 2017).

The Japanese plum industry in South Africa is based upon the export market of sweet cultivars that have large and attractive fruit. This is done by breeding procedures, especially the use of polyploidy with respect to fruit size. Polyploid breeding material is also required for crosses on the hexaploid level with the European plum. The combination of the vigour and adaptability of the Japanese plum, *P. salicina*, with the fruit quality and sugar content of certain European plums in developing a new prune cultivar which is better able to adapt in South African conditions (Hurter and Van Tonder 1963). It is believed that *P. domestica* is one of the natural alloploids that have been developed from diploid and tetraploid species (Crane and Lawrence 1952). The polyploidy levels of different plum species are given in Table 3.2.

Mutagenesis was carried out on seeds and buds in Romania by utilizing Co-60 and X-rays as a radiation source. Romanian plum cultivars like Alina and Tita were developed with X-rays of Tuleu Gras seeds. A natural bud sport, Crown, was developed from the Qing Nai tree in China (Figure 3.2). It is highly productive, and no biennial bearings were observed (Li et al. 2015).

TABLE 3.2
Ploidy of the Main Plum Species

Species	Chromosome no.
P. saliciana (and its hybrids)	2n=2x=16
P. cerasifera	2n=2x=16
P. americana	2n=2x=16
P. mariana	2n=3x=24
P. spinosa	2n=4x=32
P. domestica	2n=6x=48
P. insitia	2n=6x=48

FIGURE 3.2 New plum cultivar Crown (*Prunus salicina*) is a mutant selected from a Qing Nai plant in China. The Qing Nai fruit (A) is greenish yellow, and the fruit of Crown (B) stays yellowish in colour.

3.6 BREEDING OBJECTIVES IN SCION AND ROOTSTOCK

Different breeding centres have specific objectives for developing a variety. Late flowering and frost resistance are the main objectives in the United Kingdom. Other objectives include cold hardiness in Latvia, early bearing in Sweden, good keeping quality in Norway, prolonged ripening time in Bulgaria and Romania and self-fruitful varieties in Latvia and Romania. *Prunus domestica* is a predominant plum in which 80 per cent of breeding activities are carried out and the remaining 20 per cent was attained by *Prunus salicina* (Butac et al. 2010; Jacob 2006). Plum improvement for subtropical regions should include low chilling requirements, high tolerance to temperatures and tolerance of dwarfing rootstock to saline and stagnant soils (Kumar 2006). The development of rootstocks begins with seedlings and clonal rootstocks, the majority of which are of interspecific origin. The development of rootstock should be cheap, easy to multiply, resistant to disease and insects, tolerant to biotic and abiotic stresses, compatible with scion and influential on tree size, yield and the quality of the scion (Beckman and Lang 2003). Nowadays researchers focus on the selection of plum rootstocks for high-density planting that should be low in vigour and high in yield efficiency. One of the best examples is the Krymsk®1, which has higher efficiency in tree vigour control and precocity but is more prone to bacterial canker (Maas et al. 2011). Different breeding centres have defined the objectives on the basis of real problems, as shown in Table 3.3.

TABLE 3.3
European Plum Breeding Centres and Objectives

Country	Breeding centres	Breeder	Objectives
Serbia	Fruit Research Institute Čačak	D. Ogasanovic, N. Milosevic	Improvement of old cv. Pozegaca (size, ripening time, resistance to PPV)
Germany	University Hohenheim, Stuttgart Research Station Geisenheim	W. Hartmann M. Neumuller H. Jacob	Fruit quality PPV resistance by hypersensitivity
France	INRA, Bordeaux	R. Bernhard R. Renaud	Fruit quality PPV resistance by transgenic plants
Italy	University Bologna University Firenze University Forli Private programme	S. Sansavini E. Bellini V. Nancetti A. Liverani	Japanese and European plum breeding programme: quality, fresh consumption, resistance to PPV
Bulgaria	Fruit Growing Institute Dryanovo Fruit Growing Institute Troyan Fruit Growing Institute Plovdiv	V. Bozhkova, A. Zhivondov	Late blooming, extended ripening period, fruit quality, resistance to PPV

TABLE 3.3 *Continued*

Country	Breeding centres	Breeder	Objectives
United Kingdom	Research Station East Malling	K Tobutt R. Jones T. Laxton	Late blooming, frost resistance, PPV resistance
Latvia	Latvia State Institute of Fruit Growing	E. Kaufmane I. Gravite	Winter hardiness, vigour, precocity, fruit quality, self-fertility, early ripening
Moldova	Research Institute for Horticulture and Alimentary Technologies Chisinau	M. Pintea, A. Juraveli	Resistance to frost, fruit quality
Belarus	Institute for Fruit Growing, Minsk	M. Vasiljeva, Z. Kazlouskaya	Fruit quality, frost resistance
Czech Republic	Research and Breeding Institute of Pomology Holovousy	J. Blazek	Fruit quality, sharka tolerance
Sweden	University of Agricultural Sciences Balsgard	I. Hjalmarsson, V. Trajkovski	Short period of vegetation, fruit quality
Norway	Ullensvang Research Centre Division Njos, Lofthus	S.H. Hjeltnes	Fruit quality, storage
Romania	Research Institute for Fruit Growing, Pitesti Fruit Growing Research Station, Valcea Research Station for Fruit Growing, Bistrita	M. Butac M. Botu I. Zagrai	Improvement of old cvs. Tuleu gras, Grase romanesti, Vinete romanesti; fruit quality, PPV resistance, ripening season extension, self-fertility
China	Zhengzhou Fruit Research Institute, Chinese Academy of Agricultural Sciences (ZFRI-CAAS), Zhengzhou 450009, Henan, China	–	To select plum cultivars that possess not only high productivity and fruit storability but also high eating quality to meet the needs of both the consumer and the market

Source: Butac et al. 2013

3.7 BREEDING ACHIEVEMENTS IN PLUM

3.7.1 Scion

Plum breeding has been conducted at several institutes in different countries to develop new cultivars with desired traits. European plums are the dominant choice in producing new cultivars and have released 170 cultivars in the last several decades (Blazek et al. 2004). Some clones of plum showed outstanding fruit quality along with high degree of hypersensitivity reaction, namely, Hanita, Felsina and Fellenberg. Jojo when crossed with Haganta. Some promising cultivars produced from Jojo × Hauszwetschge differ in ripening times and could be grown worldwide (Hartmann and Neumuller 2013). In Europe, different cultivars are used as donors of certain positive characteristics of trees and fruits (Hartmann and Neumuler 2006; Jacob 2007; Blazek and Vavra 2007). Cultivars such as Cacanska Lepotica, Cacanska Najbolja and Stanley have been used for controlled cross-pollination at research institutions in Serbia. These cultivars were used for various fruit traits, e.g., Cacanska Najbolja was a donor for PPV (plum pox virus) tolerance, vigour and hardiness of trees and acceptable fruit size. Fruit size is an economically quantitative inherited trait that determines certain quality parameters, such as yield and consumer acceptance (Crisosto et al. 2004). Skin colour is considered the most important criteria when a consumer purchases fruit, as it provides information to the consumer about the fruit's taste (Tromp et al. 2005). Fruit colour varies from reddish (Roman and Romaner) to dark blue (Romanta, Pitestean and Geta). Early ripening varieties, such as Geta, Dani, Iulia, Romaner, Carpatin, Tita, Pitstean and Alina, come to the market earlier when no other fruits are available and fetch very good prices (Madalina et al. 2015). Since plum cultivars were first introduced in India they have shown a significant variation in their characters. Some cultivars, like Beauty, Frontier, Red Beaut, Tarrol, Au-Rosa and Krassivica, appear to bloom earlier, and cultivars such as Frontier, Au- Rosa and Grand Duke mature early and are heavy bearers. The Red Plum cultivar was also found better for fruit size and colour than other cultivars. These can be utilized in breeding programmes for the further improvement of plums because of their high potential for quality characteristics (Sundouri et al. 2017). Introduced plum varieties are suitable for cultivation in India because of high quality and yield potential, such as Mariposa, Au-cherry, Beauty, Monarch, Green Gauge, Methley, Frontier, Kanto-05, Santa Rosa, Red Beaut, Grand Duke and President (Kumar et al. 2016). Plum varieties Top Sugar and Valor bear significantly larger fruits, which might be important to customer's satisfaction (Suranyi 2019). Black Amber and Duarte are promising varieties concerning fruit biochemical characters and mature in mid- and late season respectively. Black Amber and Duarte cultivars cultivated under the mid-hill conditions of Himachal Pradesh. They are a great alternative to the Santa Rosa as a way to increase the harvesting season (Sharma et al. 2018). The breeding programme at Hohenheim University in Germany was started in 1980 and mainly focuses on three objectives: increase the gap in ripening time, high fruit quality, and sharka resistance. The breeding cycle can be reduced considerably by shortening the juvenility of the seedlings. However, in resistance breeding, the Stanley plum was used as a donor for sharka resistance (Hartmann 1994). By traditional breeding methods a hypersensitive variety, the Jojo, crossed with other plum varieties, can be a worthy germplasm source in the improvement of plum varieties resistant to sharka disease (Zurawicz et al. 2013). The Katinka and President cultivars are highly prone to PPV, whereas Valjevka and Vision are moderately susceptible and Jojo and Elena are resistant to PPV infection (Malinowsk et al. 2013) Stanley was considered the best variety with respect to production per hectare as compared to the Opal, Ontario, Malvazinka, Tuleu timpuriu, Althan's gage, Mirabelle de Nancy, Pacific, Bluefire and Anna Spath cultivars evaluated in Bulgaria (Ivanova et al. 2001). The Stana cultivar was evaluated in the Czech Republic and performed best with regard to highest mean yield per canopy tree volume (Blazek et al. 2018). To develop a new strategy for resistance breeding, like immunity or absolute resistance, cultivars make high yields and earlier ripening possible (Kegler et al. 1986). Hartmann (1998) utilized the hypersensitive reaction to PPV and hybrids achieved in the cultivar Ortenauer, accompanied by hypersensitive reaction, which showed field resistance at the University of Hohenheim. Now it is interesting to develop new varieties resistant to sharka, especially in prolonging the ripening period and achieving significant fruit size and high fruit quality so that we get many hypersensitive clones (Hartmann 2019). The plum breeding of the Fruit Growing Institute in Plovdiv started in 1987 with the objectives to develop cultivars tolerant to PPV from the Reine Claude group and breed cultivars with good pollinators of the Stanley, Cacanska lepotica, Pacific, President, Green Gage and Renklod Hramovih varieties used as basic parent cultivars (Zhivondov and Djouvinov 2002). The newly released cultivar Plovdivska renkloda is a superior donor parent to the Stanley in terms of ripening period and fruit weight. Its fruit colour also makes it more attractive for both producers and consumers (Zhivondov 2010). The key factor for today's breeding objectives in the plum sector is fruit-quality traits to meet market demands. In Romania, improvement in Tuleu gras, Grase romanesti and Vinete romanesti plum for fresh consumption or making other products, such as jam, marmalade and juice (Cociu et al. 2002). Black Glow and Black Sunrise are cultivars of Japanese plums that were characterized as brown to black in colour, with large fruits that ripen early. The leading American and European genotypes, along with Myrobalan plums, give rise to cold tolerance and high fertility features, so they are used in breeding programmes. Approximately 50 years ago, Romanian initiated plum breeding and mainly focused on old autochthonous cultivars, different ripening periods of cultivars and high-quality fruit (Butac et al. 2010). Plum cultivars like Valor, Bellamira, Topper, Jojo, Zhongli No. 3 and Summer Fantasia are promising because of their superior fruit quality, fruit size and sweetness (Figure 3.3).

3.7.2 Rootstocks

The influence of rootstock selection on fruit quality, yield and precocity are crucial in plum-growing areas that would differ in soil and climatic conditions (Achim et al. 2004; Botu et al. 2002). Over the last 50 years, much research has been concentrated on breeding and selecting new and improved rootstocks for plum. *Prunus salicina*, an interspecific hybrid, is resistant to cold temperatures and exhibits dwarf stature with large fruits (Zhang and Gu 2016). A few clonal rootstocks have been developed for plum which can be grown on the seedlings of Myrobalan (*P. cerasifera*). PPV a viral disease in plums which had a substantial agronomic effect and resulted in huge economic losses. GF677 (almond x peach) and Myrobalan 29C now no longer display any signs or symptoms of PPV infection (Rubio et al. 2005).

Traditional breeding is a slow process for producing new plum rootstocks that must have many desirable characteristics with minimum undesirable traits. It is more important to run rootstock breeding on plum species that will show a potential source of resistance to biotic and abiotic stresses (Bouzari 2021). The rootstocks of *Prunus* species Mariana GF8–1, Myrobalan P-1254 and Myrobalan B and their use in plum scion results in a reduction in tree size and an increase in the production of fruit.

FIGURE 3.3 New promising cultivars of Japanese plum.

Similarly, Ishtara® Ferciana (Ishtara) and Jaspi® Fereley (Jaspi) showed good performance in fruit size, yield efficiency and marketable fruit yield (Grzyb and Sitarek 2007; Hartmann et al. 2007) on the Kirke plum cultivar (Pedersen 2009). Rootstocks such as GF 677 and Myrobalan 29C are highly vigorous and promote rapid tree growth, development and bearing activity (Sottile et al. 2007). However, Penta and

Tetra exhibited low vigour, and MrS 2/5 had medium result, whereas Myrobalan 29C and GF 677 were more vigorous rootstocks. Plum scion grafted on rootstocks like Jaspi, GF 655/2, Ishtara and St. Julien A gained lower vigorous growth in a closely spaced plum orchard compared to those that were grafted on *P. divaricata*. With respect to semi-dwarf rootstocks, GF 655/2 and St. Julien A would serve as closely planted plum in orchards and the accommodation of more plum trees per unit area that may be advisable for higher productivity (Grzyb et al. 2010).

Recently, researchers have focused on developing dwarf or semi-dwarf rootstocks that can be grown in a wide range of soil conditions and allow more planting intensity in orchards. Several promising rootstocks were identified that contribute to size control and are appropriate for high-density planting, namely, Pixy, Maridon and Ferlenain. Pixy (*P. insititia*), shows high yield efficiency and resistance against silver leaf disease, a fungal disease caused by the Chondrostereum purpureum pathogen (Webster and Wertheim 1993). Krymsk 1' (VVA-1) is most important as a dwarfing and precocity rootstock for plums. On the other hand, *Prunus spinosa* (blackthorn) is an ideal rootstock because of highly resistant winter hardiness, improved drought tolerance and a high degree of compatibility with other plum cultivars (Maas et al. 2014). Rapid diffusion of the clonal Myrobalan 29C is the most popular rootstock. Almost 80 per cent of this rootstock is used in new plum scion in terms of economic growth and higher production that enhanced their rapid multiplication in the last few years (Mezzetti and Capocasa 2002).

The rootstock breeding program at SCDP Velcea produced seven plum rootstocks: Otegani 8, Miroval, Otegani 11, Rival, Corval, Oltval and Pinval have vigour from low to semi-dwarf and can be used in almost all plum-growing areas in Romania. Selections of St. Julien (*P. insititia*) rootstocks, based on excellent compatibility with different plum cultivars, have become a popular rootstock throughout Europe. The yield potential of Victoria, Czar and Oullins Golden Gage on rootstocks of Pixy and St. Julien A showed a significant difference in fruit size and soluble solids than the earlier rootstock (Webster 1980). In the previous decades, clonal rootstock propagation of plum has been used over seedling rootstocks due to problems such as lack of uniformity, compatibility problems, etc. Rival, a new clonal rootstock, provides medium vigour and good compatibility to other plum cultivars. Additionally, it is also tolerant to major pests and disease (Botu et al. 2007b). Problems like biotic stress severely affect the growth and development of rootstocks and/or scions, bringing down fruit production and fruit quality. To overcome these stresses, choose suitable tolerant rootstocks/varieties or by the use of pesticides, etc. Nevertheless, there are very few tolerant varieties that are known to overcome abiotic stresses and the detrimental effects of pesticides on human as well as soil health. In this condition, rootstock is the simplest alternative method used for multiplying fruit trees from 2,000 years ago (Nimbolkar et al. 2016). Myrobalan rootstocks are used for different plum cultivars because of its root-knot nematode (RKN) resistance, iron chlorosis, salinity and phytoplasma (Moreno 2004). The rootstocks Cadaman, GF-677, Rootpac 20 and Rootpac R, if grafted on Catherina peach, increase the proline content in both plant roots and leaves and also show that the sorbitol content in leaves increases under drought conditions (Gainza et al. 2015). GF-677 is a tolerant genotype that presented a higher expression of the P5SC gene which is involved in the accumulation of proline content (Jimenez et al. 2013). The reduction in physiological activities, such as water potential differences, photosynthetic production and transpiration from leaves increased the expression of antioxidant enzyme activity of interspecific *Prunus*. Values of plants watered after drought stress become normal, but levels of ascorbate, glutathione and H_2O_2 increase (Sofo et al. 2005). Excessive resistance to RKN has been determined in Mariana and Myrobalan rootstocks (Esmenjaud et al. 1994). This might be due to Ma genes that show resistance in a few Myrobalan clones thus suppressing reproduction of RKNs. Julior, a cross between (*P. insititia* L. x *P. domestica* L.), showed long-term hypoxia tolerance, whereas the Mariana 2624 cultivar was well suited to heavy and too wet soils (Klumb et al. 2017). The major progress made on rootstocks in recent decades is illustrated in Table 3.4.

TABLE 3.4
Important Plum Rootstocks with Their Genetic Origin

Rootstock	Genetic origin	Year of releasing	Origin	Uses
Mariana GF 8/1	Hybrid variety of *P. cerasifera* × *P. munsoniana*	1965	INRA, France	Develops a shallow root system and is resistant to cold
Myrocal®	Clonal selection of *P. cerasifera*	–	INRA, France	–
Mariana 2624	*P. cerasifera* x *P. munsoniana*	1940	Davis, California	Rootstock for almond, apricot, plum, prune tolerates wet, heavy soils
Julior®-Ferdor*	Clonal selection of *P. insititia*	1988	INRA, France	–
Ishtara®-Ferciana	(*P. cerasifera* x *P. salicina*) x (*P. domestica* x *P. persica*)	1986	INRA, France	Dwarfing rootstock for almond, apricot, peach, plum and prune
Myrobalan B.	Clonal selection of *P. cerasifera*	1936	East Malling, UK	Compatible with most cultivars and tolerates a wide range of soil types and climatic conditions
Citation®	*P. salicina* (Red Beaut plum) x *P. persica* (peach)	1983	USA	Rootstock for apricot and plum
Sharpe	Chickasaw plum (*P. angustifolia* [Marsh.]) with an unknown plum species	1974	Florida	Rootstock for plum, peach and nectarine
Otesani 8	Clonal selection into a *P. domestica* local population	1990	S.C.D.P. Valcea	Rootstock for plum and apricot cultivars
Otesani 11	Clonal selection into a *P. insititia* local population	1987	S.C.D.P. Valcea	Rootstock for plum cultivars
Miroval	Clonal selection into a *P. ceracifera* local population	1999	S.C.D.P. Valcea	Rootstock for plum and peach cultivars

TABLE 3.4 *Continued*

Rootstock	Genetic origin	Year of releasing	Origin	Uses
Rival	St. Julien H2 x *P. insititia*	2003	S.C.D.P. Valcea	Rootstock for plum and apricot cultivars
Pinval	*P. insititia* Scoldus x *P. intitia* Pixy	2005	S.C.D.P. Valcea	Rootstock for plum and apricot cultivars
Oltval	*P. besseyi* x *P. americana* x Myrobalan 2V	2005	S.C.D.P. Valcea	Rootstock for plum cultivars
Corval	Clonal selection into a *P. ceracifera* local population	2005	S.C.D.P. Valcea	Rootstock for plum and peach cultivars
Mirodad 1	Myrobalan dwarf x Adaptabil	2017	RIFG, Pitesti	Rootstock for plum cultivars

Source: Achin et al. 2010; Butac et al. 2019

INRA = Institut National de la Recherche Agronomique, France; SCDP = Statiunea de Cercetare Dezvoltare pentru Pomicultura, Valcea; RIFG = Research Institute for Fruit Growing, Pitesri

3.8 BIOTECHNOLOGICAL APPROACHES

3.8.1 In Vitro Micropropagation

Micropropagation refers to in vitro propagation of vegetative plants in a synthetic medium under aseptic conditions (Ogita 2015). Over the last ten decades, potential approaches, such as hybridization, selection, progeny evaluation and in vitro propagation of new fruit cultivars, have been developed (Okie and Hancock 2008). The selection of genotypes for commercial utilization can take several years by traditional propagation methods. Micropropagation technologies appear to be the best solution to these technical difficulties. Production of plants by this technique is true to type and large in number in a short amount of time at a cheaper price (Garcia-Gonzales et al. 2010). Plant meristem culture is a unique technique to develop virus-free plants, thus avoiding great economic losses (Gella and Errea 1998). Breeding programmes of plum vary from species to species due to their lengthy juvenile period, such as in the case of Japanese plums that can produce first flowers only after two to three years. This is one of the main drawbacks of the early evaluation, selection and release of superior hybrids (Carrasco et al. 2013). The development of Japanese plum cv. América by micropropagation had significant plant survival rates at IAA concentration (1 mg L^{-1}) for rooting during the acclimatization period (Bandeira et al. 2012). Gulf Ruby a plum cultivar was propagated though micropropagation protocol from excised mature nodes after the explants were cultivated in medium containing plant growth regulators (PGRs) when cultivated in growing medium containing PGRs for cell division and shoot elongation. Thereafter the plant exhibited a very high survival rate when subjected to greenhouse conditions (Zou 2010).

3.8.2 Genetic Transformation

Transgenesis is a genetic engineering method that allows a change in the recipient plant through introducing foreign DNA sequences isolated from a host organism (Ilardi et al. 2015). PPV causes symptoms like chlorotic rings, leaf distortion, loss of green colour and can cause severe damage on fruit quality (Usenik et al. 2015); premature leaf abscission can lead to an adverse effect on yield and market value of fruits (Sochor et al. 2012).

Symptoms of PPV were first reported in Bulgaria between 1915 and 1918, even though some reported that symptoms were observed in Macedonia as early as 1910. The viral nature of the malady became apparent in 1932 once Atanosoff named it ‘Sarka po slivite’, meaning ‘Pox of Plum’. It is currently the most important viral disease in apricots, which were first affected in Bulgaria in 1933, but it did not affect peaches until the early 1960s in Hungary (Levy et al. 2000). The PPV-resistant Honeysweet variety was developed through genetic engineering in the United States (Scorza et al. 2016). The coat protein gene-mediated resistance was due to interference of multifunctional proteins with virus infection that could affect the viral replica. The PPV-CP gene was cloned after isolation by using Agrobacterium-mediated plum transformation (Ravelonandro et al. 1992; Scorza et al. 1994). Transgenic plum lines with coat proteins have been tested for resistance for two years in controlled greenhouse conditions. Out of these transgenic plum lines only one clone (C5) showed high resistance under greenhouse conditions, but it did not express the PPV coat protein (Ravelonandro et al. 1997; Scorza et al. 2001). C5 (experimental name), or Honeysweet (commercial name), showed undetectable transgenic mRNA and indicated that this resistance was not the expressing of the CP gene. The hygromycin phosphotransferase gene (hpt) encoding for hygromycin resistance was used as a selectable marker in European plums (Fig 3.4). The sensitivity of plum tissue to hygromycin was a more efficient chemical for producing transgenic plums.

The genetic regulation of the Flowering Locus T1 (FT1) gene in European plums was successfully introduced after being isolated from *Populus trichocarpa*. Now it is a useful tool to change the dynamics of temperate plants that can be adapted to new growing areas. The expression of this gene recommended that it plays a crucial function in regulating flower induction and plant development (Srinivasan et al. 2012).

3.8.3 Molecular Breeding

Development of a new variety of plum through traditional breeding methods has some drawbacks that can be led by the heterozygous nature of fruit crops, long juvenile periods, and genetic control of self-incompatibility (Rai and Shekhawat 2014). Another disadvantage is that large land area is required for the evaluation of seedling populations and several rounds of introgressive backcrossing to achieve new offspring. The average generation time of European plums is about three to seven years. Traditional plum breeding takes almost 15–20 years to release a new crop variety for first flowering (Limera et al. 2017). The advantage of marker-assisted selection (MAS) over conventional breeding methods is that it allows the pre-selection of one or more genotypes rapidly before they can be evaluated in the field (Knapp 1998).

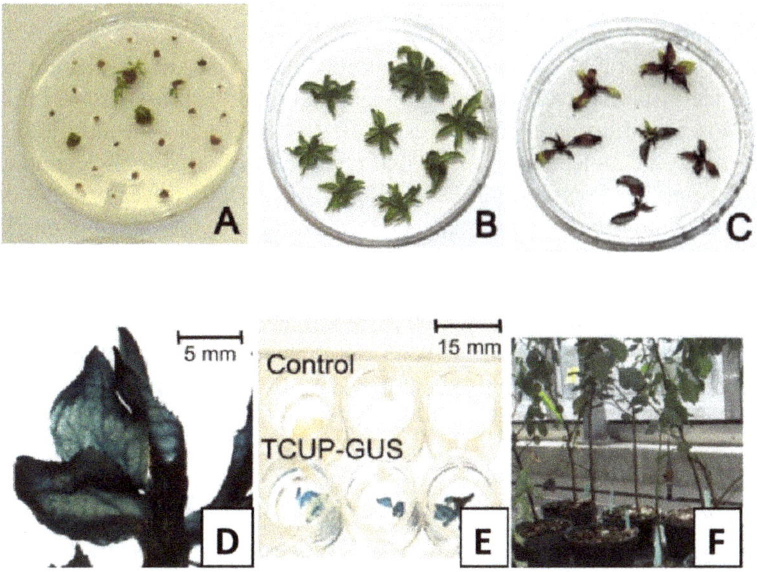

FIGURE 3.4 European plum transformation using hygromycin as the selection agent.

(A) Putatively transformed shoots form after three weeks on hygromycin selection medium.
(B) Transformed shoots grew healthily and vigorously on a selection medium containing 5 mg/L hygromycin.
(C) Untransformed shoots did not survive on hygromycin medium.
(D) GUS expression using histochemical analysis in a shoot regenerated with hygromycin selection.
(E) tCUP promoter-directed expression of the GUS reporter gene.
(F) Development of transgenic plum plants in the greenhouse.

The significance of this technique is to save time and space and to be focused on genes of interest in the development of new cultivars. DNA-based markers contain the specific advantage of microsatellites (or SSRs) over the others and are valuable tools for characterizing the Japanese plum (Ahmad et al. 2004; Carrasco et al. 2012; Klabunde et al. 2014), but the identification of fruit-quality traits has been limited (Salazar et al. 2017) since they directly reflect the genetic make-up of an organism. Single nucleotide polymorphism (SNPs) are essential markers for research in plum cultivars and genetic diversity analysis due to its high efficiency and low level of genetic variability across the genomes as compared to SSR-based markers. It has the capacity to impart genetic variation for feature breeders' programmes that can be used to improve Chinese plum varieties (Wei et al. 2021). Restricted Fragment Length Polymorphism (RFLP) was the first DNA-based genetic marker for the construction of genetic linkage maps. In plum cultivars, it is useful to efficiently determine different inherited traits within the species (Mustapha et al. 2015). AFLP may be a nice approach to decide genetic relatedness amongst plum cultivars and help the breeders to choose the genetic diverse cultivars with comparable fruit traits for further breeding programmes, whereas SSR markers are relatively low cost and are accessible for genotyping cultivars (Decroocq et al. 2004; Xuan et al. 2011). There is a need to focus on *P. salicina* and incorporate MAS so that it will enhance the performance of plum breeding towards yield and quality. *Prunus domestica*, commonly known as European plum, was a less-studied species because of the lack of accessibility of molecular markers that could be exploited for selection criteria. Random Amplified Polymorphic DNA (RAPD) and DNA finger printing are new strategies to identify and differentiate between distinctive cultivars of plum (Yu et al. 2013). Ilgin et al. (2009) characterize 14 cultivars of plum grown worldwide by using AFLP (amplified fragment length polymorphism) markers which classified the genotypes into the seven major clusters by UPGMA (unweighted pair group method with arithmetic mean) analysis. Among them, the President cultivar exhibited the most unique group, whereas maximum resemblance was determined in the Globe Sun and October Sun cultivars.

3.9 CONCLUSION

In the future, plum breeding programmes will focus on PPV tolerance, vigour, hardiness of trees and acceptable fruit size. Plum fruit size is an economically quantitative inherited trait that determines quality parameters, such as yield and consumer acceptance. These may be applied in breeding programmes for the further development of plums because of their high potential of quality characteristics. It is important to run rootstock breeding on plum species to increase biotic and abiotic stress tolerance. Recently, researchers have focused on developing dwarf and semi-dwarf plum rootstocks that can be grown in a wide range of soil conditions and allow more planting intensity in orchards. Major progress has been made on rootstocks by plum breeders worldwide, particularly for resistance to water logging, vigour control, and soil-borne disease, and have shown excellent compatibility with scion. Molecular approaches are a new strategy to identify and differentiate distinctive cultivars of plum by developed markers to help breeders in selecting diverse genotypes with comparable fruit-quality traits for further breeding programmes in a shorter period of time than traditional methods.

3.10 REFERENCES

Achim, G., Botu, I., Botu, M., Godeanu, I., Baciu, A., and Cosmulescu, S. 2004. 'Miroval'—A New Clonal Rootstock for European Type Plum Cultivars. *Acta Horticulturae* 658: 89–91.

Achim, G., Botu, L., Botu, M., Preda, S., and Baciu, A. 2010. Plum Rootstocks for Intensive Plum Culture. *Acta Horticulturae* 874: 299–303.

Ahmad, D., Potter, D., and Southwick, S.M. 2004. Identification and Characterization of Plum and Pluot Cultivars by Microsatellites Markers. *Journal of Horticultural Science and Biotechnology* 79: 164–169.

Bailey, C.H., and Hough, L.F. 1975. Apricots. In *Advances in Fruit Breeding*, edited by J. Janick and J.N. Moore, 367–383. Purdue University Press.

Bal, J.S. 2004. Genetic Resources of Plum Under Subtropical Conditions of Punjab, India. *Acta Horticulturae* 662: 147–150.

Bal, J.S., Kaur, P., and Kaur, K. 2017. Improvement of subtropical plum through hybridization. *Acta Horticulturae*. DOI:10.17660/ActaHortic.

Bandeira, J.D.M., Thurow, L.B., Braga, E.J.B., Peters, J.A., and Bianchi, V.J. 2012. Rooting and acclimatization of the Japanese plum tree, cv. América. *Revista Brasileira de. Fruticultura* 34 (2): 597–603.

Beckman, T.G., and Lang, G.A. 2003. Rootstock Breeding for Stone Fruits. *Acta Horticulturae* 622: 531–551.

Bellini, E.K., and Nencetti, V. 2002. "Dofi-Sandra": A New Early Black Japanese Plum. *Acta Horticulturae* 577: 223–224.

Bhutani, V.P., and Joshi, V.K. 1995. Plum. In *Handbook of Fruit Science and Technology: Production, Composition, Storage and Processing*, edited by D.K. Salunkhe and S.S. Kadam, 206. Marcel Dekker.

Blazek, J. 2007. A Survey of the Genetic Resources Used in Plum Breeding. *Acta Horticulturae* 734: 31–46.

Blazek, J., and Vavra, R. 2007. Fruit Quality in Some Genotypes of Plum Varieties with Tolerance to PPV. *Acta Horticulturae* 734: 173–182.

Blazek, J., Vavra, R., and Pistekova, I. 2004. Orchard Performance of New Plum Cultivars on Two Rootstocks in a Trial at Holovousy in 1998–2003. *Horticultural Science* 31: 37–43.

Blazek, J., Zeleny, L., and Krelinova, J. 2018. Productivity and Tree Performance of New Plum Cultivars from the Czech Republic. *Horticultural Science* 45 (2): 64–68.

Botu, I., Achim, G., Botu, M., Godeanu, L., and Baciu, A. 2002. The Evaluation and Classification of Growth Vieor of the Plum Cultivars Grafted on Various Rootstocks. *Acta Horticulturae* 517: 299–306.

Botu, I., Preda, S., Botu, M., and Achim, G. 2007a. New Valuable Clonal Selections for Plum in Romania. *Acta Horticulturae* 734: 333–336.

Botu, I., Preda, S., Turcu, E., Achim, G., and Botu, M. 2007b. Rival-a New Rootstock for Plum. *Acta Horticulturae* 732: 253–256.

Bouzari, N., Jolfaee, H.K., Ahmadzadeh, S., Garmaroodi, H.S., and Hosseini, S.S. 2021. Exploitation of Plum Genetic Diversity to Identify Soil-Borne Fungi Resistance Rootstocks. *International Journal of Fruit Science* 21 (1): 681–692.

Butac, M., Botu, M., Militaru, M., Mazilu, C., Dutu, I., and Nicolae, S. 2019. Plum Germplasm Resources and Breeding in Romania. *Proceedings of the Latvian Academy of Sciences* 720: 214–219.

Butac, M., Bozhkova, V., Zhivondov, A. et al. 2013. Overview of Plum Breeding in Europe. *Acta Horticulturae* 981: 91–98.

Butac, M., Zagrai, I., and Botu, M. 2010. Breeding of New Plum Cultivars in Romania. *Acta Horticulturae* 874: 51–58.

Carrasco, B., Diaz, C., Moya, M. et al. 2012. Genetic Characterization of Japanese Plum Cultivars (Prunus salicina) Using SSR and ISSR Molecular Markers. *Ciencia e Invetigacion Agraria* 39 (3): 533–543.

Carrasco, B., Meisel, L., Gebauer, M., Garcia-Gonzales, R., and Silva, H. 2013. Breeding in Peach, Cherry and Plum: From a Tissue Culture, Genetic, Transcriptomic and Genomic Perspective. *Biological Research* 46: 219–230.

Cociu, V., Botu, I., Turcu, I., Mihai, B., and Godeanu, I. 2002. The Value and Perspective of New Plum Cultivars Obtained in Romania. *Acta Horticulturae*. DOI:10.17600/577.22.

Crane, M.B., and Lawrence, W.J.C. 1952. *The Genetics of Garden Plants*, 4th ed., 30. Macmillan.

Crisosto, C.H., Garner, D., Crisosto, G.M., and. Bowerman, E. 2004. Increasing 'Blackamber' Plum (Prunus salicina Lindley) Consumer Acceptance. *Postharvest Biology and Technology* 34: 237–244.

Cullinan, F.P. 1937. Improvement of stone fruits. *Year Book* 667–672.

Das, B., Ahmed, N., and Singh, P. 2011. Prunus Diversity- Early and Present Development: A Review. *International Journal of Biodiversity and Conservation* 3 (14): 721–734.

Decroocq, V., Hagen, L.S., Fave, M.G. et al. 2004. Microsatellite Markers in the Hexaploid Prunus Domestica Species and Parentage Lineage of Three European Plum Cultivars Using Nuclear and Chloroplast Simple-Sequence Repeats. *Molecular Breeding* 13: 135–142.

Dejampour, J., Garigurian, V., and Ali Asgharzadeh, N. 2007. Evaluation of Some Morphological Characteristics and Clonal Propagation of Some Interspecific Hybrids in Prunus. *Iranian Journal of Horticultural Science and Technology* 8 (1): 43–54.

Duval, H. 1999. Prunus japonaises: Un défi à relever. *L'Arboriculture fruitière* 524: 35–41.

Ermenyi, M. 1975–77. Forrastanulmany a regeszeti korokbol szarmazo csonthejas gyümolcsleletekrol Kozep-Europaban Magyar Mezogazd. *Muzeum Kozl* 135–165.

Esmenjaud, D., Minot, J., Voisin, R., Pinochet, J., Simard, M., and Salesses, G. 1994. Inter- and Intraspecific Variability in Myrobalan Plum, Peach, and Peach-Almond Rootstock Using 22 Root-Knot Nematode Populations. *Journal of the American Society for Horticultural Science* 119: 94–100.

FAOSTAT. 2018. www.faostat.fao.org.

Faust, M., and Suranyi, D. 1999. Origin and Dissemination of Plum. *Horticultural Reviews* 23: 179–123.

Folta, K.M., and Gardiner, S.E. 2009. Genetics and Genomics of Rosaceae. In *Plant Genetics and Genomics: Crops and Models*, 6. Springer Science+Business Media, LLC. DOI:10.1007/978-0-387-77491-61.

Gainza, F., Opazo, I, Guajardo, V. et al. 2015. Rootstock Breeding in Prunus Species: Ongoing Efforts and New Challenges. *Chilean Journal of Agriculture Research* 75 (1). doi:10.4067/S0718-58392015000300002.

Garcia-Gonzales, R., Quiroz, K., Carrasco., B., and Caligari, P. 2010. Plant Tissue Culture: Current Status Opportunities and Challenges. *Ciencia e Invetigacion Agraria* 37: 5–30.

Gella, R., and Errea, P. 1998. Application of in Vitro Therapy for Ilarvirus Elimination in Three Prunus Species. *Journal of Phytopathology* 146: 445–449.

Glowacka, A., Sitarek, M., Rozpara, E., and Podwyszynska, M.P.2021. Characteristics and Ploidy Levels of Japanese Plum (Prunus salicina Lindl.). *Cultivars Preserved in Poland Plants* 10: 884. https://doi.org/10.3390/ plants10050884.

Goulet B.E., Roda, F., and Hopk, R. 2017. Hybridization in Plants: Old Ideas, New Techniques. *Plant Physiology* 173 (1): 65–78.

Grzyb, Z.S., and Sitarek, M. 2007. Evaluation of 'Jaspi' and 'Ishtara' Plum Rootstocks in Polish Climatical Conditions. *Acta Horticulturae* 734: 397–400.

Grzyb, Z.S., Sitarek, M., and Rozpara, E. 2010. Evaluation of Vigorous and Dwarf Plum Rootstocks in the High Density Orchard in Central Poland. *Acta Horticulturae* 874: 351–356.

Hartmann, W. 1994. Plum Breeding at Hohenheim. *Acta Horticulturae* 359: 55–62.

Hartmann, W. 1998. Hypersensitivity-a Possibility for Breeding Sharka Resistant Plum Hybrids. *Acta Horticulturae* 478: 429–431.

Hartmann, W. 2019. Sharka-Resistant Plum Hybrids and Cultivars from the Plum Breeding Programme at Hohenheim. *Proceedings of the Latvian Academy of Sciences* 3 (720): 226–231.

Hartmann, W., Beuschlein, H.D., Kosina, J., Ogasanovic, D., and Paszko, D. 2007. Rootstocks in Plum Growing-Results of an International Rootstock Trial. *Acta Horticulturae* 734: 141–148.

Hartmann, W., and Neumuller, M. 2006. Breeding for Resistance: Breeding for Plum Pox Virus Resistant Plums (Prunus domestica L.) in Germany. *EPPO Bullet* 36: 332–336.

Hartmann, W., and Neumuller, M. 2013. The Next Generation of European Plum Cultivars Resistant to Plum Pox Virus. *Acta Horticulturae* 985: 149–154.

Hurter, N., and Van Tonder, M.J. 1963. Colchicine Induction of Autotetraploidy in the Japanese Plum (Prunus Salicina). *African Journal of Agricultural Science* 6: 403–410.

Ilardi, V., and Tavazza, M. Biotechnological Strategies and Tools for Plum Pox Virus Resistance: Trans-, Intra-, Cis-Genesis, and Beyond. *Frontier Plant Science* 8. https://doi.org/10.3389/fpls.2015.00379

Ilgin, M., Kafkas, S., and Ercisli, S. 2009. Molecular Characterization of Plum Cultivars by AFLP Markers. *Biotechnology and Biotechnological Equipment* 23 (2): 1189–1193. DOI:10.1080/13102818.2009.10817636.

Ivanova, P., Vitanova, I., Dimkova, S., and Marinova, N. 2001. *Some Biological Characteristics of Introduced Plum Cultivars*. Proceedings form 7th IS on Plum and Prune Genetics, 235–238.

Jacob, H.B. 2006. *Plum Breeding Worldwide*, 15. Symposium on Plum of Serbia, Book of Abstracts.

Jacob, H.B. 2007. Ripening Time, Quality and Resistance Donors of Genotypes of Prunus Domestica and Their Inheritance Pattern in Practical Plum Breeding. *Acta Horticulturae* 734: 77–82.

Jimenez, S., Dridi, J., Gutierrez, D. et al. 2013. Physiological, Biochemical and Molecular Responses in Four Prunus Rootstocks Submitted to Drought Stress. *Tree Physiology* 33: 1061–1075.

Kegler, H., Fuchs, E., Gruntzig, M., and Verderevskaja, T.D. 1986. Different Types of Resistance to Plum Pox Virus. *Acta Horticulturae* 193: 201–206.

Khush, G.S. 2002. Molecular Genetic-Plant Breeder's Perspective. In *Molecular Techniques in Crop Improvement*, edited by S.M. Jain, D.S. Brar, and B.S. Ahloowalia. Kluwer Academic Publisher.

Klabunde, M., Dalbo, M.A., and Nodari, R. 2014. DNA Fingerprinting of Japanese Plum (Prunus salicina) Cultivars Based on Microsatellite Markers. *Crop Breeding and Applied Biotechnology* 14 (3): 139–145.

Klumb, E.K., Rickes, L.N., Braga, E.J.B., and Bianchi, V.J. 2017. Evaluation of Gas Exchanges in Different Prunus spp. Rootstocks Under Drought and Flooding Stress. *Revista Brasileira de Fruticultura* 39 (4): e-899. DOI:10.1590/0100-29452017899.

Knapp, S.J. 1998. Marker-Assisted Selection as a Strategy for Increasing the Probability of Selecting Superior Genotypes. *Crop Science* 38 (5): 1164–1174.

Kumar, D., Lal, S., and Ahmed, N. 2016. Genetic Diversity Among Plum Genotypes in Northwest Himalayan Region of India. *Indian Journal of Agricultural Sciences* 86 (5): 666–672.

Kumar, N. 2006. *Breeding of Horticultural Crops Principles and Practices*, 125. New Publishing House, Pitam Pura.

Layne, R.E.C. 1983. Hybridization. In *Methods in Fruit Breeding*, edited by J.N. Moore and J. Janick, 48–73. Purdue University Press.

Levy, L., Damsteegt, V., Scorza, R., and Kolber, M. 2000. Plum Pox Potyvirus Disease of Stone Fruits. *APSnet Features*. Doi:10.1094/APSnetFeature-2000-0300

Li, P., Wu, W., Chen, F., and Liu, X. 2015. Prunus Salicina 'Crown' a Yellow Fruited Chinese Plum. *HortScience* 50 (12): 1822–1824.

Limera, C., Sabbadini, S., Sweet, J.B., and Mezzetti, B. 2017. New Biotechnological Tools for the Genetic Improvement of Major Woody Fruit Species. *Frontiers in Plant Science* 8: 1418. doi: 10.3389/fpls.2017.01418

Lopez-Caamal, A., and Tovar-Sanchez, E. 2014. Genetic, Morphological, and Chemical Patterns of Plant Hybridization. *Revista Chilena de Historia Natural* 87 (16): 2–14. www.revchilhistnat.com/content/87/1/16.

Lugli, S., Correale, R., Grandi, J.M., and Sansavini, S. 2010. Breeding High-Quality Plums at Bologna University's CMVF-DCA. *Acta Horticulturae* 874: 69–76.

Maas, F.M., Balkhoven, J.M.T., Heijerman-Peppelman, G., and Van der Steeg, P.A.H. 2011. Krymsk 1 (VVA-1), a Dwarfing Rootstock Suitable for High Density Plum Orchards in the Netherlands. *Acta Horticulturae* 903: 547–554.

Maas, F.M., Balkhoven-Baart, J., and Van der Steeg, P.A.H. 2014. Selection of Prunus Spinosa as a Dwarfing Rootstock for High Density Plum Orchards. *Acta Horticulturae* 507–516.

Madalina, B., Madalina, M., Catita, P., and Mihaela, S. 2015. Evaluation of Some New Plum Cultivars for Fresh Consumption in Correlation with Consumer Preferences. *Fruit Growing Research* 31 (1): 38–43.

Malinowsk, T., Rozpara, E., and Grzyb, Z.S. 2013. Evaluation of the Susceptibility of Several Plum (Prunus domestica L.) Cultivars to Plum Pox Virus (PPV) Infection in the Field. Sharka-Like Symptoms Observed on 'Jojo' Fruit Are Not Related to PPV. *Journal of Horticultural Research* 21 (1): 61–65.

Matsumoto, K. 1977. *The Mysterious Japanese Plum*. Woodbridge Press.

Mezzetti, B., and Capocasa, F. 2002. I portinnesti del susino. *L'Informatore Agrario* Suppl. 51: 31–35.

Milosevic, T., and Milosevic, N. 2018. Plum (Prunus spp.) Breeding. In *Advances in Plant Breeding Strategies: Fruits*, edited by J.M. Al-Khayri et al., 162–215. Springer Verlag International Publishing AG, Springer Nature. www.springer.com/gp/book/9783319919430.

Moreno, M.A. 2004. Breeding and Selection of Prunus Rootstocks at the Aula Dei Experimental Station, Zaragoza, Spain. *Acta Horticulturae* 658 (658): 519–528.

Mustapha, S.B., Tamarzizt, H.B., Baraket, G., Abdallah, D., and Hannachi, A.S. 2015. Genetic Diversity and Differentiation in Prunus Species (Rosaceae) Using Chloroplast and Mitochondrial DNA CAPS Markers. *Genetics and Molecular Research* 14 (2): 4177–4188.

Neumuller, M. 2011. Fundamental and Applied Aspects of Plum (Prunus Domestica) Breeding. *Fruit, Vegetable and Cereal Science and Biotechnology* 139–156. www.globalsciencebooks.info.

NHB. 2018. *Horticultural Statistics at a Glance*. Horticulture Statistics Division Department of Agriculture, Cooperation & Farmers Welfare, Ministry of Agriculture & Farmers' Welfare.

Nimbolkar, P.K., Shiva, B., and Rai, A.K. 2016. Rootstock Breeding for Abiotic Stress Tolerance in Fruit Crops. *International Journal of Agriculture, Environment and Biotechnology* 9 (3): 375–380. DOI:10.5958/2230-732X.2016.00049.8.

Ogita, S. 2015. Plant Cell, Tissue and Organ Culture: The Most Flexible Foundations for Plant Metabolic Engineering Applications. *Natural Product Communications* 10 (5): 815–820.

Okie, W.R. 2006. Introgression of Prunus Species in Plum. *New York Fruit Quarterly* 14 (1): 9–37.

Okie, W.R., and Hancock, J.F. 2008. Plums. In *Temperate Fruit Crop Breeding*, edited by J.F. Hancock, 337–357. Springer.

Okie, W.R., and Weinberger, J.H. 1996. Plums. In *Fruit Breeding: Tree and Tropical Fruits*, edited by J. Jules and J.N. Moor, 559–607. John Willey and Sons, Inc.

Pareek, O.P., and Sharma, S. 2017. *Systemic Pomology*, 382–387. Scientific Publisher.

Pedersen, B.H. 2009. Improving Crop Load on the Plum Cultivar 'Kirke' by Utilization of New Rootstocks. *International Journal of Fruit Science* 9 (2): 136–143. DOI:10.1080/15538360902991352.

Pickering, C. 1879. *Chronological History of Plants: Man's Record of His Own Existence Illustrated Through Their Names Uses and Companionship.*

Rai, M.K., and Shekhawat, N.S. 2014. Recent Advances in Genetic Engineering for Improvement of Fruit Crops. *Plant Cell Tissue Organ Culture* 116: 1–15. DOI:10.1007/s11240-013-0389-9.

Ramsey, J., and Schemske, D.W. 1998. Pathways, Mechanisms, and Rates of Polyploid Formation in Flowering Plants. *Annual Review of Ecology, Evolution and Systematics* 29: 467–501.

Ravelonandro, M., Monsion, M., Teycheney, P.Y. et al. 1992. Construction of a Chimeric Viral Gene Expressing Plum Pox Virus Coat Protein. *Gene* 120: 167–173. https://doi.org/10.1016/0378-1119(92)90090

Ravelonandro, M., Scorza, R., Bachelier., J.C. et al. 1997. Resistance of Transgenic Prunus Domestica to Plum Pox Virus Infection. *Plant Disease* 81: 1231–1235. https://doi.org/10.1094/ PDIS.1997.81.11.1231.

Roach, F.A. 1985. *Cultivated Fruits of Britain*. Blackwell.

Rubio, M., Martinez-Gomez, P., Pinochet, J., and Dicenta, F. 2005. Evaluation of Resistance to Sharka (Plum Pox Virus) of Several Prunus Rootstocks. *Plant Breeding* 124 (1): 67–70.

Salazar, J., Pachecov, I., and Petal, S. 2017. Genotyping by Sequencing for SNP-Based Linkage Analysis and Identification of QTLs Linked to Fruit Quality Traits in Japanese Plum (Prunus salicina Lindl.). *Front. Plant Science* 476 (8): 1–14.

Sant, M.B., and Stafford, B.D. 1985. Smiling Dan Phase III Research Design. In *Structure and Function at a Middle Woodland Settlement in the Lower Illinois Valley*, edited by B.D. Stafford, M.B. Sant, and S. Dan, 23. Kampsville Archeological Center.

Scorza, R., Callahan, A, Levy, L. et al. 2001. Post-Transcriptional Gene Silencing In Plum Pox Virus Resistant Transgenic European Plum Containing the Plum Pox Potyvirus Coat Protein Gene. *Transgenic Research* 10: 201–209. https://doi.org/10.1023/A:1016644823203.

Scorza, R., Ravelonandro, M., Callahan, A.M. et al. 1994. Transgenic Plums (Prunus domestica L.) Express the Plum Pox Virus Coat Protein Gene. *Plant Cell Reports* 14: 18–22. https://doi.org/10.1007/BF00233291.

Scorza, R., Ravelonandro, M., Ornon, V. et al. 2016. 'HoneySweet' (C5), the First Genetically Engineered Plum Pox Virus-Resistant Plum (Prunus domestica L.) Cultivar. *HortScience* 51: 601–603.

Sharma, D.D., Kumar, M., Singh, N., and Bunty, S. 2018. Plant Growth and Fruiting Behaviour of Newly Introduced Plum (Prunus salicina Lindl.) Cultivars Under Mid-Hills Conditions of Himachal Pradesh. *The Pharma Innovation Journal* 7 (4): 408–413.

Singha, S., and Baugher, T. 2008. *Concise Encyclopaedia of Temperate Tree Fruit*, 1st ed., 424. CRC Press.

Sochor, J., Babula, P., Adam, V. et al. 2012 Sharka: The Past, the Present and the Future. *Viruses* 4: 2853–2901.

Sofo, A., Tuzio, A.C., Dichio, B., and Xiloyannis, C. 2005. Influence of Water Deficit and Rewatering on the Components of the Ascorbate-Glutathione Cycle in Four Interspecific Prunus Hybrids. *Plant Science* 169: 403–412.

Sottile, F., Monte, M., and De Michele, A. 2007. Rootstocks Evaluation for European and Japanese Plums in Italy. *Acta Horticulturae* 734: 375–380.

Srinivasan, C., Dardick, C., Callahan, A., and Scorza, R. 2012. Plum (Prunus domestica) Trees Transformed with Poplar FT1 Result in Altered Architecture, Dormancy Requirement, and Continuous Flowering. *PLoS One* 7 (7): e40715. doi:10.1371/journal.pone.0040715

Stacewicz-Sapuntzakis, M. 2013. Dried Plums and Their Products: Composition and Health Effects-Critical Reviews. *Food Science and Nutrition* 53: 1277–1302.

Sundouri, A.S., Verma, S.K, Sharma M.K, Kumar, A., Nazir, N., and Khalil, A. 2017. Characterization of Newly Introduced Exotic Plum Cultivars for Character Association and Genetic Improvement. *A Current Journal of Applied Science and Technology* 24 (2): 1–10.

Suranyi, D. 2019. Evaluation of Introduced Plum Varieties Under Extreme Climatic Conditions. *International Journal of Horticultural Science* 25 (1–2): 7–10.

Topp, B.L., Russell, D.M., Neumuller, M., Dalbo, M.A.W., and Liu, W. 2012. Plum. In *Fruit Breeding*, edited by M.L. Badenes and D.H. Byrne, 571–621. Springer.

Tromp, J., Webster, A.D., and Wertheim, S.J. 2005. *Fundamentals of Temperate Zone Tree Fruit Production*, 301–304. Backhuys Publishers.

Urbanovich, O.Y., Kuzmitskaya, P.V., and Kilchevsky, A.V. 2017. Identification and Genetic Diversity of Plum Cultivars Grown in Belarus. *Russian Journal of Genetics* 53 (7): 775–784.

Usenik, V., Kastelec, D., Stampar, F., and Virscek Marn, M. 2015. Effect of Plum Pox Virus on Chemical Composition and Fruit Quality of Plum. *Journal of Agriculture and Food Chemistry* 63: 51–60. https://doi. org/10.1021/jf505330.

Waugh, F.A. 1903. *Systemic Pomology*, 56–67. Orange Judd Company.

Webster, A.D. 1980. Pixy a New Dwarfing Rootstock for Plums (Prunus domestica L). *Journal of Horticultural Science* 55 (4): 425–431.

Webster A.D., and Wetheim, S.J. 1993. Comparisons of Species and Hybrid Rootstocks for European Plum Cultivars. *Journal of Horticultural Sciences* 1 (68): 861–869.

Wei, X., Shen, F., Zhang, Q., Liu, N. et al. 2021. Genetic Diversity Analysis of Chinese Plum (Prunus salicina L.) Based on Whole-Genome Resequencing. *Tree Genetics and Genomes* 17: 26.

Werneck, H.L. 1961. *Die wurzel-und kerechten Stammformen der Pflaumen in Oberosterreich (Unter Zugrundelegung der romischen Obstweihefinde von Linz Donau)*, 7–129. Naturkundliches Jahrbuch der Stadt Linz.

Williams, R.R. 1970. The Effect of Supplementary Pollination in Yield. In *Towards Regulated Cropping*, edited by R.R. Williams and D. Wilson, 6–10. Grower Books.

Wu, W., Chen, F., Yeh, K., and Chen, J. 2019. ISSR Analysis of Genetic Diversity and Structure of Plum Varieties Cultivated in Southern China. *Biology* 8 (2). doi:10.3390/biology8010002.

Xuan, H., Ding, Y., and Detal, S. 2011. Microsatellite Markers SRR) as Tool to Assist in Identification of European Plum (Prunus domestica). *Acta Horticulturae* 918: 689–692.

Yoshikawa, F.T., Ranuning, D.W., and LaRue, J.H. 1989. Rootstocks. In *Peaches, Plums and Nectarines Growing and Handling for Fresh Market*, edited by J.H. LaRue and R.S. Johnson, 9–11. Coop. Ext. University of California.

Yu, M., Chu, J., Ma, R., Shen, Z., and Fang, J. 2013. A Novel Strategy for the Identification of 73 Prunus Domestica Cultivars Using Random Amplified Polymorphic DNA (RAPD) Markers. *African Journal of Agriculture Research* 8 (3): 243–250.

Zhang, Q., and Gu, D. 2016. Genetic Relationships Among 10 Prunus Rootstock Species from China Based on Simple Sequence Repeat Markers. *Journal of American Society for Horticultural Science* 141 (5): 520–526. doi: 10.21273/JASHS03827-16.

Zhebentyayeva, T., Shankar, V., Scorza, R. et al. 2019. Genetic Characterization of Worldwide Prunus Domestica (Plum) Germplasm Using Sequence-Based Genotyping. *Horticulture Research* 6: 12. DOI:10.1038/s41438-018-0090-6.

Zhivondov, A. 2010. Plovdivska Renkloda-a New Plum Cultivar. *Acta Horticulturae* 874: 305–310.

Zhivondov, A., and Djouvinov, V. 2002. Some Results of the Plum Breeding Programme at the Fruit Growing Research Institute in Plovdiv. *Acta Horticulturae* 577: 45–49.

Zhivondov, A., and Uzundzhalieva, K. 2012. Taxonomic Classification of Plum-Apricot Hybrids. *Acta Horticulturae* 966 (966): 211–217.

Zhukovsky, P.M. 1965. Main Gene Centres of Cultivated Plants and Their Wild Relatives Within the Territory of the USSR. *Euphytica* 14: 177–188.

Zou, Y.N. 2010. Micropropagation of Chinese plum (Prunus salicina Lindl) Using Mature Stem Segments. *Notulae Botanicae Horti Agrobtanici* 38: 214–221.

Zurawicz, V.E., Pruski, K., Szymajda, M., Lewandowski, M., Seliga, L., and Malinowski, T. 2013. Prunus Domestica 'Jojo'—Good Parent for Breeding of New Plum Cultivars Resistant to Plum Pox. *Journal of Agricultural Science* 5 (9): 1–5.

4 Advances in Plum Propagation and Nursery Management
Methods and Techniques

Rehana Javid, A. Raouf Malik, M.A. Kuchay, Shaziya Hassan and Rafiya Mushtaq
Division of Fruit Science, Faculty of Horticulture, Sher-e-Kashmir University of Agricultural Sciences and Technology of Kashmir, Shalimar, Srinagar, 190025.

CONTENTS

DOI: 10.1201/9781003205449-4

4.1 INTRODUCTION

Plum (*Prunus domestica* L.), classified as a stone fruit, is one of the most important temperate fruit crops. Its cultivation is not only restricted to temperate regions, but it is also grown in the subtropics. It belongs to the *Prunus* genus of the Rosaceae family, consisting of two main species. The most common species is the European plum (*Prunus domestica* L.), and the most cultivated is the Japanese plum (*Prunus salicina* L.). European plums are hexaploid with 2n = 48, while Japanese plums are diploid with 2n = 16. The Japanese plum is regarded as a low chill plum and grows well in subtropical regions of the world. Its nutritive and high antioxidant properties have expanded its export window. Plums are classified as stone fruits because their seeds are protected by a hard covering known as the endocarp. An overview of a plum tree in the full bearing stage is shown in Figure 4.1. Plum is a rich source of carbohydrates, proteins, fat, vitamins and minerals (Bofung et al. 2002; Folta and Gardiner 2009). This fruit can be used to produce a number of products and is a staple in many countries. It can be consumed fresh or converted into juice, dried or pruned food, oil, ingredients to produce dyes, jellies, jams, etc. (Sonwa et al. 2002). Today, this stone fruit is grown all over the world. The top five plum-producing regions include China, the United States, Brazil, India and the European Union.

Propagation is important for producing a large number of plants which have been multiplied rapidly consisting of all the desirable characteristics present in the mother plant. The main and sole purpose in plant propagation is the multiplication of the superior genotypes in such a way as to preserve all the desirable characteristics of the original mother progeny. The material used in propagation is directly obtained from the certified nurseries from where the disease-free propagating material is grown in foundation mother blocks. Thus, nursery management and plant propagation together lead to the continuity of a plant or a crop species for generations. Therefore, it is high time to give attention to the nursery, nurserymen, tools and techniques used in the nursery for the production of the best-quality planting material, which in turn will

FIGURE 4.1 Plum tree in full bearing.

reflect the growth of the plant propagation industry and consequently help in the prosperity and economic build-up of the nation.

4.2 PROPAGATION

Propagation is defined as the production of a number of plants from a mother plant, a tissue, or a cell within a specific period of time with the objective to produce exactly the same plants as that of the mother plant in larger numbers. Plant multiplication is considered to be the most important occupation of present-day civilization. The origin of the fruit crops dates back to the ancient era, when man evolved different types of plants through the process of selection (Srivastava 1966). Certain fruit plants came into existence through natural selection, others through the process of hybridization, e.g., strawberries and prunes. After some time, man began to cultivate the plants through the seeds obtained from their fruits, called seedling plants. With the advent of new techniques, vegetative methods of propagation, like the use of cuttings, budding and grafting, etc., came into existence, and now vegetative means of plant propagation are commercially followed in fruit crops, keeping in view their advantages, such as the production of homozygous plants with uniform and desirable characters and a shorter juvenile phase (Sharma 2002).

Propagation not only includes the use of seeds from a mother plant, but rather incorporates both sexual and asexual methods. As far as stone fruits are concerned, propagation by seed is restricted only to the raising of the seedling rootstocks and for the production of new varieties. Compared to seed propagation, vegetative propagation is performed commercially in plum and other stone fruits for the incorporation of desired characteristics or traits. The main reason behind the use of seeds for raising seedling rootstock only is the development of undesirable traits in the progeny due to cross-fertilization and the seedling plants' long juvenile phase. Vegetative propagation provides the advantages of developing the uniform plants with only the desirable traits. Also, asexual propagation methods (like the use of cuttings) are cheap as well as easy to perform. Plum is mainly propagated by seed and vegetatively by the use of cuttings and by grafting and budding techniques (Zou 2010). The main purpose of propagating plum rootstocks is to maintain uniformity, which is possible by the application of vegetative methods that directly influence the fruit production to a larger extent. Soft wood cuttings are commonly used as a vegetative means of propagation for the Myrobalan rootstock of plum, as it has proved to be the most efficient method (Anderson et al. 2016). The first step involved in the propagation of rootstocks by cuttings is fogging, followed by biologically active substances that help in callus formation and in the development of the roots at the basal end of the cutting.

Nowadays advances in the propagation of plants through the use of biotechnological approaches have resulted in great achievements in the propagation industry. Due to the numerous advantages of in vitro propagation over more conventional propagation methods, the propagation of fruit crops, particularly stone fruits, by in vitro methods have increased progressively, mainly for the multiplication of rootstocks in nurseries (Krska and Necas 2013). The propagation of fruit crops through clonal means is an essential method of propagating most of fruit crops globally (Davies et al. 1994). Before the mass multiplication of the planting material, the methods used for in vitro propagation are verified, which is usually done by the use of standard verified conventional techniques of propagation.

4.3 ADVANCES IN THE PROPAGATION AND NURSERY MANAGEMENT OF PLUM

Although the most common method of propagation in plum is through seeds, budding and grafting—these techniques have been used for quite a long time— various advances have been made with respect to the management of various pests and diseases, the elimination of viruses like plum pox virus (PPV), clonal propagation techniques, sanitary measures, in vitro propagation methods, etc. (Preece 2003). The main propagation methods, as well as more advanced ones, are discussed next.

4.3.1 Methods of Propagation in Plum

Plum is propagated both by asexual and sexual methods, but due to plum's heterozygous nature, vegetative propagation (budding and grafting) is the preferred scientific method (Singh and Singh 2006). The most common and commercial method of propagation in plum is budding, e.g., T-budding or shield budding, either on seedling rootstocks or on clonal rootstocks. Grafting with a superior variety is another way of propagating the plum during the winter months; however, due to the higher success percentage of budding over grafting, budding is more commonly practiced in plum. Sometimes hardwood or softwood cuttings are also used for plum propagation depending on the ease of rooting. Old, unproductive plum orchards could be rejuvenated by working with newer commercial scion varieties. Besides vegetative propagation, seed propagation is also important for producing resistant seedlings and rootstocks.

4.3.1.1 Propagation by Seed

Seed produces the ungrafted plant or seedling required for crop improvement and in the production of the rootstocks or seedling population (Mayer et al. 2014). The plum seedlings are usually not recommended for commercial cultivation due to their long gestation period as well as variability leading to late bearing of the trees. Since plum seeds contain a coating that prevents germination and some seeds have other inhibitory factors that hinder their germination (Mayer and Antunes 2010), seeds must first be soaked in water, leached or stratified. Seed germination is then completed in three phases. The first phase involves the imbibition of water by the cells. This is followed by a lag phase where little to no water is taken in; however, the seed is metabolically active during this phase. After that, there occurs the emergence of radical. The processing taking place during the lag phase includes the maturation of the mitochondria, the synthesis of proteins and the metabolism of reserve food materials (carbohydrates, proteins and lipids), all of which lead to the loosening of the cell wall for the elongation of the radical. Presently, the seeds of plums are not used for commercial propagation. Seeds are only used for rootstock propagation and not for scion multiplication, as the seedlings obtained are genetically variable.

4.3.1.2 Propagation by Hardwood Cuttings

These cuttings are obtained from the mature dormant wood of the mother plant from which the leaves have been abscised. Plum propagation by hardwood and semi-hardwood cuttings, as reflected in Figure 4.2, is particularly carried out in subtropical regions of the Japan. Cuttings used in plant propagation result in the development of plants with exactly the same genetic constitution as that of the mother plant from which the cuttings are taken (Pasqual et al. 2001). The time of preparation of the cuttings is from the end of December to the end of January. The first step involved in plum propagation using cuttings is the collection of the hardwood or semi-hardwood cuttings from healthy, disease-free mother plants of medium vigor, well exposed to sunlight with equally spaced internodes.

Cuttings are taken during the dormancy period, in the month of December, when they are at least two to three years of age. Before storing the cuttings, any leaves should be removed from the cutting. Cuttings that are 20 cm long and 5 mm wide are preferred. To facilitate the rooting in the cuttings, the peel at the basal end is removed and dipped into indole-3-butyric acid (IBA) for 30 seconds at 2000 mg/L (Petri and Scorza 2010). Rooting in the cuttings can also be initiated with bottom heat treatment, which involves treating the cuttings with IBA, covering them with moist peat up to the terminal/top portion and then keeping them in bundles and placing them upright in the sand, just below the sand tubes that are filled with hot, circulating water maintained at a temperature of 18–21°C. This operation is usually performed inside a protected structure to protect the cuttings from climatic vagaries. As soon as the rooting begins in the cuttings, they are transplanted into the main field before the buds begin to grow.

The cutting to be used for plum propagation must have at least four to five buds in the dormant phase. To check the process of transpiration, any leaves or thorns must be removed. One slanting cut is made below the node and another near the top bud which helps in deciding the position for the planting of the cutting. The cut made at the basal portion of the cutting is simply to increase the area of contact between the

FIGURE 4.2 Plum propagation through cuttings.

cutting and the rooting medium, which facilitates the initiation of a large number of roots. In the nursery, the cuttings are placed in the beds in a slanted position. Treating the cuttings with biologically active substances has a positive influence on the rooting compared to untreated cuttings. The growth rate and length of the roots increase with the application of the stimulants at the base of the cuttings (Necas and Krska 2013). Furthermore, the quality of the propagating medium and the date of excision play an important role in the rooting of the cuttings. Thus, bottom heat treatment and the use of stimulants have a positive effect on the rooting of the cuttings in stone fruits.

While performing a study on the propagation of stone fruit rootstocks, Necsa and Krsk (2013) evaluated different rootstocks, namely, Pumiselect (*P. pumila* L.), VVA-1 (*P. tomentosa* × *P. cerasifera*), AP-1 (*P. cerasifera* Ehrh. × *P. persica* L.), PS-1 (probably *P. persica* L. × *P. cerasifera* Ehrh.), etc. The plum rootstocks evaluated during the study are provided in Table 4.1. The (hardwood cutting) rootstock Pumiselect showed the best results, with a 63.5% rooting percentage. The rooting potential of softwood cuttings was also evaluated with the use of rootstocks like AP-1, VVA-1, Lesiberian (*P. persica*), Isthara, MRS, etc., and the results showed VVA-1 with the highest rooting percentage, 62.7%. During the study, MY-KL-A (*P. cerasifera* L.) was used as the control.

During the experiment, two commercial and one noncommercial stimulator, i.e., Racine and Rhizopan and MP, respectively, were used. The stimulators used for the rooting of rootstocks are shown in Figure 4.3. The stimulator MP consisted of different concentrations of growth regulators, i.e., indole-3-acetic acid (IAA), IBA, 1-naphtalene acetic acid (NAA), kinetin and some other stimulating substances. Furthermore, it was concluded that any of the stimulators can improve the efficiency of rooting in the rootstocks used, as they were not statistically different.

The process of rooting using the cuttings is complex and involves the relation between a number of factors. It largely depends on the daily mean temperature, Polyphenol oxidase activity, sprouting, water content, leaf fall, phenolics and

TABLE 4.1
Average Rooting of Hardwood Cuttings in Percent

	Without bottom heat				With bottom heat			
Rootstock	**Stimulator Racine**	**Stimulator MP**	**Stimulator Rhizopon AA**	**Control variants**	**Stimulator Racine**	**Stimulator MP**	**Stimulator Rhizopon AA**	**Control variants**
Pumiselect	62.5	65.9	63.8	31.1	39	61.6	29.6	25.6
GF 665	70.1	65.3	58.5	52.2	70.9	69.1	68.5	62.2
St. Julien	62	68.3	56	55.1	65.1	69.1	61.3	65.9
MY-KL-A	44.4	46.8	63	53.1	54.4	56.1	53.9	53.1
VVA-1	51.4	48.2	44	54.1	59.9	58.6	46.1	46.9
PS-1	28	33.2	52.5	46.3	29.9	36.2	42.5	36.3
Ishtara	18.6	23.9	14	9.1	20.2	23.1	18.3	19.5
AP-1	24.4	13.9	17.3	7.5	25.5	16.9	17.9	12.3

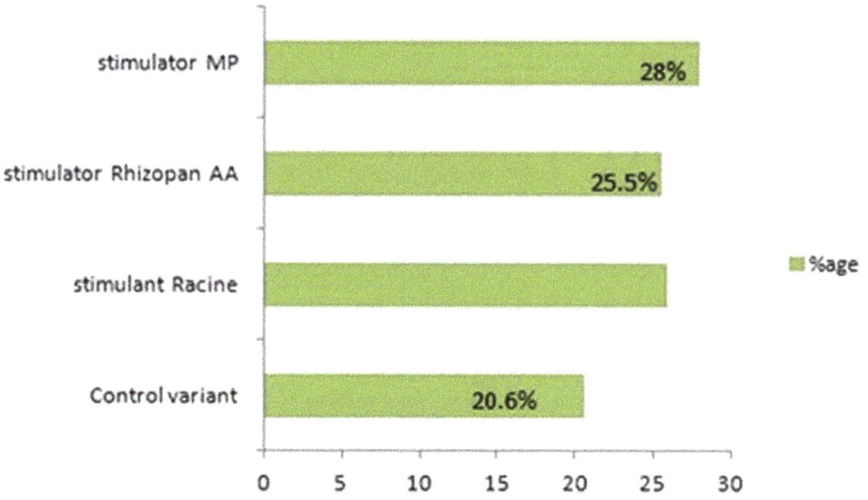

FIGURE 4.3 Effect of stimulators for the rooting of tested rootstocks—softwood cuttings (*Source*: Krska and Necas 2013).

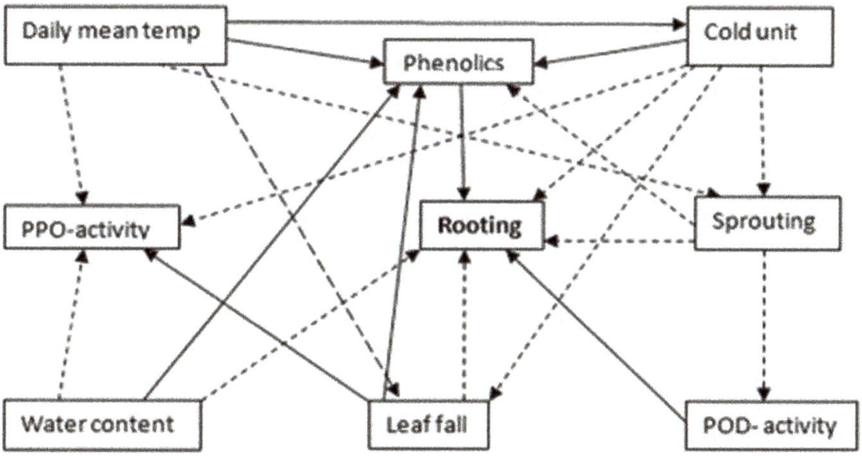

FIGURE 4.4 The possible relations between the factors influencing the rooting of plum hardwood cuttings (dotted arrows show weak correlations).
(*Source*: Szecsko et al. 2006).

Peroxidae activity, which usually vary with the changing environmental conditions. Szecskó et al. (2006), while working on the influence of factors on the rate of rooting in plum rootstocks is briefly shown in Figure 4.4. During the study, it was concluded that the rooting in plum rootstocks is strongly influenced by three factors, i.e., phenolics, daily mean temperature and PPO activity, while leaf fall and water content influenced the rooting only slightly. Another strong factor responsible for better rooting in plum rootstocks is sprouting, though it acts through the activity of the POD enzyme.

Moreover, it was revealed that the percentage of rooting in the Mariana plum was directly related to the activity of the POD enzyme system, which enhanced the process of sprouting. Hence, it is a combination of various factors that helps in the development of a complete root system in plum rootstocks; therefore, it becomes imperative to check the quality of the environment selected for the raising of the rootstocks.

4.3.1.3 Budding and Grafting

Plum are usually propagated by T-budding and tongue grafting. Budding is done during the active growing season, i.e., when the bark easily slips at the cambium region, and during active sap flow, i.e., in June, by giving a T-shaped cut in the scion (Hartmann et al. 2002). During budding, the bud from a superior scion variety is inserted under the bark of the rootstock. In grafting, the superior scion variety is grafted onto the already existing rootstock. Grafting is performed during dormancy. The bud stick used usually consists of three or more buds. Successful budding results only after the compatible matching of the cambium of both the rootstock and the scion. Hence, a grafter should be well versed in identifying the cambial layers. Only after the successful combination of cambial layers does the transport of water, minerals and manufactured food occur from the rootstock to the scion and from the scion to the rootstock, respectively.

To summarize, there are certain rules that must be followed while grafting:

1. The two plants to be grafted should be compatible; the grafting of incompatible plants is always a failure.
2. For the graft union to be successful, it is necessary that the cambial region of the stock touches the cambium region of the scion.
3. The position of the scion at the time of grafting should be upright.
4. The best time for the grafting of deciduous fruit crops, including plums, is when they are under dormant conditions.
5. One should always choose a dry, cloudy, cool weather day for grafting.
6. The aftercare of the graft union is equally important for the successful growth of the graft.

Certain criteria to be followed while selecting the scion wood:

1. The bud stick to be used for grafting must contain healthy, disease-free buds.
2. The buds present on the bud stick/scion wood must be dormant, as sprouted buds are of no purpose for grafting.
3. For grafting, always choose a one-year-old shoot that grew during the previous season.

4.3.1.3.1 Tongue Grafting

Tongue grafting is the most common and recommended method of grafting for the *Prunus* species, including the plum. The advantage of tongue grafting over other types is the greater area of contact of the cambial regions between the stock and the scion. The interlocking that occurs between the rootstock and the scion in tongue grafting creates the extra pressure required for the successful formation of the graft union. Thus, the graft union formed during the tongue grafting is the strong graft, as it allows the two grafting partners to combine or fit tightly together.

4.3.1.3.2 Preparation of the Rootstock

After choosing the stock for tongue grafting with the desired scion, place the rootstock horizontally, then with the help of a grafting knife make a slanting cut in the rootstock between the two buds, i.e., at the internodes. The rootstock preparation for tongue grafting is represented in Figure 4.5. During this technique, two cuts are made in the stock; the first cut is made at the terminal or top portion of the stock, it should be about 2.5–6 cm in length, then the second cut is made about one-third of the way down from the base of the first cut and then pulled apart.

4.3.1.3.3 Preparation of the Scion

For grafting, the rootstock and the scion should belong to the same genus, i.e., the scion as well as the rootstock should be of the genus *Prunus*. Scion preparation for tongue grafting is shown in Figure 4.6. Scion wood from the previous year with dormant buds is chosen for grafting, and the bud stick is usually 20–30 cm long. After

FIGURE 4.5 Preparation of the stock.

FIGURE 4.6 Preparation of the scion.

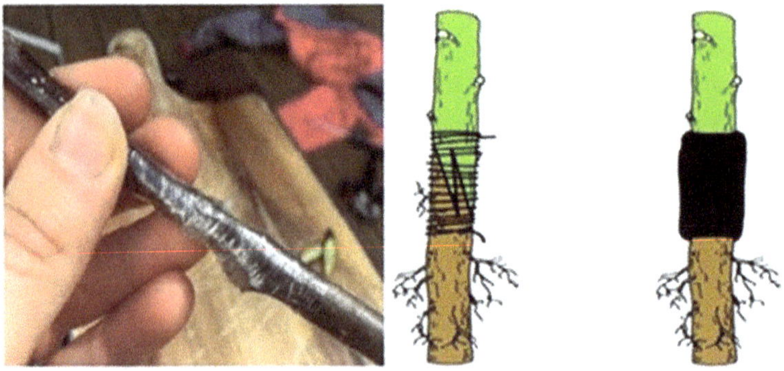

FIGURE 4.7 Stock–scion union.

the collection of the scion wood, it is wrapped with paper and placed in cold storage for future use. At the base of the scion, a similar slanted cut (like that of the first cut on the stock) is made, followed by a second cut made under the previous cut, just like that made on the stock.

4.3.1.3.4 Placing the Scion on the Stock

After creating the tongue-shaped cuts in the scion and the stock, the two partners are slipped together carefully, resulting in the tight interlocking of the tongues between the rootstock and the scion. This union between the stock and the scion is clearly shown in Figure 4.7. After the interlocking between the rootstock and the scion, the grafts are tied and waxed. Grafting tape is usually used, as it keeps the water away from the graft union, thereby preventing any sort of infection and disease occurrence.

Aftercare of grafted and budded plants:

1. The first important thing is to sanitize the equipment used for grafting or budding, i.e., the grafting knife and budding knife.
2. After the process is completed, the graft union should be wrapped with grafting tape, and the wounded portion should be covered with wax to prevent a loss of moisture from the union, thereby preventing it from desiccating.
3. The graft must be observed after every five to six days to see if the wax has cracked, and the extra shoots growing below the union should be removed, as they compete with the graft union for water and nutrients.
4. The buds located below the graft union of the rootstock should be removed regularly, as they compete with the buds of the graft union.

4.3.1.4 Micro Propagation in Plum

Apart from propagating the plum by vegetative means, it can be propagated in vitro, taking into consideration the problems associated with vegetative methods of propagation, like the requirement for a large amount of space, the dependency of the operation on a particular season, more involvement of manpower, etc. Micropropagation is preferred for producing a large number of true-to-type, disease-free plants in a shorter period of time within a smaller area; plants resulting from micropropagation are easier to transport since they are smaller in size without compromising the quality of the plant material, which in turn results in speeding up the breeding programs in plum (Topp et al. 2012). Besides the early maturing varieties in plum which don't reach maturity due to the abortion of the embryo can also be propagated through micropropagation using the embryo rescue technique or the ovule culture. The process of micropropagation basically consists of five major steps:

Stage 0: This stage involves the treatment of plants selected for micropropagation and is also called the pre-preparation stage.
Stage I: During this stage, the initiation of the explants occurs, surface sterilization is done and the mother explants are established.
Stage II: Explants are multiplied during this stage
Stage III: The process of root and shoot formation occurs in the explants.
Stage IV: Finally, the explants are hardened.

In vitro propagation involves the use of actively growing meristems which serve as the explants, i.e., shoot tips, root tips, cotyledon, axillary buds and seed explants (Minocha and Minocha 1999). Micropropagation is also used for the development of the clonal rootstocks. In vitro propagation is done using the Murashige and Skoog media with certain modifications (Murashige and Skoog 1962). Mother plants from which the explants are obtained should be from the certified agency; usually the nodal segments with a single node are used. The explants are first washed with tap water, then ethanol and, finally, with double-distilled water, followed by shade drying and then inoculation onto the growing media. The developed shoots are then transferred onto a rooting media for proper rooting. The rooting media should contain different combinations of IBA and NAA. After the roots have developed and the plantlets have acclimatized, they are transferred to the field. The developed plantlets can be micrografted at any point and can be transferred into the soil as and when required. This method of propagation is an advancement over other methods of plum propagation (seed, cuttings and grafting). The in vitro method has a great scope if coupled with the latest biotechnological tools and techniques. This method of plum propagation has revolutionized the clonal rootstock industry (Morini 2004). Moreover, it provides the option for the conservation of plum germplasm as well as for enhancing the rooting potential of the hard-to-root cuttings of plum.

For in vitro plum propagation, certain points must be taken into consideration, such as the composition of the medium used for culturing the plants. The important constituents that should be present in the propagation medium are summarized in Table 4.2. The culture medium must contain ingredients like mineral salts and sugar used as a source of carbon, gelling agents, organic supplements and growth regulators (Gamborg and Phillips 1995). Also, the most common media used for micropropagation containing all these ingredients include the MS (Murashige and Skoog 1962) and LS (Linsmaier and Skoog 1965) media; hence, it is essential to check the medium for the presence of these ingredients.

TABLE 4.2
Composition of Plant Growth Medium (MS Medium) Given by Murashige and Skoog (1962)

Compound	Amount (mg) per liter
a. Macronutrients	
NH_4NO_3	1650
KNO_3	1900
$CaCl_2.2H_2O$	440
$MgSO_4.7H_2O$	370
KH_2PO_4	170
b. Micronutrients	
$MnSO_4.4H_2O$	22.3
$ZnSO_4.7H_2O$	8.6
H_3BO_3	6.2
KI	0.83
$Na_2MoO_4.2H_2O$	0.25
$CuSO_4.2H_2O$	0.025
$CoCl_2.6H_2O$	0.025
c. Iron	
$Na_2.EDTA.2H_2O$	37.3
$FeSO_4.7H_2O$	27.8
d. Vitamins	
Nicotinic acid	0.5
Pyridoxine HCl	0.5
Thiamine HCl	0.1
e. Myo-inositol	100

Micropropagation in plum has found various applications, e.g., it can be used to develop new hybrids, for large-scale propagation, for the multiplication of transgenic plants (Faize et al. 2013) and for the elimination of diseases and viruses (Paunovic et al. 2007), thus reducing the economic losses to a large extent (Gella and Errea 1998). The technique of tissue culture or micropropagation has been developed for various cultivars in plum. Bandeira et al. (2012) successfully cultivated the Japanese plum cv. América via in vitro propagation through the use of young shoots. They further reported that the highest percentage of plant survival was obtained during the ex-vitro step using 1 mg of IAA and 1 mg/L of IAA. In the Gulf Ruby cultivar of Japanese plum, the protocol for micropropagation was established using the mature node as explants, which were then cultivated on a woody plant medium that was supplemented with certain plant growth regulators for the induction shoots. The survival of the plants was obtained after the in vitro stages inside the greenhouse with a high percentage of success (Zou 2010). For the elimination of PPV and Prunus necrotic ring spot virus (PNRSV) tissue culture through the use of meristems has been extensively used in plum (Manganaris et al. 2003).

TABLE 4.3
Effect of Type of Cytokinin on Multiplication Rate of Plum (*Prunus domestica*)

Plum cultivar	Type of cytokinin	Number of axillary shoots Cytokinin concentration (M)				Number of usable shoots Cytokinin concentration (M)			
		0	10^{-6}	$5x10^{-6}$	$2.5x10^{-5}$	0	10^{-6}	$5x10^{-6}$	$2.5x10^{-5}$
Victoria	BAP	–	0.3	7.6	6.2	–	–	2.9	1.9
	Kinetin	–	–	0.7	2.0	–	–	0.1	0.4
Brompton	BAP	–	0.3	2.6	3.1	–	–	1.2	0.9
	Kinetin	–	–	1.0	1.1	–	–	0.1	0.2

The technique of micropropagation for most plum cultivars is successfully achieved by shoot tip cultures. The hormone required for shoot multiplication is the cytokinin, and among the cytokinin the best results have been obtained with 6-Benzylaminopurine in comparison to kinetin. As studied by Abbas (1990) and shown in Table 4.3, increased concentration of BAP ($5x10^{-6}$M) increased the number of usable shoots in plum cv. Victoria and Brompton. The highest number of axillary shoots in cv. Victoria (7.6) was obtained with BAP ($5x10^{-6}$M), while the highest number of axillary shoots in cv. Brompton (about 3.1) was obtained with BAP ($2.5x10^{-5}$ M).

4.3.1.4.1 Protocol for In Vitro Micropropagation of Plum

The in vitro micropropagation protocol given by Kassaye and Bakele (2015) to produce true-to-type, disease-free plums (*P. salicina*) developed from the nodal segments includes plant growth regulators, which are essential for the induction and proliferation of shoots as well as the roots, which varied with the use of different concentrations of the growth regulators. The highest and the healthiest shoots were obtained on MS media when it was supplemented with 0.5 mg/L BAP in combination with 0.1 mg/L of IBA. As far as the rooting is concerned, the best response was obtained when the strength of the MS media was reduced to half the strength as used for shooting supplemented with 1 mg/L IBA. Hence, these concentrations are regarded as the optimum concentration for shoot and root proliferation of plum (*P. salicina*) for in vitro propagation.

4.3.1.4.2 Production of Plum Planting Material Free from Plum Pox Virus

The most devastating virus in European countries affecting the stone fruits, particularly the plum, is PPV, also called sharka disease. This virus belongs to the family Potyviridae and genus *Potyvirus* (Scorza et al. 2013) and is transmitted by aphids. It was first discovered in Bulgaria in 1900. The plum varieties infected with PPV develop many recognizable symptoms, including the development of rings of irregular lines on leaves, malformed fruits, the formation of chlorotic spots, etc., making the fruits unmarketable (Llacer and Cambra 2006). One of the most challenging issues for plum breeders is obtaining planting stock free of this virus. The technique of propagating plums that are PPV free consists of selecting plum seedlings that are free of PPV and grafting the selected scion material onto them using sterilized materials and open boxes. The selected scion is then kept inside the growth chamber for six weeks at 36°C at night and 38°C during the day. The terminal portion, or the tip, of the scion is grafted onto virus-free stock or propagated by micropropagation (tissue culture). To maintain an atmosphere of high humidity in the environment, it is covered with a plastic bag. To check whether the plants are free from viruses or not, certain tests are carried out using the indicator plants or an immunoassay test for viruses. The plants are then transferred to the field and become the certified plants for PPV. They are planted in the field at locations that are free from pests and diseases.

4.4 AIR LAYERING

Air layering is one of the methods of vegetative propagation that has been used more recently for the development of the clones in the *Prunus* species, including the plum. More importantly, this method has been employed for the development of the rootstocks (Castle 1995); however, it is not a commercial method of plum propagation. The various steps involved in air layering are shown in Figure 4.8. The first step is the selection of the mother, around the stem of which a wide strip of bark is removed for the removal of the cambium and the phloem (Cameron 1968). To facilitate rooting, the exposed part of the stem is wrapped with a material that retains moisture; the most common material used for this is the sphagnum moss, or peat moss, which is treated with the rooting hormone IBA. For the proper placement of the sphagnum

FIGURE 4.8 Technique of air layering.

moss, it is covered with either polythene film or aluminum foil. Wrapping of the polyethene or aluminum foil also maintains the temperature and makes it somewhat waterproof. After the roots develop, the layers are detached from the mother plant for transplanting.

4.5 MICROGRAFTING

Micrografting is one of the more advanced techniques of in vitro plum propagation. The purpose of micrografting is to obtain virus-free plant material (Guney 2019). In this method, virus-free shoot tips are grafted onto decapitated virus-free seedlings or rootstocks in vitro and are then used in the propagation of virus-free plants (Hartmann et al. 2002). The micrografting technique is represented in Figure 4.9. The advantage of micrografting is manyfold, e.g., to obtain virus- and disease-free plants, for virus indexing and to study the incompatibility between the scion and the rootstock; and since the plants are smaller they can be easily carried in large quantities for longer distances. Moreover, micrografting is not season dependent; it can be done throughout the year. Though expensive, it is a successful tool in the production of healthy, disease-free plants of various horticultural fruit crops including plums, almonds, apricots, apples, pears, cherries, avocados, cashew nuts, etc. More importantly, this method of propagation reduces the risk of introducing new pests and diseases during the exchange of planting material from other countries (Naverro et al. 1982), serving as an effective means of quarantine.

Rootstocks raised through micrografting have tremendous potential for the production of large-scale disease-free in vitro–grafted plants in plum and other temperate fruit crops, but it needs further investigation since it is still in its infancy.

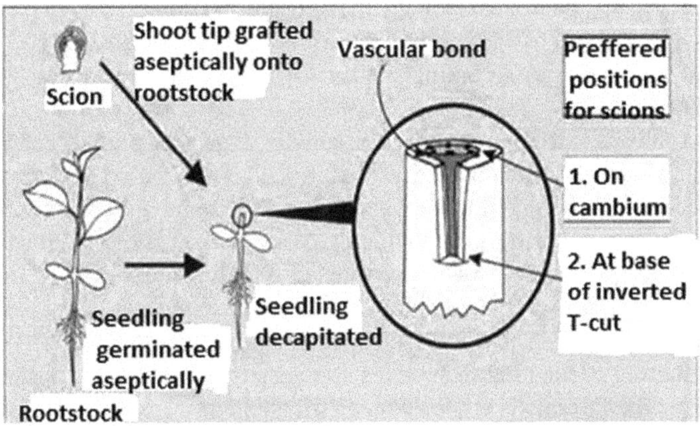

FIGURE 4.9 Shoot tip micrografting.

4.6 TOP WORKING

Top working is a propagation technique done in already existing plum trees of inferior quality. In this technique, the old mismanaged plum trees are top grafted with the new, improved varieties. This method allows the harvesting of good-quality fruits on the already existing stock without changing its root system. The method allows farmers to switch to new varieties without completely replacing the main variety, thereby saving the labor and time as well. The scion wood or the bud wood to be used for the purpose of grafting is obtained from the well-matured wood from the current season's growth during the winter. The most common method used in plum is cleft grafting (Hartmann et al. 2001). Scion is prepared by making a basal wedge cut 5 cm long. The side of the cut to be inserted into the stock should be wider so as to bear the full pressure of the split rootstock, i.e., the end of the scion with the lowest bud should be thicker than the other side. The basal cuts of the scion should be smooth and made with a knife. Both sides of the scion should press firmly against the rootstock for their entire length. Top working in plum allows the growers to be adaptable to changing market trends, as well as improved horticultural methods resulting in a high fruit yield and fruit quality. It allows the old orchards to produce new shoots by eliminating the infected branches and increasing light penetration.

4.7 USE OF CUTTINGS FOR PROPAGATION UNDER POLYHOUSE CONDITIONS

The cuttings of plum taken from a mother plant having graftable girth but devoid of roots can also be used for plum propagation. This is an advanced method that reduces the time of the plum in a nursery from almost two years to one year, with high success regarding buds and the production of saleable plants under polyhouse conditions. Negi and Upadhyay (2019) reported that the unrooted pixy plum cuttings were kept in moist sand for almost two months, i.e., December and January. These unrooted pixy were then treated with the rooting hormone IBA at different concentrations and planted in well-prepared nursery beds under tunnel-type polyhouse. The grafted stocks began sprouting within 13–15 days after grafting, with a high percentage of success.

The commercial varieties of plum are mostly grown of plum, peach or peach–plum hybrid rootstocks, including Myrobalan 29C, Mariana 2624, Nemaguard and Lovell. Recent research in plum emphasizes the development of rootstocks resistant to PPV. The rootstocks raised in the nursery should make an ideal graft with scion varieties of plum, should be precocious, disease and pest resistant, long lived and quick to grow. The growing media used in the nursery include sand, perlite or vermiculite. The rootstocks should be well spaced from each other, which is important for the production of the laterals by the scion variety grafted or budded on it, as the production of the laterals is important for the yield and productivity of the plant. There are some plum rootstocks that can be successfully propagated from the cuttings in the nursery. These rootstocks include Myrobalan 29C, Mariana GF8/1,

Myrobalan B, Tetra, Penta, Damas, St. Julien and Brompton. Also, plum rootstocks like St. Julien and Pixy can be propagated through hardwood cuttings. The cuttings are obtained in the months of November and December for planting in the nursery after being treated with the rooting hormone auxin. For the successful expression of the roots, the plum cuttings are given bottom heat treatment (Hartmann et al. 2002). After the cuttings assume the budding girth, the budding is then performed in the month of August. Moreover, propagation techniques like micropropagation, grafting and budding are employed for plum propagation in nurseries. As for budding, the percentage of success in Japanese plum is lower than for European plum, which is mainly due to the thinner bark of Japanese plum plants. Among the various budding methods, T-budding is more popular for the propagation in plum in nurseries, done in the month of June when the bark of the rootstock slips easily. The percentage of success in plum propagation with T-budding is phenomenal. In vitro propagation in plum and the use of cuttings are the easiest method of plum propagation (Okie 2006). Sansavini et al. (2004) reported that the rootstocks of plum (namely St. Julior and Ishtara) also show good graft compatibility with various peach cultivars in terms of vigor of the plant, quality of the produce and tree architecture, hence plums are propagated by budding and grafting.

The type of budding done is the shield or T-budding method and is performed in the months of July and August. Within about 1.5 years, the budded plant will be ready for sale. The type of grafting performed i is tongue or cleft grafting. Grafting is carried out when both the stock and scion are dormant. When the thickness of the stock to be used for grafting exceeds the thickness of the scion, cleft grafting is performed. Tongue grafting is the commercial method of grafting in plum, but chip budding has also proven successful.

4.8 T-BUDDING IN PLUM

The type of budding used for plum propagation is T-budding, named after the T-shaped incision made on the stock. T-budding is also called shield budding due to the shield-like appearance of the bud piece. T-budding is done when the bark of the stock slips off easily. The different steps of T-budding are shown in Figure 4.10. This type of budding is commonly used by nurserymen due to its higher success percentage and lower cost. The bud material is taken from the current-season growth of a desired vigorous plant. The buds present midway on the bud wood are used for grafting, and the top and bottom portion of the bud wood is cut off. Top buds are not recommended for budding due to their immaturity, and the buds at the bottom are usually weak and present in a cluster; hence, only the buds located in the middle of the bud wood are used. All the leaves from the stick are removed, and a length of about 30 cm is maintained for the bud wood/ bud stick. The T-shaped cut is made about 20–25 cm from the soil surface. The length of the vertical cut is 2 cm, and horizontally its length is kept at 7–8 mm. After flapping the bark with a budding knife, the bud is inserted under the bark and then wrapped with budding tape for its protection. The tape is then removed after three weeks.

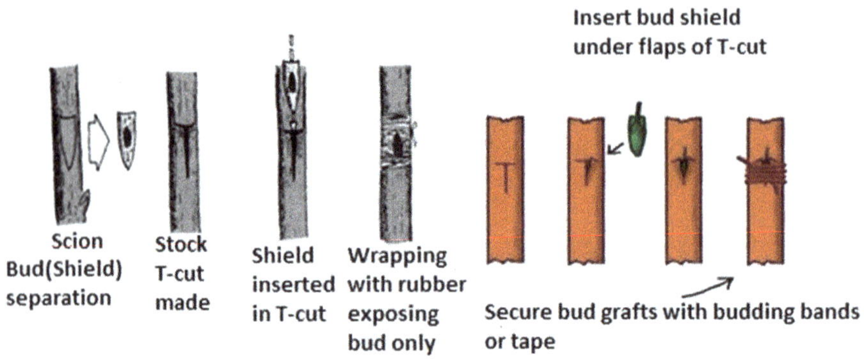

FIGURE 4.10 Steps in T-budding.

4.9 CHIP BUDDING IN PLUM

The bud stick and the stock should have the same diameter. A cut of about 2–2.5 cm long and 1/4–1/5 cm deep is made on the stock. In this cut, the bud is fixed, taking into consideration that the cambium of both the stock and the scion must be in contact with one another. After the process is completed, the graft is wrapped with polyethene or grafting tape to protect it from desiccation.

4.10 NURSERY MANAGEMENT IN PLUM

A nursery is the place where the young plantlets are grown with utmost care and maintenance for transplanting into the main field. It is the main component of the propagation of fruits crops, as the majority of the fruit crops are first raised in a nursery before being transplanted. The area selected for the establishment of the nursery should have easy access to a water supply. The area should not be waterlogged; rather, there should be proper drainage—an area with a gentle slope is preferable. For the establishment of a plum nursery, one must be well versed in plum propagation methods; furthermore, the nursery will require land, propagating structures, media of propagation, specific equipment, etc. Before planting the seed or planting material, the soil should be plowed and fumigated to prevent damage from various microorganisms. Successful nursery preparation includes preparation of the site for the nursery, preparation of the seed beds, sowing time for the seeds, the rate of seed sowing, aftercare of the plants and, finally, the transplanting of the plants to the main field. The care of plants in the nursery is an essential step for the harvesting of healthy, disease-free plants for which the cleanliness of the nursery and the material growing in the nursery are important so as to ensure that the plantlets will acclimatize to the field conditions. The first aim of developing the nursery should be as a source of obtaining healthy, disease-free planting material. The success of a nursery is reflected by the quality of the planting material it produces. Modern-day nursery fruit crops act as the backbone of the propagation industry, and nurseries are involved not only in the distribution of planting material but also in providing other services, such as training, teaching, consultation, skill and knowledge of plant propagation. Hence, nurseries need to provide the necessary infrastructure to the nurserymen to compete with the growing global fruit-propagation industry.

4.10.1 Quality of the Planting Material

The plant material to be used for raising the nursery should be taken from the elite mother plants. It should be healthy and free of any bacterial, fungal or viral infection. If the rootstocks are to be raised in the nursery, it should bear at least 4–5 healthy roots, along with some rootlets. It should have a diameter of more than 1 cm with a wider range of soil and climatic adaptability. It should also have graft compatibility with a wide range of species. Additionally, the presence of some feathers are a characteristic of good, quality plant material (Mayer et al. 2017).

4.10.2 Source of Planting Material

The material to be raised in the nursery should be obtained from a certified agency. The rootstocks or the scion wood should be obtained from the certified elite mother plants that have passed through various phases of inspection with respect to quality and insect, pest and disease symptoms. The certified material should be tagged, and only the tagged material should be supplied for propagation. The tags denote the best-quality plant material. Before the distribution of the plants from the nursery, they should be properly checked for disease and pest infestation so that the nurserymen supply only the healthy disease-free plants. Hence, there is the requirement of human resources (educated youth, NGOs, etc.), as well as the infrastructure of the nursey, for the development of quality plant material. The seedlings in the nursery are grown in nursery beds so they can be managed intensively from the very beginning, as they require greater care and attention. Nursery management of the seedlings later on provides the opportunity to select only healthy, disease-free and vigorous seedlings to be transplanted in the open field. The care of the seedlings in terms of irrigation requirements, weeding, control of insect pests and diseases and thinning helps in aiding the handing off and transplanting of the seedlings.

4.10.3 Plum Propagation in Nurseries

The seeds of the plum are usually used only for the development of the rootstocks. The most common rootstock used for European and Japanese plum is Myrobalan, which is propagated from the seeds of domestic plums. However, wild plum and wild apricot are also used as seedling rootstocks for plum. Plum is exclusively propagated by asexual methods, particularly grafting and budding. For the raising of rootstocks, the seeds are obtained from a good-quality mother plant before sowing scarification is done to break the seed coat and enhance the germination. Seeds are either treated with high temperatures, growth regulators like GA_3 or kept in running water for germination. For the development of the healthy seedlings, seeds are adequately spaced from one another. The development of the rootstocks from the seeds is a relatively cheaper and easier method of propagation compared to advanced tissue culture techniques. It is important to use only those seeds for rootstock propagation that possess characteristics like high seed germination, low rate of pathogen transfer, compatibility with scion cultivars, low vigor, as well as resistance to various insect pests and diseases (Souza et al. 2016).

Plum propagation using seed is the most profitable for nurserymen, as they are free from diseases and less expensive. The plum rootstocks raised from seed are free of viruses, which is otherwise difficult to control in the nursery. Even today, nurserymen are reluctant to raise clonal rootstocks in their nurseries because of the higher chances of disease transmittance. The seedling-raised progeny is highly heterozygous in nature, but there are certain points that need to be followed to reduce the heterozygosity to a certain extent (Iqbal and Singh 2020):

1. Use the seed obtained from a single variety for raising the progeny/rootstocks in the nursery.
2. Avoid using seeds infected with various diseases, pests and viruses.
3. If available, use the seeds obtained from the progeny of a apomictic plant.

Furthermore, propagation methods, like budding, are done in nursery-raised plum rootstocks in the same manner as discussed earlier.

4.10.4 Raising Plum Rootstocks in the Nursery and Their Management

The most commonly as well as commercially used rootstock for European plum is Myrobalan (*P. cerasifera*).

1. **Sowing of stones:** Seeds, also called stones, require about 100–120 days of ripening at 4.5–10°C. Stones should be sown directly on well-prepared beds in the months of November and December, or stratified stones may be sown in February.
2. **Seeding distance:** The seeding distance varies from crop to crop. For plum, it is advised to maintain 30 cm x 15 cm between the stones, which germinate within about 30 days.
3. **Soil management:** Myrobalan plum rootstock can do well in a wide range of soils; however, its growth response is best in light sandy soil, i.e., soil with a good proportion of sand content. Hence, for raising plum rootstock in a nursery, the preferable soil type is sandy. It is advised to avoid raising plum rootstocks in clay soil, as they will slowly decay.
4. **Water management:** Plum rootstocks require a sufficient amount of water, but they suffer if overwatered; hence, excess moisture should be avoided. Therefore, the soil must be well drained to avoid water stagnation.
5. **Transplanting:** After about 810 months, the plum seedlings should be transplanted in the well-prepared field beds, which have been irrigated, drained and applied with an ample amount of well-decamped organic manure. The seedlings should be planted 30 cm x 15 cm in rows that are 50–60 cm apart.

6. **Weed management:** It is essential to keep the floor of the nursery clean for the healthy growth of the seedlings. For this purpose, all the undesirable grasses growing in the nursery beds must be removed either manually or by using specific weedicides that will check the growth of the unwanted plants and grasses.
7. **Insect, pest and disease management:** It is important to keep a strict watch on the nursery plant materials and check for any infestation. Myrobalan plums usually suffer from root rot disease; hence, it is important to regularly check the plants for any sort of infection and take the necessary measures. Also, the nursery beds should be drenched with carbendazin (100g) and mancozeb (250 g) per 100 ml of water.
8. **Selection of mother plants for grafting or budding:** The true-to-type, healthy, disease-free, well-rooted plants are selected for propagation. The propagation (grafting or budding) on the selected rootstock is done 20 cm from the ground level after deshooting.
9. **Growing budded plants in polybags:** The well-rooted budded plants can be grown in polybags supplied with manures and fertilizers from time to time; however, on attaining further growth, the plants should be removed from polybags to prevent damage to the root system.
10. **Growing budded plants in the main field:** Finally, the well-rooted, well-grown plants are shifted in the field where manures, fertilizers, irrigation, etc. are applied as per recommended for the age of the plants.

4.11 CONCLUSION

Propagation of fruit crops is the process responsible for the prosperity of a nation, but the main purpose of this propagation industry is the production of quality plant material that acts as its seal. A number of propagation techniques are used for the multiplication of the fruit crops, including the plum, which involves the use of seed, grafting, budding, air layering, micrografting, micropropagation, etc. For the conservation of plum germplasm, propagation by seed plays an essential role. Seed propagation is also important for breeding purposes. Asexual methods of propagation in plum are used for commercial purposes and produce plants similar to the mother plant. Propagation methods used for plum propagation need refinement for the development of resistant genotypes that can cope with various climatic odds. The availability of quality planting material of plum that is free from insects, pests and diseases is the need of the hour; hence, the focus should be on the production of superior planting material, which in turn will ensure financial and economic security for growers. The tissue culture technique of propagation can prove more fruitful in producing the standard propagating material for plum. Furthermore, the best propagation technique for plum should be standardized, which can help achieve profitability and gain the trust of nursery growers.

4.12 REFERENCES

Abbas, M.C. 1990. *Micropropagation of Plum (Prunus domestica L.) with Reference to Invivo Rooting of Microcuttings*. Ph.D Thesis. Department of Agriculture, Horticulture and the Environment. Wye College, University of London.

Anderson, N.O., Hoover, E., Kostick, S.A., Tepe, E., and Tillman, J. 2016. Cutting Type and Time of Year Affect Rooting Ability of Minnesota Prunus Species. *Journal of the American Pomological Society* 70 (3): 114–123.
Bandeira, J.D., Thurow, L.B., Braga, E.B., Peters, J.A., and Bianchi, V.J. 2012. Rooting and Acclimatization of the Japanese Plum Tree, cv. America. *Revista Brasileria de Fruticultura* 34 (2): 597–603.
Bofung, C.M., Silou, T., and Mouragadja, I. 2002. Chemical characterization of Safou (Dacryodes edulis) and Evaluation of Its Potential as an Ingredient in Nutritious Biscuits. *Forest, Trees and Livelihoods* 12 (2): 105–117.
Cameron, R.J. 1968. The Leaching of Auxin from Air Layers. *New Zealand Journal of Botany* 6 (2): 237–239.
Castle, W.S. 1995. Rootstocks as a Fruit Quality Factor in Citrus and Deciduous Tree Crops New Zealand. *Journal of Crop and Horticultural Science* 23: 383–394.
Davies, F.T., Davis, T.D., and Kester, D.E. 1994. Commercial Importance of Adventitious Rooting to Horticulture. In *Biology of Adventitious Root Formation*, edited by T.D. Davis and B.E. Haissig. Plenum Press.
Faize, M., Faize, L., Petri, C., Barba-Espin, G., Diaz-Vivancos, P., Clemente-Moreno, M.J., Koussa, T., Rifai, L.A., Burgos, L., and Hernandez, J.A. 2013. Cu/Zn Superoxide Dismutase and Ascorbate Peroxidase Enhance in Vitro Shoot Multiplication in Transgenic Plum. *Journal of Plant Physiology* 170 (7): 625–632.
Gamborg, O.L., and Phillips, G.C. 1995. Media Preparation and Handling. In *Plant Cell, Tissue and Organ Culture—Fundamental Methods*, edited by O.L. Gamborg and G.C. Phillips. Springer.
Gella, R., and Errea, P. 1998. Application of in Vitro Therapy for Ilar Virus Elimination in Three Prunus Species. *Journal of Phytopathology* 146 (8–9): 445–449.
Guney, M. 2019. Development of an in Vitro Micropropagation Protocol for Myrobalan 29C Rootstock. *Turkish Journal of Agriculture and Forestry* 43 (6): 569–575.
Hartmann, H.T., Kester, D.E., Davies Júnior, F.T., and Geneve, R. 2001. *Hartmann and Kester's Plant Propagation: Principles and Practices*. Prentice Hall.
Hartmann, H.T., Kester, D.E., Davies Júnior, F.T., and Geneve, R.L. 2002. *Plant Propagation: Principles and Practices*. Prentice Hall.
Iqbal, M., and Singh, K.K. 2020. *Propagation of Temperate Fruit Crops: Innovative Agriculture and Botany*. Victorious Publishers.
Kassaye, E., and Bekele, B.D. 2015. Invitro Optimization of the Protocol for Micropropagation of Plum (P. salicina var. Methley) from Nodal Explants'. *Biotechnology International* 8 (4): 137–148.
Krska, B., and Necas, T. 2013. Propagation of Different Stone Fruit Rootstocks Using Softwood and Hardwood Cuttings. *Acta Horticulturae*. DOI:10.17660/ActaHortic.2013.985.16.
Llacer, G., and Cambra, M. 2006. Hosts and Symptoms of Plum Pox Virus: Fruiting Prunus Species. *EPPO Bulletin* 36 (2): 219–221.
Linsmaier, E.M., and Skoog, F. 1965. Organic Growth Factor Requirements of Tobacco Tissue Cultures. *Physiologia Plantrum* 18 (1): 100–127.
Manganaris, G.A., Economou, A.S., Boubourakas, I.N., and Katis, N.I. 2003. Elimination of PPV and PNRSV Through Thermotherapy and Meristem-Tip Culture in Nectarine. *Plant Cell Reports* 22 (3): 195–200.
Mayer, N.A., and Antunes, L.C. 2010. *Diagnosis of the Prunoid Seedling Production System in Southern and Southeastern Brazi'*. Embrapa Temperate Climate.
Mayer, N.A., Bianchi, V.J., and Castro, L.A.S. 2014. Rootstock. In *Pessegueiro*, edited by M.C.B. Raseira, J.F.M. Pereira, and F.L.C. Carvalho, 173–223. Embrapa.
Mayer, N.A., and Joao, B.V. 2017. Advances in Peach, Nectarine and Plum Propagation. *Revista Brasileira de Fruticultura* 39 (4): e355.
Minocha, S.C., and Minocha, R. 1999. Genetic Transformation in Conifers. In *Somatic Embryogenesis in Woody Plants,* vol. 5, edited by S.M. Jain, P.K. Gupta, and R.J. Newton, 291–312. Kluwer Academic Publishers.

Morini, S. 2004. Stato attuale della produzione di portinnesti mediante micropropagazione. *Frutticoltura, Bologna* 12 (1): 33–36.

Murashige, T., and Skoog, F.A. 1962. A Revised Medium for Rapid Growth and Bioassays with Tobacco Tissue Cultures. *Physiologia Plantarum* 15 (3): 473–497.

Naverro, L., Llacer, G., Cambera, M., Arregui, J.M., and Juarez, J. 1982. Shoot Tip Grafting In Vitro for Elimination of Viruses in Peach Plants (*Prunus persica* Batsch). *Acta Horticulturae* 130: 185–192.

Negi, N.D., and Upadhyay, S.K. 2019. Standardization of Different IBA Concentrations on Stenting Method of Propagation in Plum Cultivars Under Polyhouse Conditions. *Himachal Journal of Agricultural Research* 45 (1–2): 31–38.

Okie, W.R. 2006. Prunus Domestica—European Plum, Prunus Salicina—Japanese Plum. In *The Encyclopedia of Fruit & Nuts*, edited by J. Janick and R.E. Paull. Cambridge University Press.

Pasqual, M., Chalfun, N.J., Ramos, J.D., Vale, M.R., and Silva, C.R. 2001. Commercial Fruit Growing. In *Propagation of Fruit Plants*. Minings, UFLA/FAEPE.

Paunovic, S., Ruzic, D., Vujovic, T., Milenkovic, S., and Ievremovic, D. 2007. InVitro Production of Plum Pox Virus—Free Plums by Chemotherapy with Ribavirin. *Biotechnology and Biotecnological Equipment* 21 (4): 417–421.

Petri, C., and Scorza, R. 2010. Factors Affecting Adventitious Regeneration from In Vitro Leaf Explants of 'Improved French' Plum, the Most Important Dried Plum Cultivar in the USA. *Annals of Applied Biology* 156: 79–89.

Preece, J.E. 2003. A Century of Progress with Vegetative Plant Propagation. *HortScience* 38 (5): 1015–1025.

Sansavini, S., Bassi, D., Manucci, C., Correale, R., and Grandi, M. 2004. Nuovi portinnesti del pesco: Preliminari risultati com la nettarina Ambra e possibili alternative al GF 677. *Frutticoltura, Bologna* 12 (1): 48–56.

Scorza, R., Callahan, A., Dardick, C., Ravelonandro, M., Polak, J., Malinowski, T., Zagrai, I., Cambra, M., and Kamenova, I. 2013. Genetic Engineering of Plum Pox Virus Resistance: 'HoneySweet' Plum-from Concept to Product. *Plant Cell Tissue Organ Culture* 115 (1): 1–12.

Sharma, R.R. 2002. *Propagation of Horticultural Crops: Principles and Practices*. Kalyani Publishers.

Singh, S., and Singh, A.K. 2006. Standardization of Method and Time of Propagation in Jamun (Syzygium cumini) under Semi-arid Environment of Western India. *Indian Journal Agricultural Science* 76: 142–145.

Sonwa, D.J., Okafor, J.C., Mpungi Buyungu, P., Weise, S.F., Tchat, M., Adesina, A.A., Nkongmeneck, A.B., Ndoye, O., and Endamana, D. 2002. Dacryodes Edulis, a Neglected Non-Timber Forest Species for the Agroforestry Systems of West and Central Africa. *Forest, Trees and Livelihoods* 12 (1): 41–55.

Souza, A.G., Smiderle, O.J., Spinelli, V.M., Souza, R.O., and Bianchi, V.J. 2016. Correlation of Biometrical Characteristics of Fruit and Seed with Twinning and Vigor of Prunus Persica Rootstocks. *Journal of Seed Science* 38 (4): 332–328.

Srivastava, R.P. 1966. Research on Horticultural Crops at Chaubattia. *Indian Horticulture* 10 (1): 9–11.

Szecskó, V., Hrotkó, K., and Stefanovits-Bányai, E. 2006. Physiological Factors Influencing the Rooting of Plum Rootstocks Hardwood Cuttings. *Latvian Journal of Agronomy* 9 (2): 156–161.

Topp, B., Russell D., Neumüller, M., Dalbo, M., and Liu, W. 2012. Plum. In *Fruit Breeding, Handbook of Plant Breeding*, edited by M. Badenes and D. Byrne. Springer.

Zou, Y. 2010. Micropropagation of Chinese Plum (Prunus salicina Lindl.) Using Mature Stem Segments. *Notulae Botanicae Horti Agrobotanici Cluj* 38 (2): 214–221.

5 Flowering, Fruit Set, and Pollination of Plum

Aiza Hasnain[1], *Amna Sajid*[1], *Muhammad Shafiq*[1], *Syeda Shehar Bano Rizvi*[1], *Mukhtar Ahmed*[2], and *Muhammad Rizwan Tariq*[3]

[1] Department of Horticulture, University of the Punjab, Lahore Punjab, Pakistan

[2] Department of Higher Education (Colleges), Government of Azad, Jammu and Kashmir, Pakistan

[3] Department of Food Science, University of the Punjab, Lahore Punjab, Pakistan

CONTENTS

DOI: 10.1201/9781003205449-5

5.1 INTRODUCTION

Plums are among the essential stone fruits belonging to the *Prunus* subgenus. It has a thin, smooth outer covering, juicy flesh, and a hard pit in the middle. *Prunus* has a sandy, white wax covering that gives a leafy green appearance (Chin et al. 2014). This is also called a wax bloom. The *Prunus* class incorporates plums, almonds, apricots, cherries, and peaches. About 430 types of *Prunus* species exist in the northern temperate regions of the world. The *Prunus* subgenus, which consists of plums and apricots, is distinguished from other subgenera (cherries, peaches, and so on) because its shoots have a terminal bud, with the side buds being singular rather than bunched, the flora being assembled one to five on short stems, and the fruit having a furrow running down on one side and a stone. Plums are grown in the temperate zones of China, Romania, Pakistan, and the United States. The commercially significant plum trees are of medium size, of medium toughness, and ordinarily pruned to 5–6 meters (16–20 ft) tall. The fruits are usually medium size, between 2 and 7 centimeters (0.79–2.76 in) in diameter (Janick 1998).

Plants depend on environmental signals for their growth, development, and flowering. Floral transition is the most crucial phenomenon regulated by environmental factors (Ashraf et al. 2020). The most critical initial step for the regenerative stage, which decides the bloom power and yield limit, is the separation of vegetative primordia to conceptive primordia. The perennial natural-product fruits are continually exposed to annual environment changeability and thus require more critical thought in deciding their production. It is currently apparent that natural environmental signaling (photoperiod, light quality, vernalization, and other ecological elements, like encompassing temperature, other than supplement status and dampness stress) impacts the blooming time/blossoming control.

There are several species of the *Prunus* family, including *P. domestica*, commonly known as European plum, used in fruit cocktails and other products. Other species may include the Japanese plum and the damson plum.

5.2 FLOWERING IN PLUM TREE

A flowering plum tree has a beautiful spray of pink, red, or white flowers and purple foliage.

5.2.1 Flowering in Japanese Plum

This plum is associated with the genus *Prunus* and the family Rosaceae. Buds initiate their growth in leaf axils throughout the hot season, which is accompanied by flower opening. Flower bud initiation depends on the variety, morphological situation of the plant, weather, site, and cultural practices. After floral bud initiation, vegetative buds contain the vegetative axis and primordial flower bud. The floral bud develops on one-year-old shoot and spurs on old wood. Flower development occurs in late summer, after the last summer fruit season, and blooms in winter or early spring. Each bud is surrounded by many bud scales and more than one flower without any leaves (Guerrero et al. 2020).

Flower bud development starts after leaf fall. Dormant buds get closed and surrounded by scales. The Japanese plum is a temperate-region plant; therefore, it requires chilling hours during dormancy for 200 to 1,000 hours at temperatures below 7 °C and requires heat afterward so the bulbs can sprout by breaking the buds and enabling the tree to bloom. A rise in temperature causes biochemical and morphological changes. Flower development occurs in late winter or early spring, depending on the variety of plum and geographical location.

Flower buds swell, and brown scales appear with a rise in temperature. Single flowers appear on the stalks as the bud opens, followed by white sepals (Figure 5.1). As the flower opens, anthers are formed, followed by the development of pistils. A flower fully opens at the blooming stage. The Japanese plum is a deciduous plant in which flower buds open first and then leaf buds. Blooming depends on weather and cultivar. In Japanese plum, the blooming period is short. In warm climates, the duration of early and late flowering is longer. The flowers are small (6–25 mm in diameter) with petals opening at the top of corolla. Flowers are hermaphrodite and contain a single pistil and 20–30 stamens. The plum flowers also contain five sepals

FIGURE 5.1 Floral bud development in plum.

and five petals. The base of the stamens, petals, and sepals is inserted in tube-shaped hypanthium. Sepals and petals are of the same size at anthesis.

The development of flowers in Japanese plum is about 5–14% compared to other plums. Plants with a higher number of flowers typically produce adequate amounts of fruit.

5.2.2 Flowering in European Plum

The European plum (*P. domestica*) is another presentation in the semi-bone-dry climate of northwest India. Medium-sized European plums have a small canopy covered with bright green leaves. They give five-petaled white and highly fragrant blooms in a spectacular arrangement. It requires full sun and the season of interest is mid-spring and late summer. Its height is about 2.4–6 meters with a canopy spread of 8–20 inches, and a heavy crop of oval-shaped fruits will have dark red colored skin and golden yellow flesh. European plums are, for the most part, self-fruitful. However, Japanese and American half-breed plums typically need to cross-fertilize with a second assortment for cross-fertilization. In any case, even self-fruitful trees will deliver better if cross-pollinated with a subsequent tree.

5.3 FLOWERING PHYSIOLOGY OF PLUM

Plants depend on environmental signals for proper growth and flower development. The floral transition process in the flowering cycle is an essential phenomenon that is regulated by environmental factors (Boss and Thomas 2002; Hu et al. 2020). To begin the life cycle, a plant goes through a period of vegetative growth, which is related to the genetic potential of the species. A large number of plant species after completion of their vegetative growth and provision of their adequate requirements may start their growth and trend for flowering.

Plants start their development and blossoming on exposure to natural prompts. The fundamental initial step for the regenerative stage, which decides the bloom force and yield limit, is separating vegetative primordia from conceptive primordia. The perpetual organic product crops are continually presented to yearly environmental fluctuation and thus expect special importance in deciding efficiency. It is apparent that the various natural signs (photoperiod, light quality, etc.) impact the blooming time/blossoming control.

5.3.1 Plum Tree Categories

Plum trees are generally called European plums (*P. domestica*) and Japanese plums (*P. salicina*); depending on their development site, one variety might be superior, as it may be viable with another in the method for its yield, development, and growth responses toward the environment. Japanese plums will usually blossom in pre-spring or late winter, making them less resistant to the loss of bloom buds from frost stress, a situation that forestalls fruit set. European plums are less susceptible to cold stress since they will blossom later in the spring when ice, cold, and frost have passed. In general, European plums have a higher sugar content and need to sun-dry for longer than other varieties.

5.3.2 Interspecies Plum Hybrids

Trees that result from cross-pollination between two varieties consolidate the properties of both plums and apricots (*P. armeniaca*) and can be acceptable for home plantation, according to the University of California Master Gardeners of Inyo & Mono Counties. These trees are designated "pluots" or "plumcots" (Bastias et al. 2020).

5.4 FLOWERING AND FRUITING CHARACTER OF PLUM

Flower bud differentiation was studied by electron microscope technique. First, morphological evidence of flower bud initiation was evident about 60 years after full bloom, and fruit distribution of reproductive structures was described according to the age of the wood. The total number of buds increases with the age of the wood.

5.5 PHYSIOLOGY OF FLOWERING

Plants develop flowers after they have gone through a specific period of vegetative development. Stages included in bud development might consist of:

- Differentiation of flower bud
- Development of blossom buds
- Opening of blossom or flower

The partition of blossom buds is the crucial turning point in this communication, and the plant's cycle changes from vegetative to a conceptive turn of events. This trade is called blooming.

5.5.1 Events in the Bud Leading to Flowering

Three phases were proposed to happen during the change of buds to blooming: induction, evocation, and initiation.

5.5.1.1 Induction

The sequence in which the flower is produced is called flower induction. Searle (1965) characterized the physiological conditions that start in the tissues by external stimuli, such as photoperiod, water pressure, and stress (Searle 1965).

5.5.1.2 Evocation

The progression step at which the shoot apical meristem has received the botanical or inflorescence stimulus and is irreversibly dedicated to the flower bud primordial early stage is called flower evocation. These cycles are fundamental for the arrangement of flower primordia occurring in the apex (Metzger 1995).

5.5.1.3 Initiation

In this interaction, the evoked bud becomes conspicuous as a flower bud and is dedicated to a regenerative turn of events called its reproductive growth. It clears the expanding and straightening of the developing point while simultaneously developing lobes.

5.6 GENE FLOW IN *PRUNUS*

Prunus species are significant business fruit products, and decorative trees have been developed extensively worldwide. *Prunus*, particularly cherries and plums, show high gene flow capacity. (Cici and Van Acker 2010).

5.7 FACTORS THAT AFFECT FLOWERING IN PLUM

Some self-incompatible varieties of plum, even when suitably interplanted with pollinators, may sometimes be very irregular croppers. A study on the behavior of pollinating insects and flower structure was made to throw light on the causes of this irregularity. It was observed that hive bees never collect nectar and pollen on the same visit and that they are more efficient as pollinators of plum blossoms when collecting pollen. Although present in smaller numbers, wild bees and blowflies are very efficient cross-pollinators because of their wandering habits and the fact that they usually contact stigma even when collecting nectar. Some varieties appeared to be much more attractive than others to hive bees, about four times as many bees being observed on Utility plums as on Jefferson ones, although both were in full flower. Measurements of the amount of nectar secreted by blossoms of several varieties indicated that a nectar abundance is one of the factors that determines the number of bees attracted to a tree. High nectar-producing types continue to produce it in large quantities, likely to attract bees for a more extended period than low nectar-producing varieties. These plums are also more attractive to bees collecting pollen. The unreliable cropping of the President plum is shown to be due not only to low nectar yield but also the stigma, which projects so far above the stamens that hive bees can collect both pollen and nectar without touching it. Therefore, those hive bees are of practically no importance in the cross-pollination of this variety. There is probably a large population of wild bees in the surroundings. The same applies to the Jefferson plum. Old Greengage, on the other hand, has an abundance of nectar, and is achieved by cross-fertilization, however, it is irregular because the stigmas become covered with incompatible pollen from the same tree so rapidly that compatible pollen brought to the stigmas later is likely to have a reduced chance of germination.

Developing organic fruit trees means having a lot of beloved natural product. Although planting in your yard might seem to be a brilliant idea, there are several factors regarding developing fruit trees that need to be considered when figuring out what kind of tree will fit best in your yard. Depending on the type of tree you choose and the climatic conditions, some fruit trees need more distance in between each one, some have chilling requirements, and many require cross-fertilization to set natural fruit alongside a radiant area with adequate sunlight.

5.7.1 Sunlight

Bloom trees need a place where they are exposed to six hours of sunlight consistently during their growth and development stages. Sunlight is necessary for excellent fruit quality and quantity as well. During the first two developing seasons, young plants require an adequate amount of sun for their bark.

5.7.2 Chill Requirement

To develop natural product-bearing trees, it should be understood which types of trees can be sufficiently grown in your available space and have a flexible climatic range. Many trees require a particular proportion of frost environment (chill requirement) annually to sprout and grow well.

Every plum type needs a particular amount of combined long chilling periods—temperatures above freezing yet below 45ºF—to crack open torpidity and permit dynamic development to start again. While chill hours don't need to be successive, extremely bright days and midwinter heat waves can have negative impacts. In an environment where winter temperatures are once in a while under 45°F, plum can be grown with low chilling requirements (Simpson 2021).

5.7.3 Chilling for Plum Crops

Plum requires chilling below 45°F from November through February. Conceded leafing-out, delayed or extended sprouting, or diminished sum or nature of natural item set may result from an insufficient chilling period. While chill hours shouldn't be progressive, incredibly bright days and midwinter heat waves can impact the effects of chilling. In a climate where winter temperatures are only occasionally below 45°F, you should think about growing a plum with low chilling necessities, such as the Methley.

5.7.4 New "Low-Chill" Varieties

Many Japanese plums need 700–1,000 hours in chilling temperatures to break lethargy and reverse improvement. The European plum variety Methley, at 250 hours, is far below the Japanese average. For a highly extended period, plant analysts have made species known as "low-chill" varieties that widen the degree of conditions and suitable temperatures. In addition to the Methley, the new assortment of low-chill plums fuses the Greatness, Burgundy, and "St. Nick Rosa varieties, all with chill-hour necessities somewhere in the range of 150–300 hours.

5.7.5 Climate

Some natural product trees are more frost resistant and safe than other fruit trees depending on their climatic requirements. Most trees experience damage due to temperatures below the freezing point, or 32°F. At the same time, some plums, cherries, and apples are resistant and tolerant to fluctuations in temperature well below freezing.

5.7.6 Distance

Distance and the application of fertilizers are critical for fruit trees. Some trees may grow to a width of 40 feet and spread over 30–40 feet, so you need to be sure you have enough space for your desired fruit tree or find a semi-dwarf assortment that is half to three-fourths the size of an ordinary tree.

5.7.7 Plum Flowering

Plums are deciduous stone fruit trees and incorporate two standard varieties: Japanese plums (*P. salicina*) and European plums (*P. domestica*). All types of plums sprout from pre-spring to late winter and fruit, for the most part, from May through September, contingent upon species, cultivar, and environment. Japanese plums are famous in the U.S. Department of Agriculture (USDA) Plant Hardiness Zones 4–10, while European plums flourish in zones 3–9, contingent upon the cultivar. Most, yet not all, plums are not self-fruitful, and trees require cross-fertilization to set natural product fruit, so breeders will need to establish at least two trees (Allman 2019).

5.7.8 Vernalization in Plum

Vernalization is a pattern of going dormant, or the resting stage, in cool temperatures, which helps specific plants set up for the following year. Plants that have vernalization requirements need to be introduced to long periods of cold temperatures. The temperatures and lengths of chilling depend on the plant species. So, gardeners need to pick plant varieties that fit their current circumstances for the best development design and to get the best fruiting and blooming. After vernalization, these plants are ready for blossoming. In years or locales in which the cold does not last long enough, these plants will yield a weak harvest, or sometimes they might not bloom or produce fruit at all.

As per the home plantation data at the University of California, Davis, in the cold weather months, a tree's interior cycles are in a condition of rest, known as torpidity, because of the presence of growth inhibitors. Development won't happen much under ideal temperature conditions. This keeps the trees from starting to develop during abnormal times of warm weather just to become harmed by normal cold temperatures later on.

Dormancy is broken when adequate cold temperatures separate the development inhibitors inside the tree. This is called vernalization, chilling, or winter cool. A particular number of aggregate long stretches of chilling (temperatures between 32°F and 45°F) are needed to break lethargy, which changes from one variety to another. When a variety has accomplished the appropriate number of long stretches of chilling, dynamic development resumes in the spring, yet only after trees are introduced to warm enough temperatures for average growth do cycles start. The majority of Northern California gets somewhere in the range of 800–1,500 hours of vernalization each winter. Southern California may only get 100–400 hours. The number of hours below 45°F is a good record of the sufficiency of winter chilling. The chilling prerequisites of choosing calm tree products of the soil communicated as the number of hours < 45°F expected to break torpidity are observed in December and January. The trees start proving to be fruitful four to six years after planting. Plums likewise need winter chill, pruning, and the right environment to deliver a decent yield (Hickman 2004).

5.8 STAGES OF FLOWERING IN PLUM

5.8.1 Dormancy to Green Clusters

Plum growth starts from the dormant stage when the branches are devoid of fruit or buds. The bud swells when buds begin bulging from the branch. In the green cluster stage, multiple flowers within each bud are visible as green points, although the flowers are not yet open.

5.8.2 White Bud to Petal Fall

When the individual flower petals in the bud begin to appear, the white bud stage has been reached. The first bloom follows when the first few flowers completely uncurl, showing white petals and yellow centers. The time when all or nearly all of the tree's flowers are fully open is the complete bloom stage. Eventually, the flowers fall to the ground in a set called petal fall, leaving behind a bud remnant known as a shuck.

5.8.3 Shuck Split to Pit Hardening

The plum tree is in the shuck split stage when the green, immature fruits emerge from the former flowering site. The fruits continue growing until reaching their total diameter of 1–2 inches from this stage. Inside the fully grown but still green fruit, the plum pits take shape in a location known as pit hardening. Since the state of the pit is not detectable from outside the fruit, growers will cut green plums open and check the holes.

5.8.4 Pale Green Fruit to Market

Once the plum pits harden, the fruits change from bright green to a more yellow hue. This is the pale green fruit stage. When the purple begins showing, it reaches the coloring fruit stage. Finally, when the plums are purple, they are ready for harvest and sent to market. After picking, plums can be stored for two to three weeks if kept in temperature- and humidity-controlled conditions.

5.9 HORMONES

Ethylene has been viewed as the vital controller of maturing fruit products for some time. The late proof showed that auxin additionally assumes a significant part during fruit aging, yet the idea of connection between two chemicals has stayed hazy. To comprehend the distinctions in ethylene- and auxin-related practices that may uncover how the two chemicals interface, two plum (*P. salicina* L.) cultivars with generally fluctuating organic product improvement and maturing ontogeny were considered. Exogenous auxin significantly speeds up plum advancement and aging, demonstrating that this chemical is effectively associated with the maturing system. Further, it can be exhibited that the varieties in auxin affectability between plum cultivars could be somewhat because of PslAFB5, which encodes a TIR1-like auxin receptor. Two unique PslAFB5 alleles were recognized, one (Pslafb5) dormant because of replacing the moderated F-box amino corrosive build up Pro61 to Ser. The early-maturing cultivar, EG, showed homozygosity for the inert allele. The late cultivar V9 showed a PslAFB5/afb5 heterozygous genotype. Our outcomes feature the effect of auxin in invigorating organic product advancement, particularly the aging system and the potential for differential auxin affectability to change significant fruit formative cycles.

5.10 GENES THAT INDUCE FLOWERING IN PLUM

The Flowering Locus T1 (FT1) quality from *Populus trichocarpa*, heavily influenced by the 35S advertiser, was changed into the European plum (*P. domestica* L). Transgenic plants communicating more elevated levels of FT bloomed and delivered natural products in the nursery within 1–10 months. FT plums did not enter lethargy after cold or brief day medicines, yet field-established FT plums remained tough down to −10°C (Srinivasan et al. 2012). The plants likewise showed pleiotropic aggregates abnormal for plum, including bush sort development propensity and panicle blossom design (Srinivasan et al. 2010).

The blooming and fruiting phenotypes were viewed nonstop in the greenhouse but were restricted to spring and fall in the field. The example of blossoming in the area is related to lower everyday temperatures. This temperature impact was accordingly affirmed in development chamber studies. The pleiotropic phenotype associated with FT1 expression in plum recommends a primary job of this gene in plant development and advancement. This shows the potential for a single transgene occasion to notably influence the vegetative and regenerative growth and improve a financially significant temperate woody perennial crop. It is recommended that FT1 be a valuable instrument to adjust delicate plants to changing environments and adjust these yields to new developing regions.

5.11 VARIETIES OF FLOWERING PLUM TREES

Plum trees (*Prunus spp.*) with white or pink blooms convey the brilliance and concealing to yards and scenes. Filling in USDA Plant Hardiness Zones 8–10 and various districts worldwide, plum trees consistently produce nearly nothing, berry-like natural items, or drupes. Whether or not deciduous or evergreen, their foliage is generally green, purple, bronze, or red. So, pick trees that fit best with the development and concealing arrangement (Silver 2021).

5.11.1 White Flowers

Elaborate plum tree arrangements attract birds and other wildlife that produces White sprout–making plum trees fuse the Pigeon, Chickasaw, and Flatwoods plum varieties. Pigeon (*Coccoloba diversifolia*) plum trees grow up to 25 feet tall and 35 feet wide. It produces faint green, circular, endured leaves that do not change tone in fall; they also have white blooms and small, round, and complete purple fruits. Chickasaw (*P. angustifolia*) plum trees grow to 25 feet tall and wide with thorny branches on short trunks. Their deciduous green leaves do not change their tone in the fall, but their white blooms, which sprout through winter in warm conditions, radiate a fragrant smell. Chickasaw plum trees produce fruits that are from 1/2 to 1 inch wide. Flatwoods (*P. umbellata*) plum trees sprout close to nothing, white blossom bunches in the spring and winter, and sweet-to-tart purple fruits. Flatwoods trees are deciduous, and their green leaves become yellow in harvest time (Table 5.1).

5.11.2 Pink and Pink-White Flowers

Pink and pink-white blooms sprout on some plum cultivars with diversely tinted leaves. For example, the Storm cloud (*P. cerasifera*)—or cherry plum—sprouts ruby leaves that are dark to reddish-purple. Its pinkish-white blooms sprout in spring and are followed by purple fruits, each around 1 inch wide. Its foliage becomes purple in the fall.

TABLE 5.1
Types of Flowering in Plum Trees with White Leaves

Species name	Scientific name	Flower size	Common name	Description
Pigeon	*Coccoloba diversifolia*	About 30 feet tall	Pigeon, Seagrape, doveplum, tietongue	Broadleaf-evergreen tree white-green flowers
Chickasaw	*Prunus angustifolia*	12–20 feet tall	Cherokee plum, sand plum, mountain cherry, Florida sand plum, sandhill plum	Scaly-black bark, reddish side branches, small oval-green leaves, fragrant small white blossoms with reddish-orange antlers
Flatwoods	*Prunus umbellata*	12–20 feet tall	Hog plum	Oblonglance-shaped leaves, smooth reddish brown to black bark, showy white flowers in early spring, flowers have five petals and multiple stamens and are borne in single or groups of five

TABLE 5.2
Types of Flowering in Plum Trees with Pink and Purple Leaves

Species name	Scientific name	Flower size	Common name	Description
Thunder cloud Purple Leaf	*P. cerasifera*	15–25 feet tall	Cherry plum, Myrobalan plum	Deciduous tree, purple leaves, pink flowers, 1 inch long
Blireana	*P.* × *blireana*	15–25 feet tall	Purple-leafed plum	Deciduous tree, bronze or purple-red leaves, showy dark-pink flowers in winter and spring
Pissard	*P. cerasifera Atropurpurea*	15–25 feet tall	Purple cherry plum, purple leaf plum, Newport cherry plum	Deciduous tree, alternate leaf arrangement, pinkish-white flowers that bloom in January and February
Krauter Vesuvius Purple Leaf	*P. cerasifera Krauter Vesuvius*	15–20 feet tall	Cherry plum	Dark reddish-purple leaves, pink bloom grows as a dense, upright-rounded tree

The Picard (*P. cerasifera atropurpurea*) plum has pink and white flora that begin blooming in the spring. Its leaves become astonishingly red immediately, dark to purple through mid-summer, and become green-bronze by the season's end. Pissard plum trees produce small purple fruits, and their fall tone is purple (Table 5.2). Krauter Vesuvius (*P. cerasifera*) has the darkest foliage of sprouting plum trees—a dull purple color. It may produce fruit, though unpredictably.

5.11.3 Dwarf and Semi-Dwarf

Semi-dwarf blooming cultivars consist of Purple Pony, which attains a height of 12 feet and delivers profound purple leaves and single, pale-pink blossoms, and Enormous Cis, which reaches a height of 14 feet and has large, purple-colored leaves, white flowers, and small, dark purple fruits.

Dwarf blooming plum trees grow up to a height of 8–10 feet, and semi-dwarf blends grow to only a few feet tall. For instance, the Dwarf Natal Plum (*Carissa macrocarpa*) is an evergreen ground-cover plant that ends up being 12–18 inches high yet spreads 4–8 feet wide. It has green leaves and forked spines. Its white, star-framed blossoms sprout in spring. The Red-Leaf plum (*P. x cistena*) is 6–10 feet tall and 6–8 feet wide. The tree produces single white sprouts and small, faint purple fruits.

5.11.4 Fruiting

Many sprouting plum varieties produce fruits of several tones, surfaces, tastes, and sizes. The American goose plum (*P. americana*) has dwarf bark and thorny branches and grows up to 25 feet tall and 20 feet wide. Its tiny, white blooms and ruddy yellow fruits attract birds and other wildlife. The Mexican (*P. Mexicana*) combination of plum trees grow to around 30 feet tall and 25 feet width. They structure shades of pale leaves that are fleecy on one side and shiny on the other. Their white blooms sprout in spring, and their tart-sweet fruits are eaten fresh or made into jams. Deciduous leaves on Mexican plum trees become a splendid shade of orange in the fall.

5.12 FRUIT SETTING IN PLUM

Ovary retention when influenced by the stimulus of pollination is called as setting fruit or fruit set. Pollinators play a vital role in providing stimuli for the fruit set. Ovary growth continues until the flower matures sexually and stops at anthesis or just before pollination. If pollination occurs, growth resumes. Ovaries remain receptive to pollen for a specific period of time, and if they remain unpollinated they undertake senescence or abscission (Srivastava 2002). Temperature affects fruit set strongly, but extreme temperatures may induce seed abortion. The most favorable temperature for fruit set is about 25°C, which is associated with speedy pollen tube growth and quick fertilization (Table 5.3). At 20°C, the pollen tube grows slowly and leads to late and reduced fertilization (Srivastava 2002).

TABLE 5.3
Fruit Setting in Plum Species

European plum			Japanese plum	
Group	Fruit set	Frequency of cultivars	Group	Fruit set
Low	<10%	10.3%	Low	<5%
Intermediate	10–20%	22.4%	Intermediate	5–10%
High	20.1–40%	54.0%	High	>10%
Very high	>40%	10.3%		

5.13 INITIAL AND FINAL FRUIT SET IN PLUM CULTIVAR POZNA PLAVA AS AFFECTED BY DIFFERENT TYPES OF POLLINATION

Plum cultivar Pozna plava is exposed to various sorts of fertilization in agroecological states of Serbia. Under self-fertilization and cross-fertilization conditions (with Čačanska najbolja and Prezenta as pollinizers). The advancement, term, bounty, blooming time cross-over of Pozna plava, Čačanska najbolja, and Prezenta plum cultivars were observed. Pozna plava is a late-sprout cultivar with medium blooming bounty and medium in vitro dust germination (25.62%). An organic fruit set under open-fertilization conditions was higher in the second year of examination because of positive climate conditions during the blooming time frame (Table 5.4). Real outcomes demonstrate that Pozna plava has a low last fruit set with Čačanska najbolja and Prezenta as pollinators (Milošević and Milošević 2018).

Fruit set and yield are clearly subject to genotype inside Rosaceae (plums). Nonetheless, even after sufficient fertilization, a couple of blossoms form into fruits, demonstrating that some different elements are innate to the actual bloom. Insufficient fruit set in plums might be because of a hereditary inclination for unusual incipient organism sac improvement and low-temperature conditions during and after blossoming time that result in weak dust tube development (Jaumien 1968).

Rounplava is one of the most recent maturing plum cultivars with high-quality fruits. Notwithstanding the great fruit production, development of this plum in this space has been restricted, essentially because of a lacking organic fruit set.

Rounplava trees produce huge quantities of blossoms; however, the fruit set rate is extremely low. Improved creation of these plums would be possible by planting appropriate pollinator plants. This examination researched the similarity of two dust contributors for Pozna plava considering dust germination in vitro and the impact on underlying and last fruit set. The outcomes have significant ramifications for the major foundation for this yield.

The decline of a fruit set after self-fertilization could be because of another impact, like pistil-to-pistil variety in ovule treatment, post-zygotic dismissal, or incipient organism early terminations, which open new windows for future examination. In the subsequent year, fruit sets were somewhat higher than after self-fertilization. These outcomes demonstrate that it is essential to explore the impact of these pollinators and some more on the fruit set of Pozna Plava to get a suitable foundation for this harvest.

5.14 IMPROVING FRUIT SET IN PLUM

Plums bloom in spring, and half of the blossoms ought to be. Getting reliable yields in plums can be trying due to:

- A lower-than-required degree of pollination
- A lack of a viable pollinizer
- Frost during blooming
- Excessive levels of nitrogen in the dirt
- Biennial bearing

When getting steady yields in plums, there is freedom to develop fruit sets and yields (Defieldsja 2019).

5.14.1 Pollination

Plums are one of the most punctually blossoming trees. However, cool and wet sprouting climates can adversely affect honeybee movement. A few investigations propose that some honeybee species, like the blue plantation bricklayer honeybee, are more qualified to pollinate fruit trees. These lone honeybees are local to Canada and the United States and can assist with expanding fertilization in plantations as the new age of honeybees arises in the coincidence of apple and peach sprouts in late winter.

Research has shown that blossoms of weeds and wild species might be more interesting to bumblebees than the blossoms of tree fruits. Producers ought to oversee different weeds and blooming plants that are near plum plantations to guarantee they do not meddle with plum fertilization. It is better to cut the plantation floor before blossoming to decrease rivalry with dandelions.

Most stone fruit cultivators are aware of the significance of shielding pollinators from poisonous pesticides (Fantinato et al. 2018; Tison et al. 2016). With more than 23 local species in Ontario and economically accessible settling boxes, these honeybees are an optimal pollinator for late winter crops. Local honeybees can be urged to settle by leaving exposed patches of earth, old empty stems, and old wood as likely environments. Bumblebees are additional decent late winter pollinators. Business producers ought to have at least one hive of bumblebees per section of land of plum trees (Council 2007).

5.14.2 Pollinizer

Numerous plum assortments require cross-pollination to set natural products. A pollinizer assortment should be planted near the principal plum cultivar, ideally in a 1:8 dissemination of pollinizer to primary cultivar. There ought to be something like two pollinizers for self-sterile plum cultivars, with one pollinizer starting to blossom two days prior and the other two days after self-sterile cultivar.

5.14.3 Frost

Cold temperatures during sprouting can harm the blooms, bringing about diminished yields. Wind machines and other ice alleviation methodologies might assist with reducing the effect of the chilly temperatures on products.

5.14.4 Nitrogen

The examination has shown that excessive nitrogen levels in the dirt can adversely affect yields. Gather soil tests like clockwork and leaf tests every year to further develop manure applications.

5.14.5 Biennial Bearing

Plums are biennial bearing, and a guard crop one year regularly brings about a need for a shine crop the following. There are a few alternatives that attempt to relieve these issues. Studies have shown that fruit diminishing can bring about a steadier yield from one year to another. Another investigation has shown that applying Gibberellic acid and an auxin at four and a half months after petal fall has been displayed to bring about expanded fruit set in Victoria plums (Erogul and Sen 2015). Nonetheless, Gibberellic acid not enlisted for this utilization in Canada.

5.15 AMERICAN PLUM AS CROSS-POLLINATOR

The American plum and its varieties cross-fertilize with the Japanese plum; both grow in USDA Plant Hardiness Zones 5–9. The American plum is developed for ornamental purposes in a natural environment and for its fruit. It is helpful as a pollinator for Japanese plums, as they blossom in the same season.

5.15.1 Cross-Pollination

Self-fertile or self-pollinating trees are defined as trees whose blossoms set fruit when they are pollinated from flowers on the same tree, a tree of the same cultivar, or a tree of the same species. According to the University of Missouri Cooperative Extension Service, cross-pollination transfers pollen between different species or varieties.

5.15.2 Cultivars

Cultivars of American plums include Hawkeye, Blackhawk, De Soto, Fairlane, Alderman, Weaver, Stoddard, Tecumseh, Toka, and Waneta. The Japanese plum cultivar Santa Rosa cross-pollinated with American plums does well in the Mediterranean, requiring little winter cold to bloom and set fruit.

5.15.3 Japanese Plums

Japanese plums sprout early and are not appropriate for regions with late ices. They have a low chill necessity—some will prove fruitful with just 250 chill hours, making them reasonable for hotter environments. Their fruits arrive in various tones, going from profound purple to brilliant yellow. The Methley variety is a self-ripening Japanese plum bearing small, ruddy purple fruits, and the St. Nick Rosa delivers large, ruddy red fruits. Other self-prolific assortments include the Catalina and Early Orleans, with dark blue skin, and the Magnificence, with blush red skin.

5.15.4 European Plums

Most European plums have dull purple skins with white wax, causing them to seem blue. Once in a while, they all have golden tissue and a hint of red. Most will endure lower temperatures than the Japanese, frequently challenging Zone 4 winters. There are a lot more self-productive European plums than Japanese plums. A couple of the most well-known assortments are the Stanley and the Italian, both with purple skin and golden tissue, and the Green Gage, with greenish-tan fruits grown from the ground tissue.

5.16 CONCLUSION

There are many native species of plum, depending on their growth habitat. Plums go through various stages of flower bud initiation, provided with adequate conditions. Generally, the development of flowers in Japanese plums is low compared to other varieties. The physiology of flowering involves bud formation, including the differentiation of the flower bud, the development of the flower bud, and the opening of the flower. Step-by-step events occur during bud formation with the flow of genes. Plum shows a high capacity for gene transfer, with a broad span of genetic diversity. The *Prunus* species gives a vast and complicated ability for pollen grains.

Ethylene is considered a key regulator of ripening in climacteric fruits. We demonstrate that auxin sensitivity variations between plum are due to PSIAFB5, which encodes auxin receptors. Transgenic plants expressing a high level of F1 are produced in the greenhouse in one month.

The pleiotropic aggregates related to FT1 expression in plum recommend a vital job of this quality in plant development and advancement. This examination exhibits the potential for a solitary transgene occasion to extraordinarily influence the vegetative and regenerative growth and progress of a commercially significant mild perpetual woody harvest. We propose that FT1 may be a helpful instrument to adjust calm plants to changing environments and additionally to adjust these yields to new developing regions.

5.17 REFERENCES

Allman, M. 2019. *How Old Is a Plum Tree Before It Bears Fruit?* https://homeguides.sfgate.com/old-plum-tree-before-bears-fruit-60024.html.

Ashraf, S., Ahad, S., Wani, M.Y., and Kumar, A. 2020. Physiology of Flowering in Apple and Almond: A Review. *International Journal of Current Microbiology & Applied Science* 9 (9): 1912–1929.

Bastias, A., Oviedo, K., Almada, R., Correa, F., and Sagredo, B. 2020. Identifying and Validating Housekeeping Hybrid Prunus spp. Genes for Root Gene-Expression Studies. *PLoS One* 15 (3): e0228403. doi: 10.1371/journal.pone.0228403.

Boss, P.K., and Thomas, M.R. 2002. Association of Dwarfism and Floral Induction with a Grape 'Green Revolution' Mutation. *Nature* 416 (6883): 847–850.

Caprio, J.M., and Quamme, H.A. 1999. Weather Conditions Associated with Apple Production in the Okanagan Valley of British Columbia. *Canadian Journal of Plant Science* 79 (1): 129–137.

Chin, S.W., Shaw, J., Haberle, R., Wen, J., and Potter, D. 2014. Diversification of Almonds, Peaches, Plums and Cherries—Molecular Systematics and Biogeographic History of Prunus (Rosaceae). *Molecular Phylogenetics and Evolution* 76: 34–48.

Cici, S.Z.H., and Van Acker, R.C. 2010. Gene Flow in Prunus Species in the Context of Novel Trait Risk Assessment. *Environmental Biosafety Research* 9 (2): 75–85.

Council, N.R. 2007. *Status of Pollinators in North America.* National Academies Press.

Defieldsja. 2019. *Improving Fruit Set in Plums*. https://onfruit.ca/2019/05/16/improving-fruit-set-in-plums/.

Erogul, D., and Sen, F. 2015. Effects of Gibberellic Acid Treatments on Fruit Thinning and Fruit Quality in Japanese plum (Prunus salicina Lindl.). *Scientia Horticulturae* 186: 137–142.

Fantinato, E., Del Vecchio, S., Silan, G., and Buffa, G. 2018. Pollination Networks Along the Sea-Inland Gradient Reveal Landscape Patterns of Keystone Plant Species. *Scientific Reports* 8 (1): 1–9.

Guerrero, B.I., Guerra, M.E., and Rodrigo, J. 2020. Establishing Pollination Requirements in Japanese Plum by Phenological Monitoring, Hand Pollinations, Fluorescence Microscopy and Molecular Genotyping. *JoVE (Journal of Visualized Experiments)* 165: e61897.

Hickman, G.W. 2004. *Home Fruit Trees in Mariposa County, Backyard Horticulture*. http://cemariposa.ucanr.edu/newsletters/Chilling_Hours_for_Home_Fruit_Trees36985.pdf.

Hu, J., Liu, Y., Tang, X., Rao, H., Ren, C., Chen, J., Wu, Q., Jiang, Y., Geng, F., and Pei, J. 2020. Transcriptome Profiling of the Flowering Transition in Saffron (Crocus sativus L.). *Scientific Reports* 10 (1): 1–14.

Jackson, J.E., and Hamer, P.J.C. 1980. The Causes of Year-to-Year Variation in the Average Yield of Cox's Orange Pippin Apple in England. *Journal of Horticultural Science* 55 (2): 149–156.

Janick, J., ed. 1998. *Horticultural Reviews*, vol. 22. John Wiley & Sons.

Jaumień, F. 1968. The Causes of Poor Bearing of Pear Trees of the ariety 'Doyenne du Comice'. *Acta Agrobotanica* 21: 75–106.

Metzger, J.D. 1995. Hormones and Reproductive Development. In *Plant Hormones*, 617–648. Springer.

Milošević, T., and Milošević, N. 2018. Plum (Prunus spp.) Breeding. In *Advances in Plant Breeding Strategies: Fruits*, 165–215. Springer.

Searle, N.E. 1965. Physiology of Flowering. *Annual Review of Plant Physiology* 16 (1): 97–118.

Silver, T. 2021. *Varieties of Flowering Plum Trees*. https://homeguides.sfgate.com/varieties-flowering-plum-trees-50541.html.

Simpson, M. 2021. *The Chilling Requirements for a Methley Plum Tree*. https://homeguides.sfgate.com/chilling-requirements-methley-plum-tree-55867.html.

Srinivasan, C., Callahan, A., Dardick, C., and Scorza, R. 2010. *Expression of the Poplar Flowering Locus T1 (FT1) Gene Reduces the Generation Time in Plum (Prunus Domestica L.)*. Paper presented at the 28th International Horticultural Congress.

Srinivasan, C., Dardick, C., Callahan, A., and Scorza, R. 2012. Plum (*Prunus domestica*) Trees Transformed with Poplar FT1 Result in Altered Architecture, Dormancy Requirement, and Continuous Flowering. *Plos One*. https://doi.org/10.1371/journal.pone.0040715.

Srivastava, L.M. 2002. *Plant Growth and Development: Hormones and Environment*. Elsevier.

Tison, L., Hahn, M.L., Holtz, S., Rößner, A., Greggers, U., Bischoff, G., and Menzel, R. 2016. Honey Bees' Behavior Is Impaired by Chronic Exposure to the Neonicotinoid Thiacloprid in the Field. *Environmental Science & Technology* 50 (13): 7218–7227.

6 Nutrition and Orchard Manuring Practices of Plum Trees

S. Parameshwari[1], *Jobil J. Arackal*[2], *Sithara Suresh*[2] *and Imtiyaz Ahmad Malik*[3]

[1] Department of Nutrition and Dietetics, Periyar University, Salem 636011, Tamil Nadu, India

[2] Department of Food Technology, Saintgits College of Engineering, Pathamuttom, Kottayam—686532, Kerala, India

[3] Agriculture Production & Farmers Welfare Department, Government of UT of J&K, India

CONTENTS

DOI: 10.1201/9781003205449-6

6.1 INTRODUCTION

A sweet fruit that grows on the plum tree, whose botanical name is *Prunus domestica*, is a drupe commonly known as a plum. The characteristic feature of plums is that they have a terminal bud and non-clustered solitary side buds with a fleshy fruit that has an edible outer part and a groove on one side that is enclosed in a smooth stone-like shell where the seed is enclosed. The term 'plum' is also used in a generalized manner for the species of the genus *Prunus*, which also includes bird cherries, peaches, and cherries. The fruits of plum trees are defined as smooth-skinned stone fruit (Naser et al. 2018).

The plum tree has multiple benefits. Almost all parts of the tree are found to be useful in one way or other. Ecologically speaking, plum trees provide flowers to insects that pollinate, while the fruits are consumed by animals. Plums are quite healthful for humans and also have a very pleasant taste. Additionally, plum trees can serve as ornamental trees where the color of the species varies based on their delicate and attractive flowers. Moreover, a number of plum varieties are developed based on their growing attributes and other factors (Bender and Bender 2005).

All over the world, plums are grown in temperate climate zones (Somogyi 2005). Plums have quite an interesting history in that plum cultivation has been going on since time immemorial compared only to the apple. The earliest data reveal that plums find their origin in China in 470 BCE, while European plums were discovered over 2,000 years ago, somewhere closer to West Asia or East Europe (Bhutani and Joshi 2005). Plums seem to have entered the United States in the 17th century and slowly through the entire world. Today, plum cultivation is going on through the entire world in all temperate climate countries. *Prunus domestica* was bred first in Europe, while *Prunus americana* originated in North America. The cherry plum was cultivated in South Asia, and the damson plum, or *Prunus salicina*, has its origin in western Asia (Birwal et al. 2017). Table 6.1 shows the estimate of global plum production.

TABLE 6.1
The Estimate of Global Plum Production

Position	Country	Production in millions of tons (MT)
1.	People's Republic of China	6,788,107
2.	Romania	842,132
3.	Serbia	430,199
4.	United States of America	368,206
5.	Iran	313,103
6.	Turkey	296,878
7.	India	251,389
8.	Chile	229,951
9.	Morocco	205,222
10.	Ukraine	198,070

*Top 10 major plum producers in the world in 2018, as per the estimated data from the Food and Agriculture Organization of the United Nations

In India, plums are grown mainly in the Kullu and Solan districts of Himachal Pradesh, while low-chill varieties are grown in Punjab, Haryana and Eastern Uttar Pradesh.

6.2 NUTRIENT COMPOSITION OF PLUM FRUIT

Plum, similar to all fruits, is a very good source of sugar with a small amount of starch. The acids present in plum include quinic acid and malic acid, which impart the characteristic acidic taste of fruits. Plum also contains a significant amount of vitamin C. Plum is a rich source of antioxidants, with phenolic substances such as catechin, caffeic acid, chlorogenic acid, rutin and phenolic acid, which give it its specific astringent characteristics. The major anthocyanins include cyanidin-3-rutinoside and peonidin-3-rutinoside. The fruit's seed contains amygdalin, a glucoside. Compounds of phenols which are found in fruits and vegetables are known for their health-promoting and antioxidant effects, thus aiding in disease prevention. The total antioxidant capacity (TAC) and total phenolic content (TPC) of plums have been under study for several years and have proved to be excellent (Kazim 2012). Tables 6.2 and 6.3 show the nutritional composition and phenolic components present in plums (Augur et al. 2004).

TABLE 6.2
Nutrient Composition per 100 Grams of Plums

Nutrient	Amount
Moisture	86.9
Protein	0.7 g
Fat	0.5 g
Minerals	0.4 g
Fibre	0.4 g
Carbohydrate	11.1 g
Energy	52 K. cal
Calcium	10 mg
Potassium	12 mg
Iron	0.6 mg

**Source*: Nutritive Value of Indian Foods, National Institute of Nutrition, Hyderabad

TABLE 6.3
Phenolic Components Present in Plums (Augur et al. 2004)

Acids	Quantity (mg/kg)
Neochlorogenic acid	85–1300 mg/kg
Chlorogenic acid	13–430 mg/kg
Cryptochlorogenic acid	956 mg/kg

6.2.1 Health Benefits of Plums

- As plums contain isatin, sorbitol and dietary fibre, the functioning of one's digestive system is regulated, thus relieving constipation.
- The vitamin C present in plums provides immunity against infectious diseases and thereby scavenges the free radicals.
- It has a moderate amount of beta-carotene content, preventing lung and oral cancer.
- Carotenoids, such as cryptoxanthin, lutein and zeaxanthin, are present in plums and act as antioxidants.
- Minerals, including fluoride, potassium and iron, are present in plums and help in the maintenance of bodily functions.
- Plums have moderate amounts of B-complex vitamins, which help in the metabolism of carbohydrates, proteins and fats.
- Plums provide 5% RDA levels of vitamin K, which is important in blood clotting, and recent findings reveal that it helps reduce the risks of Alzheimer's disease in the elderly.
- Studies show that plums can prevent macular degeneration, protect against heart disease and reduce damage to neurons (Birwal et al. 2017).

6.3 CURRENT CULTIVATION PROCEDURES

Plums are generally cultivated under rain-fed conditions in places which receive an average annual rainfall of 100–125 cm. Moreover, fruitfulness can be obtained through winter chilling, otherwise it will remain vegetative. An important developmental phase is dormancy, which can allow the trees to survive unfavorably low winter temperatures. This dormant phase can be achieved only through winter chilling. The period of winter chilling varies depending on the varieties; European plums can be grown in places where the chilling is below 7°C for 800–1,500 hours, while Japanese plums only need 100–800 hours. The elevation of European plums can be 1,525–2,745 ft above mean sea level, while Japanese plums can grow from subtropical plains to an elevation of 1,525 ft above mean sea level. Indian plums are grown in both plains and sub-mountain areas of Punjab with fewer than 300 chilling hours and a rainfall of 100–125 cm distributed throughout the growing season. Wind-break trees should be planted to protect plum orchards and to prevent damage through high wind velocity. Areas that are free from spring frost are conducive for plum cultivation. Clay soil is better for European plums, while lighter soils are preferred by Japanese plums (Teskey and Shoemaker 1978). In general, plums can survive well in deep, well-drained loamy soils with a depth of 699 mm and soil pH between 5.5 and 6.5. Good drainage that is free of alkalinity and salinity conditions along with good-quality irrigation water is of prime importance (DAFF 2010). A water deficit can lead to the failure of crops commercially. Soil erosion can be prevented if plums are planted on gentle slopes. Waterlogging, poorly drained soil with excessive salts, should be avoided. Soil profile pits can be carried out to inspect soil to determine the best method for preparing the soil.

6.4 PROPAGATION AND ROOTSTOCK

Rootstocks are used for plums, though seedlings are used for almonds, peaches, and Japanese apricots. Shield budding is the technique used in India, where wild apricot is used as rootstock. The clonal selections for plums include Clone GF-43 (*Prunus domestica*), GF-655/2, Montizo 20 and Montpol 21 (*Prunus institia*) (Rehalia 2004). Seedling or clonal rootstock is commonly followed in the autumn or spring for T-budding or chip budding. Propagation of these can be done through leafy, softwood and hardwood cutting under intermittent mist (Bhutani and Joshi 1995). When there is a dormant stock and scion, propagation through cleft or tongue grafting is done. Rootstock seedlings can be raised by keeping the seeds under alternate layers of moist sand for a period of 1–3 months of stratification, depending on the species at temperatures of 3–5 degrees centigrade to break the rest period before they germinate.

6.5 PLANTING

Usually, plums are planted in the first fortnight of January in solid blocks with a distance of 6 m x 6 m as a filler tree in mango, litchi, and pear orchards. For instance, in India, in the orchard of Satluj Purple, Kala Amritsari is used as pollinizer, which implies that the pollinizer plants are planted alternately to improve the pollination in a specific fruit orchard. Through this practice, a quality fruit with a high yield can be expected.

6.6 ESSENTIAL NUTRIENTS FOR THE GROWTH OF PLUMS

As many as 12 essential nutrients are required for normal growth and optimum production of fruit trees in varying amounts. Regarding fruit-bearing trees, the macronutrients include nitrogen, potassium, phosphorouscium, sulphur and magnesium. The micronutrients are iron, manganese, boron, zinc, copper, chlorine and molybdenum (Chadha 1998).

6.6.1 Nitrogen

Nitrogen and potassium are found to be the primary nutrients required by plum plants (Marschner 1995). But ironically, excessive nitrogen can cause the plant to lose its colour, quality, flesh firmness and sweetness, and the plants can also become more prone to post-harvest diseases (Cuquel et al. 2011). Studies show that a deficiency of nitrogen can result in poor yield and a shorter span of leaf maintenance as well as shorter reserves accumulation for posterior cycle. The leaves can also be paired with decreased shoot growth and reduced flower numbers because of a nitrogen deficiency. Trees which are deficient in nitrogen are more prone to bacterial canker (Olson et al. 1982). In the case of a severe deficiency, the leaves might appear scorched and may face early defoliation (LaRue and Gerdts 1973). But symptoms alone cannot identify nitrogen deficiency; rather, tissue analysis is essential to confirm the issue (Niederholzer et al. 2012).

Nitrogen excess can lead to excessive shoot growth in both plums and prunes, which can increase the cost of pruning and shades, which might force the fruit to be developed in a lower canopy, which can delay maturity and decrease the sugar content. There can be a decrease in the colour and quality of the fruits, and the trees will be more vulnerable to brown rot (Pope 2014).

6.6.2 Potassium

The productivity, quality and potential of storage are negatively affected if there is a potassium deficiency. Poor to lack of colour leading to a dirty look was also seen with potassium deficiency (Johnson and Uriu 1989). Additionally, the drying ratio is increased and the size of the prune and soluble solids are reduced (Niederholzer et al. 2012). The most exposed part of the tree is more affected by potassium deficiency, and hot and dry weather triggers such a condition. Midsummer is the most common season for potassium deficiency, and generally it is more noticed in the upper canopy where the leaves turn pale and a tan colour is developed with scorching in the margins (Johnson and Uriu 1989). Hot temperatures and heavy cropping may result in the scorching and dropping of the entire leaf, leading to bark sunburn. This makes the tree prone to a cytospora canker infection, leading to limb and scaffold death (Niederholzer et al. 2012). Severe deficiencies can lead to limb 'dieback' through the tree top (Lilleland 1932). Potassium-deficient trees have bare, defoliated leaves after harvest. There might be yield reduction in serious potassium deficiency, and if it is untreated it might even kill the tree (Lilleland 1932). Japanese plums are more resistant to cytospora infection and potassium deficiency thanEuropean plums (DeVay et al. 1974).

6.6.3 Phosphorous

Phosphorous is recycled from the leaves before they fall (Strand 1999). The leaves of phosphorous-deficient trees are smaller than normal, dark green in colour, with a leathery texture and may fall easily. The young shoots, leaves and stems develop a purplish colour. More pronounced symptoms are seen in actively growing shoots and as vegetative growth slows down, the symptoms decrease or disappear. Overall, the fruits are smaller in size and ripen earlier and have poor flavour. Ultimately, there is also a reduction in yield (Shear and Faust 1980).

6.6.4 Zinc

Zinc deficiency is more common in plum orchards that are grown in sandy soils. There is limited availability of zinc in soils which have a greater pH than 6.5 or heavily manured lands. Zinc deficiency was found in 72–84% of the samples in a study conducted in Sacramento Valley in 2003–2004 (Olson et al. 2004). Obvious zinc deficiency is seen in the spring when the deficiency is severe. Due to zinc deficiency, there is a delay in leaf bud opening with small yellowish leaves which form rosettes, or tuft, and this condition is called 'little leaf' (Uriu 1981). There is only a slight reduction in the size of the leaf when there is a slight deficiency. Leaves that are affected might have wavy margins and dormant flower buds might fall (DeBuse 2012). The entire yield might be lost in cases of severe deficiencies, and dieback might result (Niederholzer et al. 2012).

6.6.5 Magnesium

Magnesium is essential for greener trees, to prevent leaf fall, and to fight chlorosis. Lack of magnesium can result in decreased photosynthesis, leading to smaller leaves which have a lighter colour and poor flowering with a decreased fruit size and fruit drop. Magnesium deficiency can be made worse by sandy acidic soils which are rich in potassium. Soil which has higher potash applications can face a magnesium deficiency. Prolonged cold and wet periods can lead to magnesium deficiency. Symptoms of interveinal chlorosis and bright yellowing can develop primarily at the margins and tips of leaves and can spread back to the main vein. The chlorosis thus formed can have a characteristic herringbone pattern of chlorosis which becomes obvious during fruit fill. Leaves curl up and fall prematurely in severe instances.

6.6.6 Copper

Copper is essential for vigorous and healthy growth, prevention of premature leaf fall and increase in yield. Generally copper deficiencies are rare in stone fruit, as copper sprays are used to correct diseases such as bacterial canker. Copper deficiency can be aggravated by organic, chalky, sandy soils with increased nitrogen applications. Symptoms of copper deficiency include chlorosis, where the older leaves become dark green and progress to yellowish green to yellow between veins. Leaf scorch is seen along the margins in severely affected plants and may have wavy margins and leaf drop (Johnson and Uriu 1989).

6.6.7 Iron

Iron is essential to prevent premature leaf fall, to produce healthy green trees and to fight chlorosis. Iron deficiency is worsened when the pH of the soil is high, the soil is calcareous and waterlogged and contains high copper, zinc and manganese levels. Iron deficiency can lead to chlorotic yellowing and bleaching affecting new leaf growth, though the leaf veins are green. The severity of the condition is aggravated towards the youngest leaves at the shoot tip. Severe deficiency can lead to leaf scorching (Johnson and Uriu 1989).

6.6.8 Boron

Though boron is a micronutrient, it is essential for bud development, improvement in flowering and fruit set, better skin finish and prevention of storage disorders (Shear and Faust 1980). Boron deficiency is generally found in the fruit or tree wood, with lesser incidence towards leaves, unless the deficiency is severe. Boron deficiency can be made worse by sandy, alkaline soil which is low in organic matter and high levels of nitrogen and calcium. A symptom of boron deficiency is wood cracking, where the new wood becomes rough and blistered while the old wood cracks and splits (Johnson and Uriu 1989).

6.6.9 Calcium

Calcium deficiency is quite rare in trees, as most soils contain abundant calcium. As the ions are adsorbed or dissolved in the soil, there is generally less calcium deficiency. Deficiency symptoms are rarely seen, except sometimes there is reduced shoot growth because of shortened internodes with dieback and defoliation. Chlorotic patch is seen often as leaves develop before abscission (Johnson and Uriu 1989).

6.6.10 Sulphur

Sulphur is seen abundantly in soil from lipids, amino acids and proteins which are broken down by microorganisms to inorganic sulfates (Chapman 1973). Sulphur deficiency is rare, and if it occurs, the symptoms are similar to that of nitrogen deficiency except that in sulphur deficiency the symptoms occur on young leaves, as sulphur cannot be remobilized from older leaves. The deficiency can be rectified easily by applying ammonium sulfate, elemental sulphur or gypsum (Chadha 1998).

6.6.11 Manganese

Manganese is required in a lesser quantity, though the manganese content varies from one soil to another (Chadha 1998; Mengel and Kirby 1982). Interveinal chlorosis, which extends from midrib to margin, occurs due to a manganese deficiency. Wide bands are seen around the major veins, which remain green, giving it a herringbone pattern. When compared with other plants, the symptoms are seen on older basal leaves. Though the entire tree looks pale, the terminal leaves are found to be green. In case of severe deficiency, the size of the leaf, shoot growth and yield are affected seriously. Deficiency can be corrected by the application of foliar sprays of about 3–5 lb/100 gal (0.4 to 0.6 kg/100 l) of manganese sulfate. Sometimes the deficiency can be due to high pH of the soil; this can be rectified by lowering the soil's pH.

6.6.12 Molybdenum

Molybdenum deficiency is rarely seen, as plum plants require the least amount of this nutrient. Molybdenum content, similar to manganese, varies from soil to soil. Molybdenum deficiency symptoms are similar to those of nitrogen deficiency. Though symptoms have not been noticed, studies on induced soil deficiency of molybdenum showed dwarf leaves, diffuse mottling of leaves and irregular areas of dead tissues (Chadha 1998; Shear and Faust 1980).

6.7 FERTILIZERS AND MANURES

The amount of fertilizer needed is governed by a number of factors, including soil type, climate, fertility, tree age, planting density, crop load and shoot growth. The tree's manure and fertilizer requirements are established after leaf and soil

TABLE 6.4
Manure and Fertilizer Schedule for Plums and Peaches

Age of tree (year)	FYM (kg)	N (9 g)	P_2O_5 (g)	K_2O (g)
1	10	70	35	100
2	15	140	70	200
3	20	210	105	300
4	25	280	140	400
5	30	350	175	500
6	35	420	210	600
7 and above	40	500	250	700

**Source*: Unit 2—Peach and Plum -egyankosh

analysis. The plum needs a sufficient amount of nutrients to ensure optimum development of fruit quality. Soil fertility, soil type, topography, tree age, cultural practices and crop load all influence manure and fertilizer application. The amount of fertilizer required varies by region. Farmyard manure (FYM) needs to fill the quantity of phosphorus and potassium and must be applied in December and January. The first partial nitrogen dose is sprayed before flowering in the spring, and the second half is applied a month later. Under rain-fed conditions, in a single application 15 days before bud break, nitrogen fertilizers should be applied. FYM, together with the full doses of P_2O_5 and K_2O, should be used between December and January. Manure and fertilizer should be sprinkled and lightly mixed into the soil surface beneath the tree's spread. To reduce nutrient leaching, band application of fertilizers is preferred over broadcasting in areas with significant rainfall and steep slopes Chadha 1998). Table 6.4 shows the manure and fertilizer schedule for plums and peaches.

The requirement of fertilizer varies according to the region and mostly depends on the climate, plant species and age, fertility, type and pH of soil, rootstock and moisture. The rough estimate is that a hectare of plum orchard removes 4.3–6.5 kg P_2O_5, 30.1–30.9 kg N and 24.2–38 kg K_2O annually, while only 1.2 kg N, 0.4 kg P_2O_5 and 1.1 kg K_2O are removed by bearing trees. The plums grown in the plains of Punjab require 200 g of nitrogen, 75 g of phosphorous and 200 g of potassium per tree with a minimum dose of FYM at 35 kg per tree, which is applied in December along with the full dose of phosphorous and potassium and half of the nitrogen one month before flowering and another half after a month of maximum yield. In Meghalaya, for a fruit-bearing tree of 10–13 years of age, a dose of about 200–205 g P, 100g N and 20–100g K per tree per year is the optimal requisite for maximum yield. In Himachal Pradesh, 50 g N, 60 g K, 25 g P along with well-decomposed FYM is what is recommended. Again, nitrogen is applied generally in split doses, once before flowering and once a month later. In plums which grow during the summer months, when a zinc deficiency is observed, 3 kg of zinc sulfate in 500 l of water can be sprayed. Misshapen fruits with corky spots resulting in fruit cracking is a sign of boron deficiency, which can be rectified by spraying 0.1% boric acid in the month of June. Studies report that plum trees treated with biofertilizers (60 g each/tree basin)

TABLE 6.5
FYM, Urea Super Phosphate and Muriate of Potash to Plum Tree

Age (years)	FYM (kg/tree)	Dose per tree (g) urea	Dose per tree (g) superphosphate	Dose per tree (g) muriate of potash
1–2	6–12	60–120	95–190	60–120
3–4	18–24	180–240	285–380	180–240
5–6	30–36	300–360	475–570	300–360
6 and above	36	360	570	360

* *Source*: Punjab Agricultural University, Ludhiana

+ NPK 75% + green manure (sunhemp at 25 g seeds/tree basin) showed the maximum annual shoot growth, tree height and volume and fruit set and yield (Johnson and Uriu 1989). Manure and fertilizers are applied to plum trees based on their age.

FYM, along with muriate of potash and super phosphate should be applied in December. Nitrogen fertilizer should be split into half and applied in spring before flowering and again a month after fruit set. In case of a zinc deficiency, a foliar spray of 0.6% zinc sulfate solution made from 3 kg zinc sulfate and 1.5 kg unslaked lime in 500 liters of water should be sprayed (Chadha 1998; Bhutani and Joshi 1995). A mixture of biofertilizers, organic fertilizers and chemical fertilizers can help in the prevention of plum tree diseases. Table 6.5 shows the FYM, urea super phosphate and muriate of potash to plum trees.

6.8 CONCLUSION

Plums are fruits which are grown universally, require minimum tending with high yield and are cultivable in almost all climates. Proper soil management, propagation techniques and periodical application of fertilizers and manure can give a very good plum fruit yield. In many mountains we can see plums growing as a wild tree. Since it can grow so well in a wide range of environments, if properly tended, it will provide a good yield.

6.9 REFERENCES

Auger, C., Al-Awwadi, N., Bornet, A., Rouanet, J.M., Gasc, F. et al. 2004. Catechins and Procyanidins in Mediterranean Diets. *Food Research International* 37: 233–245.

Bender, D.A., and Bender, A.E. 2005. *A Dictionary of Food and Nutrition*. Oxford University Press.

Bhutani, V.P., and Joshi, V.K. 1995. Plum. In *Handbook of Fruit Science and Technology: Production, Composition, Storage and Processing*, edited by D.K. Salunkhe and S.S. Kadam, 206. Marcel Dekker Inc.

Bhutani, V.P., and Joshi, V.K. 2005. Plums, Production, Composition, Storage and Processing, in: D.K. Blazek: A Survey of the Genetic Resources Used in Plum Breeding. *Acta Horticulture* 734: 31–45.

Birwal, P., Deshmukh G., Saurabh S.P., and Pragati S. 2017. Plums: A Brief Introduction. *Journal of Food, Nutrition, and Population Health* 1: 1.

Chadha, T.R. 1998. *A Text Book of Temperate Fruits*. ICAR.
Chapman, H.C., ed. 1973. *Diagnostic Criteria for Plants and Soils*. Department of Soils and Plant Nutrition, University of California.
Cuquel, F.L., Motta, A.C.V., Tutida, I., and De Mio, L.L.M. 2011. Nitrogen and Potassium Fertilization Affecting the Plum Postharvest Quality. *Revista Brasileira de Fruticultura* 33: 328–336.
DeBuse, C. 2012. *Prune Orchard Nutrition*. Sutter-Yuba Counties Pomology Notes, November.
Department of Agriculture, Forestry and Fisheries (DAFF). 2010. *Plums Production Guidelines*, 35. DAFF, Republic of South Africa.
DeVay, J.E., Gerdts, M., English, H., and Lukezic., F.L. 1974. Controlling Cytospora Canker in President Plum Orchards of California. *California Agriculture* 28: 12–14.
Johnson, R.S., and Uriu, K. 1989. Mineral Nutrition. In *Peach Plums and Nectarines: Growing and Handling for Fresh Market*, edited by J.H. La Rue and R.S. Johnson, 252. University of California, Cooperative Extension, 3331.
Kazim, G., and Onur, S. 2012. Variation in Total Phenolic Content and Antioxidant Activity of *Prunus cerasifera Ehrh: Selections from Mediterranean region of Turkey, Scientia Horticulturae* 134: 88–92.
LaRue, J.H., and Gerdts, M. 1973. *Growing Plums in California*. California Experiment Station Extension Service, Circular 563.
Lilleland, O. 1932. Experiments in K and P Deficiencies with Fruit Trees in the Field. *Proceedings of the American Society for Horticultural Science* 29: 272–276.
Marschner, H. 1995. *Mineral Nutrition of Higher Plants*, 2nd ed., 889. Academic Press.
Mengel, K., and Kirby, E.A. 1982. *Principles of Plant Nutrition*. International Potash Institute.
Naser, I., Al-Hamad, E.T., and Olbinado, E.E. 2018. Evaluation of Super High Density Planting of Stone Fruits Cultivars at TADCO, Tabuk, Saudi Arabia. *Research & Reviews: Journal of Agriculture and Allied Sciences* 7 (2): 41–72.
Niederholzer, F.J.A., Buchner, R.P., Southwick, S.M. 2012. Nutrition and Fertilization. In *Prune Production Manual*, edited by R.P. Buchner, 151–168. University of California, Division of Agriculture and Natural Resources, Publication 3507.
Olson, W., Andris, H., Buchner, R., Holtz, B., Klonsky, K., Krueger, W., and Walton, J. 2004. *Integrated Prune Farming Practices (I.P.F.P.): A Six Year Summary*. Report submitted to the California Dried Plum Board.
Olson, W., Ramos, D., Yeager, J., Uriu, K., and Pearson, J. 1982. *Efficient Nitrogen Application Timing in Prune Production*. Report submitted to the California Dried Plum Board.
Pope, K. 2014. *Potassium Nutrition- Maintaining Optimum Levels by Replacing What You've Lost*. Glenn County Orchard Facts, October.
Rehalia, A.S., and Tomar, C.S. 2004. Plum and Prune. In *Recent Trends in Horticulture in the Himalayas—Integrated Development under the Mission Mode*, edited by K.K. Jindal and R.C. Sharma, 128. Indus Publishing.
Shear, C.B., and Faust, M. 1980. Nutritional Ranges in Deciduous Tree Fruits and Nuts. In *Horticultural Reviews*, edited by J. Janick, vol. 2, 142–163. The AVI Publishing Company, Inc.
Somogyi, L.P. 2005. *Plums and Prunes Processing of Fruits Science and Technology*, 2nd ed., 513–530. CRC Press.
Strand, L.L. 1999. *Integrated Pest Management for Stone Fruits*. University of California, Division of Agriculture and Natural Resources, Publication 3389.
Teskey, B.J.E., and Shoemaker, J. 1978. *Tree Fruit Production*, 3rd ed., 358. AVI.
Uriu, K. 1981. Soil and Plant Analysis and Symptomology for Diagnosis of Mineral Deficiencies and Toxicities. In *Prune Orchard Management*, edited by D.E. Ramos, 89–97. University of California Division of Agricultural Sciences Special Publication 3269.

7 Recent Development in Plum Harvesting and Handling

Tridip Boruah[1], *Debasish Das*[1], *Gargee Dey*[1] *and Imtiyaz Ahmad Malik*[2]

[1] PG Department of Botany, Madhab Choudhury College, Barpeta-781301, Assam, India

[2] Agriculture Production & Farmers Welfare Department, Government of UT of J&K, India

CONTENTS

DOI: 10.1201/9781003205449-7

7.1 INTRODUCTION

Plum is one of the oldest cultivated plants; our ancestors started its cultivation more than 2,000 years ago. Plum is cultivated in many countries of the world, including China, Serbia, India, the United States, the United Kingdom, Romania, Turkey, Italy, Bosnia and Hungary. China holds the first position and Serbia is in the second position regarding the area of land occupied by plum trees. According to a recently published document, approximately 1,752,675 hectares of China and 158,000 hectares of Serbia are covered by plum trees (Matković 2015). Unsurprisingly, China produces 56.2% of the world's total plum production, which is 6,022,744 tons per year; the second largest producer of plum is Romania, with a production of 424,068 tons per year. Slovakia produces 25.45 tons of plum per hectare area, which is the most production in a unit area (Matković 2015). Plum is very underrated in the Indian subcontinent, and as a result it is a relatively new crop for Indian cultivators. Several states of India, including Uttar Pradesh, Himachal Pradesh, Jammu and Kashmir and the states of the Northeast region, came forward to cultivate different species of plum. A few varieties of plum that are cultivated in India are the Santa Rosa, Satsuma, Titron, Burbank, Grand Duke, Kelsey, Mariposa and Jamuni plums (Bhatt and Tyagi 2021). The Santa Rosa plum needs 115–120 days to reach maturity, and it is the most widely cultivated species in hilly regions. In Himachal Pradesh 1.31 tons of plum fruits are produced per hectare, in Jammu and Kashmir the number is 0.74 tons per hectare, in Uttar Pradesh it is 2.5 tons per hectare and in Northeast India it is 1.18 tons per hectare, as illustrated in Figure 7.1 (Papademetriou et al. 1999).

Plum fruit contains a wide range of beneficial biochemical compounds. Fresh plum fruit contains water; carbohydrates; fats; proteins; various types of sugars (glucose, sucrose, fructose, sorbitol); many dietary fibres (pectin, cellulose, hemicelluloses,

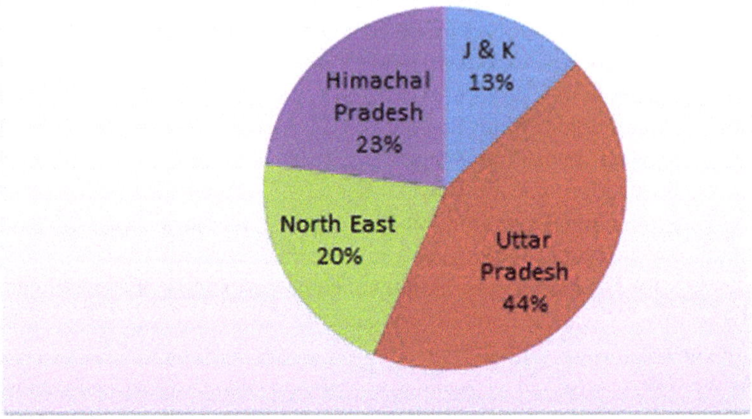

FIGURE 7.1 Major contributing states to the total plum production in India (Papademetriou et al. 1999).

lignin); amino acids, including aspartic acid; minerals (iron, calcium, phosphorus, sodium, potassium, magnesium, zinc, copper, manganese, boron); vitamins (vitamin A, folate, pyridoxine B6, pantothenic acid, niacin, riboflavin, thiamin, ascorbic acid); different carotenoids (lutein, alpha-carotene, beta-carotene) and organic acids, such as malic acid and quinic acid (Stacewicz-Sapuntzakis et al. 2001). A recent report revealed 113 different types of flavonoid compounds in the plum, out of which 55 are flavone and flavonol compounds, 9 are anthocyanins, 3 are isoflavonoids, 33 are proanthocyanin, 13 are dihydroflavonol and 1 is dihydrochalcone. Apart from this various types of phenolic compound, including seven cinnamic acid compounds, seven benzoic acid derivatives, three comarolquinic acid derivatives, two feruloylquinic acid derivatives, two shikimic acids, two types of abscisic acid, ellagic acid and two propionic acid compounds, are found extensively in *Prunus cerasifera*, *Prunus domestica*, *Prunus salicin* and *Prunus spinosa* (Liu et al. 2020). The presence of these important biochemical compounds makes plum-harvesting methods more sensitive because the quality of this compound and its presence or absence is completely determined by the harvesting time and the methods associated with it (Kucuker et al. 2014). The concentration of acidic compounds in plum is severely affected by harvesting date and acid concentration (mainly malic acid). In Japanese plums yield decreases gradually if harvested late or a few days after ripening (Singh et al. 2009). Acidity is a big factor for customer satisfaction in the case of plums, so it should not be compromised at any cost. The plum fruit which is harvested at the right time contains a large concentration of ascorbic acid, which is a very important nutrient present in plums, but early harvested plums contain a very low quantity of major phenolic compounds, such as anthocyanins, which ultimately degrades the quality of plum as a whole (Yousef et al. 2016).

The most popular and widely cultivated plum species are *Prunus domestica* (European plum) and *Prunus salicina* (Japanese plum). Both have enormous economic significance in the international market (Lozano et al. 2009). These fruits are popularly known as stone fruits due to their hard pits. Plum fruits are processed in various ways after harvesting to produce different food products for commercial use on an industrial scale throughout the world. Plum fruits produce a wide range of products, including jellies and jams, fruit juice and alcoholic drinks (Okie and Hancock 2008). A few important yeast strains (*Candida oleophila* and *Pichia fermentans*) with significant commercial value were already isolated from *Prunus domestica* and are primarily used in the production of acetic acid and oxalic acid (García-Fraile et al. 2013). Looking at the enormous potential of plum fruit in the near future, this chapter will try to cover all the relevant aspects of plum harvesting and handling, which will help the future researchers of this field to synchronize their studies in a more efficient, scientific and practical way.

7.2 METHODS OF PLUM HARVESTING AND HANDLING

Plum was commonly planted in gardens in different countries of Europe as early as the 1st century CE (Okie and Hancock 2008). The most ancient and conventional method of fruit harvesting is handpicking (Li et al. 2011). Plum producers follow a few conventional, mandatory and efficient rules to obtain better-quality

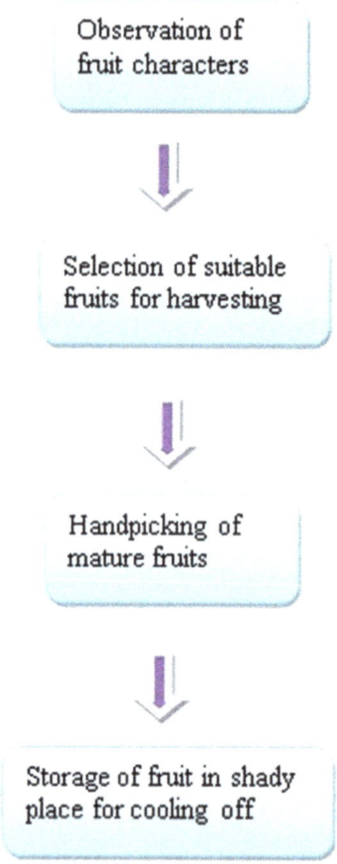

FIGURE 7.2 Different steps involved in the conventional method of harvesting plum.

fruits and to increase the storage capacity are, including harvesting fruits at the proper time of maturity and preserving the physical structure of the fruit as much as possible (Prasad et al. 2018). A schematic diagram is provided in Figure 7.2 for a better understanding of the process of the conventional methods of plum fruit harvesting.

There are a few characteristics of plum fruits usually observed before and during harvesting through conventional methods, including the firmness of the fruit's flesh and the age of the fruit from complete flowering; most plum species cultivated in India need 90 ± 3 days for complete maturity, but late cultivars need up to 135 days (Shamrao 2020). Traditionally, plums are also harvested by observing skin colour; the black skin colour of Japanese plums has high demand in the consumer market, while green and yellow plums are mostly sold in Asia. Large fruit size, regular shape of the fruits, firm flesh and colour are all important for customer satisfaction (Okie and Hancock 2008). Farmers tend to look for various aspects of the fruits through morphological observation to identify the perfect time of harvesting because it was noted that fruit harvested before maturity never contains the texture and flavour of ripe fruit and fruit harvested after maturity was easily damaged (Crisosto et al. 1995).

The most common conventional method of plum harvesting is picking the fruits by hand. Fruits are usually harvested with a stalk to prevent any type of injury on the fruit's skin because plum fruits are very soft when they reach maturity. Plums are harvested in baskets made of soft material to avoid any kind of injury to the delicate fruit. Harvested fruits are kept in a cool place because heat can damage the fruits or change their texture and can make them inconsumable (Shamrao 2020). A study revealed that handpicked fruits are better in quality than those are collected with the help of mechanical means, as only 5–10% of fruits are damaged during handpicking; however, handpicking is a much more time-consuming process (Mika et al. 2012). In India, different types of soft-bodied plants (mainly grasses) are used to make the harvesting baskets soft for better and smoother collection. In Punjab, plum fruits are harvested in May. To supply plum fruits to local markets, fruits are collected in the morning to provide customers with the freshest fruits, and for markets in locations farther from harvesting areas, fruits are harvested halfway to maturity so they are ready when they reach the customers (Thind 2019).

The advancement of modern robotic or automatic harvesting methods raised the bar for a larger collection of fruit in a shorter amount of time. But it was evident that mechanically harvested fruits are more damaged, resulting in 74% perfectly harvested fruit compared to 95% for the conventional handpicking methods (Kader 1983). The design of a mechanical plum-harvesting mechanism can be divided into two main components—software design and system hardware design. The success of mechanical harvesting is comparatively less because it is a new technology and is currently in development. The rate of plum harvesting through mechanical means is not satisfying in the commercial situation because although it can be done quickly, it contributes to more damaged fruits (Brown and Sukkarieh 2021).

Handling plum fruit is a very sensitive process because the quality of the product should be maintained during handling. The first step of handling begins just after the harvesting. The harvested plum fruits are sifted in a cool and dark environment to remove the heat they absorb from their cultivated climate; this is done to increase the life of the fruits (Shamrao 2020). Plum fruits can survive from one to six weeks if a proper handling process is followed (Manganaris et al. 2008). Fruits are selected based on their physical characteristics, and the damaged, immature or over-mature fruits are discarded from the fresh fruits. Fruits are assigned different grades according to their quality. After the selection and separation of unsuitable fruits, the useable fruits are packed by layers into wooden boxes with many layers of paper used for protection. Then the box is sealed and labelled; the labels carry data about the plum variety, quality, price, etc. Recent evidence suggests that the storage life of plum fruits can be increased with cold treatment, so these boxes are kept in cold storage (Thind 2019). Fruits can be injured in various ways during handling, packaging and transportation; to prevent this plum cultivars everywhere take careful precautions when harvesting plum fruit (Crisosto et al. 1995).

7.3 IMPACT OF HARVESTING MECHANISMS ON THE QUALITY OF PLUM FRUIT

The quality of the plum fruit depends on the harvesting, handling and storage of the product. It has been observed that both immaturity and over-maturity significantly reduce the quality of the fruit (Crisosto et al. 1995). Over-mature fruits are more delicate and will not be able to withstand the post-harvesting processes (Mitchell et al. 1991), so they undergo a shorter post-harvesting and marketing time, whereas immature fruits will be incapable of ripening and will lose water more rapidly, contributing to a loss of flavour. Mostly the fruits are damaged due to the tree limbs and friction generated from other fruits. In the case of manual harvesting, the fruit selection depends on the standard acceptance of understanding; the technique of removing the fruit should exert proper force to reduce plugging-related damages, as plum is a delicate fruit that requires rigid pails (Prussia and Woodroof 1986). The unwanted and harmful materials are separated both manually and mechanically, as they may cause ruptures in the cell wall, along with bacterial and fungal contamination. Some of the injuries that can occur during handling include cuts, abrasions, compression, bruising and other wounds (Crisosto and Valero 2008). Bruising occurs due to excessive compression in a stack from overloading; stone fruits, like plums, increases their resistance to bruising in low temperatures, leading to higher-quality fruit (Mitchell et al. 1991). The timing of herbicide and fungicide also plays a critical role in the quality of plum; if fungicides are applied too close to the harvesting time, it may lead to black staining in the plum (Cheng and Crisosto 1994). Vibration or abrasion can also damage the fruit while transporting or conveying; the top layer of the container undergoes a maximum amount of damage because of the amplitude and frequency of friction during transportation. The quality of the product can be achieved by sorting the fruits in a uniform manner, such as maturity, size, colour, weight and firmness. Various reports have shown that mechanical harvesting methods show increased accuracy in sorting mature fruit in an orchard than manual harvesting methods and also has high harvest rates (Prussia et al. 1986). The mature fruits are selected based on the force required during detachment rather than the colour of the fruit.

7.4 POST-HARVESTING MANAGEMENT OF PLUM

The generalized concept of post-harvesting is the process of removing the crop immediately after harvesting. Soon after the crop is removed from the plant body it starts to deteriorate, and therefore post-treatment includes cooling, cleaning, sorting and packing the crop to increase the shelf life and maintain the quality of the crop. This is also known as post-harvesting management. The main purpose of post-harvesting management is to keep the fruit as intact as possible. The Japanese plum ripens first, European varieties ripen later. The Japanese plum must be consumed fresh, but European plums can be consumed either fresh or dried (Topp et al. 2012). Plum fruits undergo many biochemical changes during ripening as well as in storage, such as anthocyanin accumulation, which leads to colour development, softening of the cell wall, decrease in fruit acidity and production of aroma compounds (Manganaris et al. 2008). Anthocyanin and carotenoid are mainly responsible for the surface colour of the fruit. The main carbohydrates found in fresh plums are glucose, sucrose, fructose and a sugar alcohol called sorbitol; hence the soluble solid content (SCC) is a quality-determining factor (Kim et al. 2003). Titratable acid (TA) has a major role in consumer acceptance; the main acid present in plum fruit is malic acid, which gradually decreases during ripening if not properly handled after harvesting (Crisosto et al. 2007). If the SCC content in the fruit ranges from 10% to 11.9% with a low TA percentage (i.e. <0.6%), it is more likely to be accepted by consumers (Crisosto et al. 2007). Cell wall modification due to components like cellulose and hemicellulose has not been extensively studied yet, but some cell wall–degrading components like poygalacturonase, galactosidase and methylesterase have been identified in harvested plum fruits (Iglesias-Fernandez et al. 2007). Ethylene plays a key role in the ripening of plum, which can be treated exogenously by chemicals that contain ethylene. The auxin concentration in plum fruit regulates several ethylene-dependent pathways; it also mediates the disassembly of cell wall–related transcripts (El-Sharkawy et al. 2016). The production and accomplishment of ethylene can be affected by 1-Methylcyclopropene (1-MCP), leading to a delay in the ripening process as well as extending its storage life (Watkins 2006). The rapid softening of the fruit during post-harvesting leads to economic losses during marketing, so extending the storage life at a low temperature of 0°C is generally recommended (Crisosto et al. 2007). Plum fruits are very delicate, and due to improper preservation they may get infected by brown rot (*Monolinia spp.*), and powdery mildew caused by *Sphaerotheca pannosa* and *Podosphaera triacty* (Khan et al. 2018). Some of the most extensively used post-harvesting techniques are discussed next.

7.4.1 Pre-cooling

Pre-cooling is performed after harvesting to remove the excess absorbed heat of the plum fruit, which leads to a reduction of important metabolic activity, like ethylene reduction, during the storage period (Dincer et al. 1992; Thompson et al. 2008). Pre-cooling treatments, such as air forced cooling and hydro cooling, are much more beneficial for the quick removal of heat. The forced cooling method is a commercial cooling technique where cold air flows quickly through the container with great force; velocity and temperature play a key role in this technique (Dehghannya et al. 2010; Rao 2015). The hydro cooling technique is another effective method where the fruits experience an agitated bath of cold or ice water, hence increasing surface heat transfer.

7.4.2 Heat Treatment

In terms of heat treatment, it has been recorded that hot water dips at 45°C for 10 minutes can slow down the physiological changes due to mechanical damages in plum fruit (Serrano et al. 2004). Due to the heat treatment in plum, an amplification in cell wall–bound spermidine, causing stability and firmness in the cell wall, was also observed; the slowing down of fruit ripening takes place because the inactivation of the heat deals with enzyme degradation. This heat treatment also reduces chilling injuries (Valero et al. 2002). The heat treatment controls fungal decaying and prevents insect contamination by killing any pests. The fruits are sometimes treated with calcium, followed by a heat treatment that directly lowers the physiological activities of ethylene and also reduces the respiration rates, thus increasing the life of the plum in storage (Daillant-Spinnler et al. 1996).

7.4.3 Cold Storage

Cold storage is generally employed to expand the storage life of the plum. It also maintains fruit quality during long-distance transport, but since plums are sensitive to low temperatures, they can only be stored in cool temperatures for a limited amount of time before chilling injuries appear (Manganaris et al. 2008). The main disorders in plum due to chilling injuries are flesh translucency or breakdown of gel, browning of the flesh, lack of juiciness, failure to ripen, red pigment accumulation and loss of flavour, but if the atmosphere is controlled with 3–5% CO_2 and 1–2% O_2 modification during the storage and shipment, then the fruit firmness and ground colour are maintained, and decay is reduced (Crisosto et al. 2004). The fruits are treated with 1-MCP by treating it first with water or a buffer and then releasing the 1-MCP gas in an enclosed area to reduce the fruit's metabolic activity as well as the likelihood of chilling disorders (Candan et al. 2008).

7.4.4 Controlled Atmosphere Storage and Edible Coatings

In this type of storage, the level of carbon dioxide and oxygen is monitored in a closed, airtight chamber. In modified atmosphere packaging (MAP) such an environment is created. The packaging of the fruit with polymeric MAP reduces the overall weight and makes it suitable for transport (Khan et al. 2018). Edible coatings are an external layer applied to the surface of the fruit to create a partial barrier to minimize the loss of moisture from the surface by enhancing the waxy cuticle. Chitosan, alginate and carboxymethyl cellulose are examples of edible coatings which provide a barrier without hampering the quality of the fruit (Dhall 2013). The alginate coating reduces the mass loss and showed some bioactive compounds, like carotenoid and anthocyanin, when the fruit was stored for 20 days at 5°C (Valero et al. 2013).

7.5 ROLE OF FORCED-AIR COOLING IN THE HANDLING OF PLUM FRUIT

During the post-harvesting period, the fruit undergoes mechanical damage, causing a massive decrease in the quality of the product. Various treatment methods are adopted to maintain the quality of the fruit and maintain its susceptibility (Ericsson and Tahir 1996). The pre-cooling method is used to increase the shelf life of the fruit and to remove the absorbed heat. Among the various pre-cooling methods, the forced-air cooling technique has proven to be the most effective (Kader 2002). The rapid decrease in the temperature of the fruit from its core temperature inhibits the development of pathogens and reduces metabolic activities, like respiration and water loss (Zhou et al. 2015). In the forced-air cooling technique, the air flows through the container to cool the pack. Depending on the velocity and pressure of air-cooling arrangements, the forced-air cooling method has three variations: the air is circulated at a high velocity inside the refrigerator, commonly called tunnel-type cooling; the forced air is moved through a tunnel in bulk, called forced-air serpentine cooling; or the air is forced through the produce that is packed in containers in a process known as cold wall forced-air cooling (Rao 2015). Velocity and temperature play an essential role in the forced-air cooling method; the high velocity of cooled air may lead to chilling injuries and water loss in the stone fruit, thus producing low-quality fruit (Delele et al. 2013). One of the most important roles of the normal cooling method is to slow the rate of the respiration of the fruit. The heat generated by the fruit during respiration in the form of sugars, fats and proteins inside the cell get oxides, and this loss of nutrients leads to a decrease in the quality of the product (Liberty et al. 2013). However, the forced-air cooling technique is a clean and simple method that is faster and more efficient than most of the other techniques. One of the disadvantages of this method, though, is that it is not uniform; however, by adjusting the air velocity, pressure and temperature it can be an energy-efficient method as well (Han et al. 2018). After many experiments, it has been concluded that the internal average temperature of plum decreases from 24.54 ± 0.15 °C to 3.1°C in an average of 65 minutes (Martínez-Romero et al. 2003). The forced-air cooling technology plays a key role during the handling of harvested plums. The pre-cooled fruits have higher steadfastness; increased weight loss; well-maintained colour index, although a slight variation can be noticed after 10–15 days in storage; lower pigment degradation and a reduction in respiration rates in both damaged and non-damaged fruits (Li et al. 2019). The forced-air cooling method delays the onset production of ethylene when treated with 1-MCP, protecting it from rapid ripening, but the SSC remains unaffected (Minas et al. 2013). Sometimes due to the non-uniform supply of forced air in the room, the fruits near the fan get much more chilled air and soon develop chilling injuries that can be maintained by treating them with 1-MCP (Candan et al. 2011).

Depending on the time of the pre-cooling treatment, the pattern of respiration is different. The fruits that were damaged before the pre-cooling treatment had a higher rate of respiration during storage than the fruits that were damaged after pre-cooling, leading to a higher loss in weight and reduction in fruit firmness, rupture of the tissue of the fruit and leakage in carbohydrate reserves in break tissue in the former type of the plum fruit (Zainal et al. 2019). There is no significant change in the production of oxygen in the fruit during the treatment; although during the storage of the fruit at a low temperature, the production of carbon dioxide (CO_2) significantly decreases depending on the time of pre-cooling (Li et al. 2019). According to a successful experiment enumerated in Figures 7.3 and 7.4, the weight loss in the control fruit is 0.52 ± 0.09% after four days; fruits that were mechanically damaged before pre-cooling showed 3.05 ± 0.33%, the plum in which the damage was observed after pre-cooling was 1.20 ± 0.05% (Martínez-Romero et al. 2003).

In the same experiment, it was also concluded that the firmness of the controlled fruit was higher compared to any sort of mechanical damage, i.e., 1.27±/0.02 N, fruit that was mechanically damaged during storage had firmness of 1.20 ± 0.05 N, fruits that were damaged before pre-cooling had firmness of 0.71±/0.07 N. On a concluding remark, we can justify that the plum fruit is liable to perish quickly, and hence they

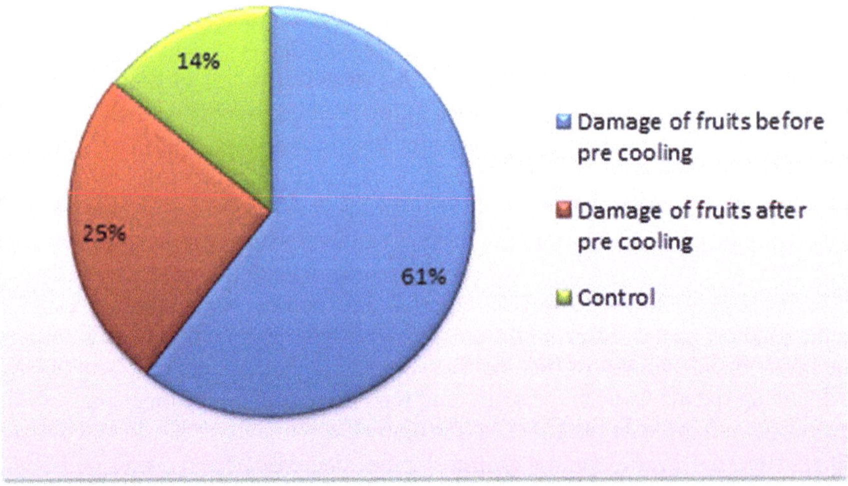

FIGURE 7.3 Percentage of plum weight loss after four days in storage at 1°C.

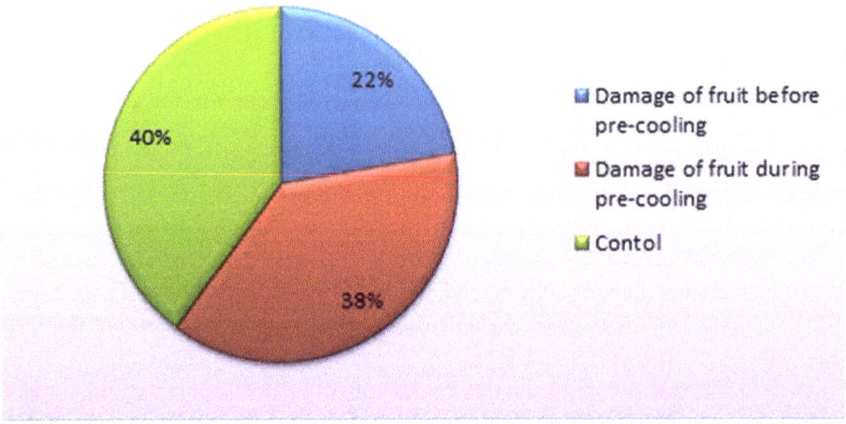

FIGURE 7.4 Flesh firmness (N) of plum fruits after four days in storage at 1°C.

require pre-cooling treatment to maintain their quality in terms of market value by increasing their storage life.

7.6 POST-HARVESTED PLUMS AND ASSOCIATED FOOD PRODUCTS

Plum fruits are used for the production of various plum products which are consumed across the globe. Various products, such as dry plum fruit, plum juice or drinks, alcoholic beverages, plum jam and canned plum, are produced from plums

(Satora et al. 2017). Some of the techniques and methods involved in the production of plum products are described next.

7.6.1 Production of Dry Plum

Dried plums have high demand among consumers, and different methods are used to increase the quality of dried plums to obtain a considerable market value. Drying helps in the preservation of fruit quality even if the fruit sustains a small injury because the water content in dried fruits is much less, or even absent, compared to fresh fruit (Molnár et al. 2016). Drying is effective for storing fruits for a long time; this method was used by our ancestors to preserve different kinds of fruits. Ancient people tried drying by applying heat generated from firewood, but modern technological equipment has allowed us to improve the drying process. For drying, plums are harvests at the late ripening period, and structural shapes are observed before selecting. Fruits without any mechanical injury are usually collected for drying. The availability of water content in plum fruit is counted by drying it into a uniform weight. To remove the oily layer above the plum, fruits are boiled in water for a very short time, followed by a mild heat treatment. Then fruits are dried by applying hot air at a temperature between 44°C and 74°C. The heat is applied in two stages. In the beginning a comparatively low temperature is applied for more than five hours, and later a higher temperature is applied. To obtain uniform dryness in all fruits, fruits are removed from the drying equipment many times to get rid of excess heat from the fruits; this also helped in the appearance of a moist coat in dried plum fruits. This drying process has been very effective in preserving the nutrient quality of plum fruits (Kurmanov et al. 2015).

7.6.2 Production of Plum Powder

Plum is a very soft fruit with a high quantity of water, resulting in its short storage life. But in plum powder, the water content is removed completely, so the shelf life of the fruit increases. The dehydration of plum fruits serves as an effective technique for reducing microbial activity during the process of producing plum powders. The powdery form of these fruits is used extensively for medicinal purposes and to add flavour and texture to other edible material. Fruits are collected in maturity, and the skin is usually removed and fruit cut in the centre before the drying process starts. A freeze dryer is commonly used in the drying process where a temperature less than –50°C is employed. The vacuum-drying process is also used to remove water from plum fruits. In a vacuum drier, high temperature and pressure are used to make plum powder. Convective dryers use two different kinds of temperature simultaneously applied to the plum in a controlled air velocity (Shukitt-Hale et al. 2009). Three different kinds of plum powders are collected from a microwave vacuum dryer; here plum fruits are heated in a glass container and the temperature of the drying plum is observed on an infrared camera. Rotation of the equipment is used to prevent overheating of the plum. In the combined drying process, fruits are dried at two temperatures, one after another, until the desired moisture content is obtained. These are some of the methods employed extensively to produce dry plum, which is the most important step in the production of plum powder (Michalska et al. 2016). These dried products are then converted to powder using a grinder (Sharma et al. 2011).

7.6.3 Production of Plum Wine

Wines are one of the most economically important products that are produced from the plum. Wines are produced from various carbohydrate sources, but the production of wine from plum provides more flavour and texture, so it is in high demand among consumers. Fresh plum fruits have a shelf life of a maximum of five to six days, but plum wine has a very long shelf life; it can be used even after 10 years from its production date. One of the biggest advantages of wine is that older wines have higher demand and more economic value. For the industrial production of plum wine, two strains of yeast (*Saccharomyces cerevisiae* and *Schizosaccharomyces pombe*) are commonly used in industrial production. Fresh plum fruits of different species are collected and boiled by adding a low quantity of water; boiling helps in the removal of upper skinny layers and the inner hard stony layer with the help of a pulper (Miljić and Puškaš 2014). The remaining parts of the plum fruits are then transferred to a cold place with a temperature less than 10°C and kept until the beginning of the next step (Srinivasan et al. 2012). Yeast is generally collected from various research and horticulture institutes and cultured in an agar-based culture medium. Yeast culture is added to the stock of the plum fruits to start the fermentation process. Substances like sugar (sucrose) and honey are added for manipulation of flavour, aroma and taste. The fermentation of plum is normally performed in a temperature range between 24°C and 26°C. After the completion of the fermentation process, plum wine is filtered and kept in a glass apparatus before mixing it with different concentrations of fruit juice (Joshi et al. 2014).

7.6.4 Production of Plum Bar/Leather

Plum fruits can be stored for a maximum of six days after harvesting, and after few days they start decaying due to the activity of different microorganisms. To prevent the waste of plum fruit, different types of products are developed from them. Plum fruit bars are one of the most valuable substances produced from plum fruits. They have good industrial demand because they are less costly and need very few supporting substances (Sun-Waterhouse 2011). Plum bars lack water content, resulting in good storage life, and have nutritional value similar to fresh plums (Singh et al. 2019). Traditionally, plum fruit bars are produced by solar drying without mixing any other substance or treating with other substances to improve the quality. These fruit bars can be used as a replacement for biscuits and sweet dishes (Madhav and Parimita 2016).

For the production of plum bars, plum fruits are first cleaned with water then separated into two halves; the inner stony part is also removed during this process. The plum piece is then put into a hot air oven on a cookie sheet and the uncut side of the fruit is kept in the bottom. The oven is heated close to 200°C for close to 19 minutes; during this period fruits are subjected to heat for a very short amount of time to prevent heat damage to the fruits. The plums are then transferred to a blender, which converts them into a fine paste. The paste is then put in a big container and sugar is mixed in and then boiled for more than four minutes. The plum paste is then poured into a tray and treated with buttery substances to add a greasy effect. The temperature of the oven is maintained close to 59°C for one day. After a day the leathery substance in the tray is cut into bars of different sizes and packaged for commercial use (Leonzio 2018).

7.6.5 Production Technology of Plum Jam

Plum fruits contain various kinds of essential biomolecules that are very beneficial for human health. To provide these nutrients in a product with a longer shelf life than fresh fruits, plums can converted into jam, which is similar to jelly in texture. Plum jam is a semi-solid plum product that has very good market demand. In conventional methods of jam production, large amounts of sucrose are added to improve the flavour, but this makes the jam less healthful. So now less harmful natural sweetening agents, such as honey, plant syrup, stevia, sorbitol and fructose, are added instead (Edwards et al. 2016). *Prunus domestica* is the variety usually used for plum jam. The production of plum jam involves many steps. First, the fruits are washed and cut in half. Then the fruits are mixed with sweeteners and boiled; the fruits are boiled at more than 90°C for about 15 minutes and continuously stirred. Pectin is also added during the heating process, which gives the jam its semi-solid texture, as well as citric acid. After heating, jams are stored in glass pots and the accumulated heat of the product is removed completely during storage. The drying process is done with heat generated from a fire until the water content of the plum jam is reduced to less than 50%. In industrial production, jam is also examined to observe the presence or absence of harmful microorganisms before further processing. If plum jam is produced under low pressure, it requires lower temperatures than normal, which is close to 60°C. The heating process is traditionally performed in an open vessel, like a stockpot (Stamatovska et al. 2017).

7.6.6 Production of Plum Juice

Plum fruit is very rich in nutrients; it contains different types of vitamins, minerals, organic acids and carbohydrates (Birwal et al. 2017). Plum juice is a favourite drink in the countries of the European Union because it retains more nutritional value than other plum products. The production of plum juice contains many intermediate steps. Fruit is collected at its perfect ripening period and kept below –20°C. The stony part of the plum fruit is removed with a knife while the fruit is still frozen and then the fruit is converted into mash by a perforated disc mill. Enzymes are added to the mash at a temperature close to 100°C for more than 50 minutes. Then the plums are pressed and the juices are collected from the juicer (Zbrzeźniak et al. 2015). The juice can be extracted in the microwave as well with a temperature close to 70°C. In the microwave, fruits are treated with heat generated with a fixed power, and water is not applied to the plums. The hot juices are then collected in the equipment tubes. Juices are collected without any break in between as far as possible; the process is culminated with the appearance of steam and it gives the final product when allowed to cool at room temperature (Cendres et al. 2012).

7.7 METHODS FOR ESTIMATION AND MAINTENANCE OF THE QUALITY OF PLUM AFTER HARVESTING

The plum fruit has high antioxidant potential; it is also perishable and has a short storage life, hence new preservation methods have been introduced to extend its shelf life and proper distribution in the market. Mainly innovative progression in harvesting and handling methods has been introduced to maintain the physicochemical and bioactive components of plum. The chromatic characteristic of plum depends on many factors, such as growth conditions, ripening stages, storage technology and, most importantly, the cultivars (Gonçalves et al. 2007). The intense homogeneity of skin texture and bright, shiny skin colour can be achieved by providing a perfect storage temperature that determines the quality of the product (El-Ramady et al. 2015). Plum fruit has high organoleptic attributes, like SSC, TA and anthocyanin, along with five different types of sugar, i.e., glucose, fructose, maltose, sucrose and sorbitol (Usenik et al. 2008). The levels of these different sugars vary, which mainly depends on the cultivars and the environmental conditions of the orchard (Petkovsek et al. 2010). According to some experiments, during the growth and development of the fruit in 14 different stages, concerning their size and colour, it has been concluded that these sugars gradually increase during the maturation period (Guillén et al. 2013). The sugar level can be estimated in a stone fruit through a rapid and convenient method by using a portable digital refractometer (Díaz-Mula et al. 2012). It is normally determined with the help of high-performance liquid chromatography (HPLC) and is generally equipped with a refractive index detector and Ca^{2+} or with N_2 column; for quantification, a pure curve of sugar is used (Kelebek et al. 2011). The organic compound is one of the most essential compounds present in the stone fruit because the characteristic tart or sour taste of the fruit is a result of organic acids (McLellan et al. 2004). Many organic acids, such as amino acids and malic acid, are found in plum, but malic acid is the most abundant (Serrano et al. 2005). The level of malic acid increases during the development of the fruit, but an excess of malic acid may lead to fermentation during storage. The titratable acid (TA) content in plum fruit can be measured by using titration against 0.1N NaOH, and the pH is maintained up to 8.1 (Serradilla et al. 2013). Individual acids like shikimic acid, furamic acid malic acid can be estimated by the HPLC method (Kelebek and Selli 2011). Anthocyanin is an important component that contributes to the redness in the fruit's skin colour and is an indicator of maturity (Ndou et al. 2019). The anthocyanin content in plum fruit is calculated with the help of the pH-differential method, and individual estimation of anthocyanin is done by the HPLC method, along with a detector that monitors at 520 nm and 530 nm (Serradilla et al. 2013; Hayaloglu and Demir 2016). Some phenolic components like flavonoids and phenolic acids, such as chlorogenic acid, neo-chlorogenic acid and p-coumaroyl quinic acid, can be detected with the help of the HPLC method (Lara et al. 2020). Even though the stems of plums are not consumed, they play a crucial role in determining the freshness of the fruit. Both plum fruit and stem require a good amount of water to maintain its freshness; when the water content in the plant body is not sufficient, the stem starts to lose its original appearance which is a direct indicator of the health of the plum fruit present in this stem. The firmness of the fruit is maintained by providing an optimum temperature of 0.5–2°C for the conservation of the plum fruit (Wang et al. 2016). To prevent the diseases, the fruit must be treated with proper fungicides and pesticides during post-harvesting and handling (Adaskaveg et al. 2005). Numerous factors are responsible for the production of plum, and to maintain good quality fruits, new tools and techniques are continually emerging (Clark et al. 2014). However, there is a need for improved, cost-efficient techniques that will result in fewer damaged products.

7.8 CONCLUSION

Plum fruit has already established itself as an important food product in foreign markets, although it is still in the emerging stage in the Asian market, including on the Indian subcontinent. Researchers and economists across the globe have already labelled it as the 'next big thing' in the fruit markets of South-east Asia. Looking at the enormous potential of plum fruits, it is getting extremely important to find a way to preserve these delicate and sensitive fruits through effective post-harvesting methods. Although a lot of work has been carried out to understand the efficiency of various already existing post-harvesting handling techniques, there is a big scope of improvement that should be monitored closely by future researchers of this field.

7.9 REFERENCES

Adaskaveg, J., Förster, H., Gubler, W., Teviotdale, B., and Thompson, D. 2005. Reduced-Risk Fungicides Help Manage Brown Rot and Other Fungal Diseases of Stone Fruit. *California Agriculture* 59 (2): 109–114.

Bhatt, G.D., and Tyagi, P. 2021. Production of Fruits in India. *The International Journal of Modern Agriculture* 10 (2): 2173–2180.

Birwal, P., Deshmukh, G., Saurabh, S.P., and Pragati, S. 2017. Plums: A Brief Introduction. *Journal of Food, Nutrition and Population Health* 1 (1): 1–5.

Brown, J., and Sukkarieh, S. 2021. Design and Evaluation of a Modular Robotic Plum Harvesting System Utilizing Soft Components. *Journal of Field Robotics* 38 (2): 289–306.

Candan, A.P., Graell i Sarle, J., and Larrigaudière, C. 2008. Roles of Climacteric Ethylene in the Development of Chilling Injury in Plums. *Postharvest Biology and Technology* 47 (1): 107–112.

Candan, A.P., Graell i Sarle, J., and Larrigaudière, C. 2011. Postharvest Quality and Chilling Injury of Plums: Benefits of 1-Methylcyclopropene. *Spanish Journal of Agricultural Research* 9 (2): 554–564.

Cendres, A., Chemat, F., Page, D., Le Bourvellec, C., Markowski, J., Zbrzezniak, M., Renard, C.M., and Plocharski, W. 2012. Comparison Between Microwave Hydrodiffusion and Pressing for Plum Juice Extraction. *LWT-Food Science and Technology* 49 (2): 229–237.

Cheng, G.W., and Crisosto, C.H. 1994. Development of Dark Skin Discoloration on Peach and Nectarine Fruit in Response to Exogenous Contaminations. *Journal of the American Society for Horticultural Science* 119 (3): 529–533.

Clark, S., Jung, S., and Lamsal, B. 2014. *Food Processing: Principles and Applications*. John Wiley & Sons.

Crisosto, C.H., Crisosto, G.M., Echeverria, G., and Puy, J. 2007. Segregation of Plum and Pluot Cultivars According to Their Organoleptic Characteristics. *Postharvest Biology and Technology* 44 (3): 271–276.

Crisosto, C.H., Garner, D., Crisosto, G.M., and Bowerman, E. 2004. Increasing 'Blackamber' Plum (*Prunus salicina* Lindell) Consumer Acceptance. *Postharvest Biology and Technology* 34 (3): 237–244.

Crisosto, C.H., Mitchell, F.G., and Johnson, S. 1995. Factors in Fresh Market Stone Fruit Quality. *Postharvest News and Information* 6 (2): 17–21.

Crisosto, C.H., and Valero, D. 2008. Harvesting and Postharvest Handling of Peaches for the Fresh Market. In *The Peach: Botany, Production and Uses*, edited by D.R. Layne and D. Bassi. CAB International.

Daillant-Spinnler, B., MacFie, H.J.H., Beyts, P.K., and Hedderley, D. 1996. Relationships Between Perceived Sensory Properties and Major Preference Directions of 12 Varieties of Apples from the Southern Hemisphere. *Food Quality and Preference* 7 (2): 113–126.

Dehghannya, J., Ngadi, M., and Vigneault, C. 2010. Mathematical Modeling Procedures for Airflow, Heat and Mass Transfer During Forced Convection Cooling of Produce: A Review. *Food Engineering Reviews* 2 (4): 227–243.

Delele, M.A., Ngcobo, M.E.K., Getahun, S.T., Chen, L., Mellmann, J., and Opara, U.L. 2013. Studying Airflow and Heat Transfer Characteristics of a Horticultural Produce Packaging System Using a 3-D CFD Model. Part I: Model Development and Validation. *Postharvest Biology and Technology* 86: 536–545.

Dhall, R.K. 2013. Ethylene in Post-Harvest Quality Management of Horticultural Crops: A Review. *Research & Reviews: A Journal of Crop Science and Technology* 2 (2): 9–24.

Díaz-Mula, H.M., Serrano, M., and Valero, D. 2012. Alginate Coatings Preserve Fruit Quality and Bioactive Compounds During Storage of Sweet Cherry Fruit. *Food and Bioprocess Technology* 5 (8): 2990–2997.

Dincer, I., Yildiz, M., Loker, M., and Gun, H. 1992. Process Parameters for Hydrocooling Apricots, Plums, and Peaches. *International Journal of Food Science & Technology* 27 (3): 347–352.

Edwards, C.H., Rossi, M., Corpe, C.P., Butterworth, P.J., and Ellis, P.R. 2016. The Role of Sugars and Sweeteners in Food, Diet and Health: Alternatives for the Future. *Trends in Food Science & Technology* 56: 158–166.

El-Ramady, H.R., Domokos-Szabolcsy, É., Abdalla, N.A., Taha, H.S., and Fári, M. 2015. Postharvest Management of Fruits and Vegetables Storage. In *Sustainable Agriculture Reviews*, edited by E. Lichtfouse. Springer.

El-Sharkawy, I., Sherif, S., Qubbaj, T., Sullivan, A.J., and Jayasankar, S. 2016. Stimulated Auxin Levels Enhance Plum Fruit Ripening, but Limit Shelf-Life Characteristics. *Postharvest Biology and Technology* 112: 215–223.

Ericsson, N.A., and Tahir, I.I. 1996. Studies on Apple Bruising: II. The Effects of Fruit Characteristics, Harvest Date and Precooling on Bruise Susceptibility of Three Apple Cultivars. *Acta Agriculturae Scandinavica B-Plant Soil Sciences* 46 (4): 214–217.

García-Fraile, P., Silva, L.R., Sánchez-Márquez, S., Velázquez, E., and Rivas, R. 2013. Plums (*Prunus domestica* L.) Are a Good Source of Yeasts Producing Organic Acids of Industrial Interest from Glycerol. *Food Chemistry* 139 (1–4): 31–34.

Gonçalves, B., Silva, A.P., Moutinho-Pereira, J., Bacelar, E., Rosa, E., and Meyer, A.S. 2007. Effect of Ripeness and Postharvest Storage on the Evolution of Colour and Anthocyanins in Cherries (*Prunus avium* L.). *Food Chemistry* 103 (3): 976–984.

Guillén, F., Díaz-Mula, H.M., Zapata, P.J., Valero, D., Serrano, M., Castillo, S., and Martínez-Romero, D. 2013. Aloe Arborescens and Aloe Vera Gels as Coatings in Delaying Postharvest Ripening in Peach and Plum Fruit. *Postharvest Biology and Technology* 83: 54–57.

Han, J.W., Zhao, C.J., Qian, J.P., Ruiz-Garcia, L., and Zhang, X. 2018. Numerical Modeling of Forced-Air Cooling of Palletized Apple: Integral Evaluation of Cooling Efficiency. *International Journal of Refrigeration* 89: 131–141.

Hayaloglu, A.A., and Demir, N. 2016. Phenolic Compounds, Volatiles, and Sensory Characteristics of Twelve Sweet Cherry (*Prunus avium* L.) Cultivars Grown in Turkey. *Journal of Food Science* 81 (1): C7–C18.

Iglesias-Fernández, R., Matilla, A.J., Rodríguez-Gacio, M.C., Fernández-Otero, C., and de La Torre, F. 2007. The Polygalacturonase Gene PdPG1 Is Developmentally Regulated in Reproductive Organs of *Prunus domestica* L. subsp. insititia. *Plant Science* 172 (4): 763–772.

Joshi, V.K., Gill, A., Kumar, V., and Chauhan, A. 2014. Preparation of Plum Wine with Reduced Alcohol Content: Effect of Must Treatment and Blending with Sand Pear Juice on Physico-Chemical and Sensory Quality. *Indian Journal of Natural Products and Resources (IJNPR)[Formerly Natural Product Radiance (NPR)]* 5 (1): 67–74.

Kader, A.A. 1983. Influence of Harvesting Methods on Quality of Deciduous Fruits. *Horticultural Science* 18 (4): 409–411.

Kader, A.A. 2002. *Postharvest Technology of Horticultural Crops*, vol. 3311. University of California Agriculture and Natural Resources Publications.

Kelebek, H., and Selli, S. 2011. Evaluation of Chemical Constituents and Antioxidant Activity of Sweet Cherry (*Prunus avium* L.) Cultivars. *International Journal of Food Science & Technology* 46 (12): 2530–2537.

Khan, A.S., Singh, Z., and Ali, S. 2018. Postharvest Biology and Technology of Plum. In *Postharvest Biology and Technology of Temperate Fruits*, edited by S. Parvez, I.A. Wani, M. Shah, S. Mir, and M. Mir. Springer.

Kim, D.O., Jeong, S.W., and Lee, C.Y. 2003. Antioxidant Capacity of Phenolic Phytochemicals from Various Cultivars of Plums. *Food Chemistry* 81 (3): 321–326.

Kucuker, E., Ozturk, B., Celik, S.M., and Aksit, H. 2014. Pre-Harvest Spray Application of Methyl Jasmonate Plays an Important Role in Fruit Ripening, Fruit Quality and Bioactive Compounds of Japanese Plums. *Scientia Horticulturae* 176: 162–169.

Kurmanov, N., Shingissov, A., Kantureyeva, G., Nurseitova, Z., Tolysbaev, B., and Shingisova, G. 2015. Research of Plum Drying Process. *In CBU International Conference Proceedings* 3: 494–495.

Lara, M.V., Bonghi, C., Famiani, F., Vizzotto, G., Walker, R.P., and Drincovich, M.F. 2020. Stone Fruit as Biofactories of Phytochemicals with Potential Roles in Human Nutrition and Health. *Frontiers in Plant Science* 11: 13–23.

Leonzio, G. 2018. State of Art and Perspectives About the Production of Methanol, Dimethyl Ether and Syngas by Carbon Dioxide Hydrogenation. *Journal of CO2 Utilization* 27: 326–354.

Li, J., Fu, Y., Yan, J., Song, H., and Jiang, W. 2019. Forced Air Precooling Enhanced Storage Quality by Activating the Antioxidant System of Mango Fruits. *Journal of Food Quality* 1: 1–12.

Li, P., Lee, S.H., and Hsu, H.Y. 2011. Review on Fruit Harvesting Method for Potential Use of Automatic Fruit Harvesting Systems. *Procedia Engineering* 23: 351–366.

Liberty, J.T., Okonkwo, W.I., and Echiegu, E.A. 2013. Evaporative Cooling: A Postharvest Technology for Fruits and Vegetables Preservation. *International Journal of Scientific & Engineering Research* 4 (8): 2257–2266.

Liu, W., Nan, G., Nisar, M.F., and Wan, C. 2020. Chemical Constituents and Health Benefits of Four Chinese Plum Species. *Journal of Food Quality* 1–17.

Lozano, M., Vidal-Aragón, M.C., Hernández, M.T., Ayuso, M.C., Bernalte, M.J., García, J., and Velardo, B. 2009. Physicochemical and Nutritional Properties and Volatile Constituents of Six Japanese Plum (*Prunus salicina* Lindl.) Cultivars. *European Food Research and Technology* 228 (3): 403–410.

Madhav, K., and Parimita, K. 2016. Studies on Development of Tomato Leather Prepared for Geriatric Nutrition. *Journal of Nutrition & Food Sciences* 6 (1).

Manganaris, G.A., Vicente, A.R., and Crisosto, C.H. 2008. Effect of Pre-Harvest and Post-Harvest Conditions and Treatments on Plum Fruit Quality. *CAB Rev. Perspectives in Agriculture, Veterinary Science, Nutrition and Natural Resources* 3 (9): 1–9.

Martínez-Romero, D., Castillo, S., and Valero, D. 2003. Forced-Air Cooling Applied Before Fruit Handling to Prevent Mechanical Damage of Plums (*Prunus salicina* Lindl.). *Postharvest Biology and Technology* 28 (1): 135–142.

Matković, M. 2015. Possibilities of Plum Cultivation in the Republic of Serbia. *Agricultural Economics* 62 (4): 1045–1060.

McLellan, M.R., and Padilla-Zakour, O.I. 2004. Sweet Cherry and Sour Cherry Processing. In *Processing Fruits: Science and Technolog*, edited by D.M. Barrett, L.P. Somogyi, and H.S. Ramaswamy. CRC Press.

Michalska, A., Wojdyło, A., Lech, K., Łysiak, G.P., and Figiel, A. 2016. Physicochemical Properties of Whole Fruit Plum Powders Obtained Using Different Drying Technologies. *Food Chemistry* 207: 223–232.

Mika, A., Wawrzyńzak, P., Buler, Z., Konopacka, D., Konopacki, P., Krawiec, A., Białkowski, P., Michalska, B., Plaskota, M., and Gotowicki, B. 2012. Mechanical Harvesting of Plums for Processing with a Continuously Moving Combine Harvester. *Journal of Fruit and Ornamental Plant Research* 20 (1): 29–42.

Miljić, U.D., and Puškaš, V.S. 2014. Influence of Fermentation Conditions on Production of Plum (*Prunus domestica* L.) Wine: A Response Surface Methodology Approach. *Hemijska Industrija* 68 (2): 199–206.

Minas, I.S., Crisosto, G.M., Holcroft, D., Vasilakakis, M., and Crisosto, C.H. 2013. Postharvest Handling of Plums (*Prunus salicina* Lindl.) at 10 C to Save Energy and Preserve Fruit Quality Using an Innovative Application System of 1-MCP. *Postharvest Biology and Technology* 76: 1–9.

Mitchell, F.G., Mayer, G., Saenz, M., Slaughter, D., Johnson, R.S., Biasi, B., and Delwiche, M. 1991. *Selecting and Handling High Quality Stone Fruit for Fresh Market.* 1991 Research Reports for California Peaches and Nectarines, California Tree Fruit Agreement.

Molnár, Á.M., Ladányi, M., and Kovács, S. 2016. Evaluation of the Production Traits and Fruit Quality of German Plum Cultivars. *Acta Universitatis Agriculturae et Silviculturae Mendelianae Brunensis* 64 (1): 109–114.

Ndou, A., Tinyani, P.P., Slabbert, R.M., Sultanbawa, Y., and Sivakumar, D. 2019. An Integrated Approach for Harvesting Natal Plum (*Carissa macrocarpa*) for Quality and Functional Compounds Related to Maturity Stages. *Food Chemistry* 293: 499–510.

Okie, W.R., and Hancock, J.F. 2008. Plums. In *Temperate Fruit Crop Breeding*, edited by J.F. Hancock, 337–358. Springer.

Papademetriou, M.K., Herath, E.M., George, A.P., Dorji, P., and Zailong, L. 1999. *Deciduous Fruit Production in Asia and the Pacific.* RAP Publication (FAO).

Petkovsek, M.M., Slatnar, A., Stampar, F., and Veberic, R. 2010. The Influence of Organic/Integrated Production on the Content of Phenolic Compounds in Apple Leaves and Fruits in Four Different Varieties Over a 2-Year Period. *Journal of the Science of Food and Agriculture* 90 (14): 2366–2378.

Prasad, K., Jacob, S., and Siddiqui, M.W. 2018. Fruit Maturity, Harvesting, and Quality Standards. In *Preharvest Modulation of Postharvest Fruit and Vegetable Quality*, edited by M.W. Siddiqui, 41–69. Academic Press.

Prussia, S.E., and Woodroof, J.G. 1986. Harvesting, Handling, and Holding Fruit. In *Commercial Fruit Processing*, edited by S.E. Prussia and J.G. Woodroof. Springer.

Rao, C.G. 2015. *Engineering for Storage of Fruits and Vegetables: Cold Storage, Controlled Atmosphere Storage, Modified Atmosphere Storage*. Academic Press.

Satora, P., Kostrz, M., Sroka, P., and Tarko, T. 2017. Chemical Profile of Spirits Obtained by Spontaneous Fermentation of Different Varieties of Plum Fruits. *European Food Research and Technology* 243 (3): 489–499.

Serradilla, M.J., del Carmen Villalobos, M., Hernández, A., Martín, A., Lozano, M., and de Guía Córdoba, M. 2013. Study of Microbiological Quality of Controlled Atmosphere Packaged 'Ambrunés' Sweet Cherries and Subsequent Shelf-Life. *International Journal of Food Microbiology* 166 (1): 85–92.

Serrano, M., Guillén, F., Martínez-Romero, D., Castillo, S., and Valero, D. 2005. Chemical Constituents and Antioxidant Activity of Sweet Cherry at Different Ripening Stages. *Journal of Agricultural and Food Chemistry* 53 (7): 2741–2745.

Serrano, M., Martínez-Romero, D., Castillo, S., Guillén, F., and Valero, D. 2004. Role of Calcium and Heat Treatments in Alleviating Physiological Changes Induced by Mechanical Damage in Plum. *Postharvest Biology and Technology* 34 (2): 155–167.

Shamrao, B.S. 2020. Production Technology of Peach, Plum and Apricot in India. In *Prunus*, edited by A. Küden. IntechOpen.

Sharma, K.D., Sharma, R., and Attri, S. 2011. Instant Value Added Products from Dehydrated Peach, Plum and Apricot Fruits. *Indian Journal of Natural Products and Resources* 2 (4): 409–420.

Shukitt-Hale, B., Kalt, W., Carey, A.N., Vinqvist-Tymchuk, M., McDonald, J., and Joseph, J.A. 2009. Plum Juice, but Not Dried Plum Powder, Is Effective in Mitigating Cognitive Deficits in Aged Rats. *Nutrition* 25 (5): 567–573.

Singh, A., Sonkar, C., and Shingh, S. 2019. Studies on Development of Process and Product of Plum Fruit Leather. *The International Journal of Food Sciences and Nutrition* 4 (5): 74–79.

Singh, S.P., Singh, Z., and Swinny, E.E. 2009. Sugars and Organic Acids in Japanese Plums (*Prunus salicina* Lindell) as Influenced by Maturation, Harvest Date, Storage Temperature and Period. *International Journal of Food Science & Technology* 44 (10): 1973–1982.

Srinivasan, C., Dardick, C., Callahan, A., and Scorza, R. 2012. Plum (*Prunus domestica*) Trees Transformed with Poplar FT1 Result in Altered Architecture, Dormancy Requirement, and Continuous Flowering. *PLoS One* 7 (7): e40715.

Stacewicz-Sapuntzakis, M., Bowen, P.E., Hussain, E.A., Damayanti-Wood, B.I., and Farnsworth, N.R. 2001. Chemical Composition and Potential Health Effects of Prunes: A Functional Food?. *Critical Reviews in Food Science and Nutrition* 41 (4): 251–286.

Stamatovska, V., Karakasova, L., Babanovska-Milenkovska, F., Nakov, G., Blazevska, T., and Durmishi, N. 2017. Production and Characterization of Plum Jams with Different Sweeteners. *Journal of Hygienic Engeneering and Desing* 19: 67–77.

Sun-Waterhouse, D. 2011. The Development of Fruit-Based Functional Foods Targeting the Health and Wellness Market: A Review. *International Journal of Food Science & Technology* 46 (5): 899–920.

Thind, B.S. 2019. *Phytopathogenic Bacteria and Plant Diseases*. CRC Press, Taylor & Francis.

Thompson, J.F., Mitchell, F.G., and Rumsay, T.R. 2008. *Commercial Cooling of Fruits, Vegetables, and Flowers*. UCANR Publications.

Topp, B.L., Russell, D.M., Neumüller, M., Dalbó, M.A. and Liu, W. 2012. Plum. *In Fruit Breeding*, edited by M.L. Badenes and D.H. Byrne. Springer.

Usenik, V., Fabčič, J., and Štampar, F. 2008. Sugars, Organic Acids, Phenolic Composition and Antioxidant Activity of Sweet Cherry (*Prunus avium* L.). *Food Chemistry* 107 (1): 185–192.

Valero, D., Díaz-Mula, H.M., Zapata, P.J., Guillén, F., Martínez-Romero, D., Castillo, S., and Serrano, M. 2013. Effects of Alginate Edible Coating on Preserving Fruit Quality in Four Plum Cultivars During Postharvest Storage. *Postharvest Biology and Technology* 77: 1–6.

Valero, D., Pérez-Vicente, A., Martínez-Romero, D., Castillo, S., Guillen, F., and Serrano, M. 2002. Plum Storability Improved After Calcium and Heat Postharvest Treatments: Role of Polyamines. *Journal of Food Science* 67 (7): 2571–2575.

Wang, J., Pan, H., Wang, R., Hong, K., and Cao, J. 2016. Patterns of Flesh Reddening, Translucency, Ethylene Production and Storability of 'Friar' Plum Fruit Harvested at Three Maturity Stages as Affected by the Storage Temperature. *Postharvest Biology and Technology* 121: 9–18.

Watkins, C.B. 2006. The Use of 1-Methylcyclopropene (1-MCP) on Fruits and Vegetables. *Biotechnology Advances* 24 (4): 389–409.

Yousef, A.R., Ahmed, D.M., and Sarrwy, S. 2016. Effect of Different Harvest Dates on the Quality of Beauty and Japanese Plum Fruits After Ripening. *International Journal of Chemical and Technology Research* 9 (7): 8–17.

Zainal, B., Ding, P., Ismail, I.S., and Saari, N. 2019. Physico-Chemical and Microstructural Characteristics During Postharvest Storage of Hydrocooled Rockmelon (*Cucumis melo* L. reticulatus cv. Glamour). *Postharvest Biology and Technology* 152: 89–99.

Zbrzeźniak, M., Nordlund, E., Mieszczakowska-Frąc, M., Płocharski, W., and Konopacka, D. 2015. Quality of Cloudy Plum Juice Produced from Fresh Fruit of *Prunus domestica* L.—the Effect of Cultivar and Enzyme Treatment. *Journal of Horticultural Research* 23 (2): 83–94.

Zhou, H., Ye, Z., and Su, M. 2015. Effects of Low Temperature and Forced-Air Precooling on Precooling Performance of Different Varieties of Peach Fruits. *Storage and Process* 15 (1): 16–19.

8 Diseases, Pests, and Disorders in Plum

Diagnosis and Management

Parthasarathy Seethapathy[1], *Rajadurai Gothandaraman*[2], *Thiribhuvanamala Gurudevan*[3], *and Imtiyaz Ahmad Malik*[4]

[1] Department of Plant Pathology, Amrita School of Agricultural Sciences, Amrita Vishwa Vidyapeetham, Coimbatore—642109, Tamil Nadu, India

[2] Department of Plant Biotechnology, CPMB&B, Tamil Nadu Agricultural University, Coimbatore—641003, Tamil Nadu, India

[3] Department of Plant Pathology, Tamil Nadu Agricultural University, Coimbatore—641003, Tamil Nadu, India

[4] Agriculture Production & Farmers Welfare Department, Government of UT of J&K, India

CONTENTS

DOI: 10.1201/9781003205449-8

8.1 INTRODUCTION

Prunus is an economically imperative genus of deciduous or evergreen flowering trees consisting of more than 430 species, mostly native to the temperate regions, and taxonomically placed within the Rosaceae family. The *Prunus* species are economically valued for their drupe fruit, as a decorative ornament, and for timber. Plum is a notable affiliate of the *Prunus* genera; its fruits are delicious, temperate, stone drupes cultivated widely in China, France, Germany, India, Italy, Japan, Russia, Romania, Spain, and the United States. Although numerous Old World and New World similar fruit species are phylogenetically called plums, only two species are *Prunus domestica* L. (hexaploid European plums or prune plums) and *Prunus salicina* Lindl.

(diploid Chinese or Japanese plums) are widely grown all over the world (Jayasankar et al. 2016). In India, *P. salicina* was first introduced into orchards in Mashobra, Himachal Pradesh (Shamrao et al. 2020), and it grows widely in the Himalayan states of the country. Plums and their interspecific hybrids are the richest sources of nutritious carbohydrates, proteins, minerals, vitamins (A, B, C, E, K), soluble fibres, and health-promoting compounds, like anthocyanins, antioxidants phenols, and sorbitol. Also, plums are the most preferred fruit for human consumption, with well-known cooling effects, and are considered the best option for reducing jaundice (Ahmed et al. 2010).

Naturally, plants have evolved elegantly to sense, succeed, sustain, or combat changing environmental conditions (Atkinson and Urwin 2012). But, as a consequence of global warming and climate catastrophe, several biotic and abiotic stresses collide, limiting crop production. Abiotic stress environments in trees can elevate the host vulnerability to pathogenic organisms, insects, and nematodes and lower resistance/tolerance capacity (Pandey et al. 2017). Abiotic stresses also influence the spread and occurrence of pests and pathogens in several fruits crops (Mittler 2006). Prolonged drought can cause plant pathogen infection on fruit crops. Increases in temperature and humidity in the atmosphere can have detrimental effect son fungal, bacterial, phytoplasma, viral, and nematode disease resistance in fruit crops (Atkinson and Urwin 2012). A plant-parasitic nematode infestation drastically alters plant water intake and can even exacerbate or inverse the effects of abiotic stress on plants. Simultaneously, pests and pathogens, on the other hand, may induce susceptible plant responses to physiological stresses. Therefore, both biotic and abiotic stresses impact severe consequences, especially when co-occurring. The yield losses due to diseases and pests are more extreme than abiotic factors to the economy and fruit production worldwide. Cultivated plums worldwide have encountered an increased number of biotic and abiotic stresses, including over 65 diseases affected by pathogenic fungi, oomycetes, bacteria, and viruses; 20 herbivorous pests; and 4 nematodes. Understanding and diagnosing the symptoms of biotic and abiotic stresses is thus critical in establishing management practices to ensure sustainable and quality fruit production. Plant disease and pest management in plums continues to concern, owing to increased customer demand for fresh; hygienic; and pest-, disease-, and chemical residue–free fruits. In this chapter, the authors discuss several biotic and abiotic problems, but not all diseases and pests. Readers must keep in mind that some of the diseases addressed might be more economically significant in some regions than in others.

8.2 PLANT DISEASES

Plant diseases play a detrimental role in plum cultivation by reducing the value and yield of plum fruits. Past and recent cases show the widespread destruction that disease outbreaks can cause. Some disease problems with plums are an issue everywhere plums are grown, whereas others, while significant in their specific locations, are not nearly as problematic globally. The major issue in European plum production is sharka disease, caused by an infection of plum pox virus (PPV); more than 100 million *Prunus* sp. in European countries suffer from this disease, and predisposed trees can exhibit an 80–100% decrease in yield. The development of sharka on leaves, flowers, fruits, and seeds in European plums and the severity of the disease led to the establishment of the Sharka International Working Group in the 1970s, which facilitates the collaboration of regional research between nations. Quarantine guidelines and restrictions were enforced between nations trading *Prunus* germplasms, which slowed the spread of diseases and pests in infestation-free regions (Levy et al. 2000). Likewise, most of the plum diseases are deleterious in the productivity of fruits. Still, it can be effectively prevented by a proper diagnosis of visible symptoms on the infected parts at earlier stages, which are caused by pathogenic oomycetes, fungus, bacteria, phytoplasma, and viruses. The diagnosable symptoms of plums are usually the consequences of a pathogen, and morphological changes include chlorotic and necrotic lesions, abnormalities, alterations, or impairments to plant tissue and cells due to an intrusion of the regular plant's metabolism. All visible and underground parts of plums are subject to attack by numerous pathogens. The delay in precise disease diagnosis and management may result in massive losses in fruit production and tree health, which causes poor fruit quality and economic losses to producers. Symptomatically, the expression of a disease will display at a late stage of infection and/or internal colonization of an infectious pathogen. Well-studied pathogens allow diagnosticians to detect it rapidly and precisely establish consistent management strategies. Proficient symptomatic detection, exact diagnosis of the pathogen, and well-timed management with organic or inorganic sources play an influential role in keeping plums free from pathogens.

8.2.1 Fungal Diseases

8.2.1.1 Wilt

The fungus *Verticillium* spp. has been reported to infect over 430 plant species comprising herbaceous annuals, perennials, and woody plants and causes a serious vascular wilt disease. This host range is expanding annually, and most plant species are susceptible to infection by *Verticillium* spp. (Deketelaere et al. 2017). The first report of *Verticillium* wilt on European plum trees was described in Pavia, Italy (Pollaci 1933). Then, the fungal species *Verticillium dahliae* Kleb., known to cause wilt symptoms on plum orchards in Hastings, England, was found (Smith 1965). In Moldavia, similar wilt symptoms were reported in plants affected by *Verticillium albo-atrum* Reinke and Berth as severe outbreaks in irrigated orchards (Gavrilenko and Kropis 1978). The soil -nvading *V. albo-atrum* and *V. dahliae* are prevalent in temperate and subtropical countries and are infrequent in tropical countries, especially in irrigated regions. Plum wilt is most likely to occur in the first three to ten years of cultivation. The diagnostic symptoms of *Verticillium* wilt are of two types depending on the pathogenic species and environment. A severe infection can lead to the rapid development of chlorosis, stunting, early defoliation, and wilting, leading to acute wilt; conversely, a delayed exhibition of minor symptoms, including reduced branch growth and sparse and twisted foliage from which the tree recuperates, is known as chronic decline. In mature trees, chronic infection is more common. Most accurate symptoms develop over a single growing season or over several years; leaves initially show light green, milder yellow, and brown, later becoming withered and curled (Parthasarathy 2021). Symptoms occasionally develop from the base of the branch upwards. Leaves may abscise or remain connected to the twigs, resulting in the eventual death of trees. When infection occurs at an early stage, blossoms and fruits fall off; if it matures later, fruits become withered, mummified, and abscised. The infected tree may completely disfigure, and sapwoods may exhibit xylem staining with abundant streaks of brown to black discoloration on the trunk and main branches (Hiemstra 1998). This characteristic indication gives it the common name 'blackheart'. Blackheart is only a severe problem in young trees that are three to six years old. *Verticillium* sp. survives as chlamydospores, resting mycelium, and microsclerotia in dying or dead plant tissues. *V. albo-atrum* also settled and persists as saprophytic in the uppermost soil surface, sometimes for years, by forming a microsclerotia structure in the dead and vanishing tissues of the infected trees. Dust storms and irrigation or rainwater disseminate the microsclerotia between the trees or orchards. High populaces of below-ground-feeding migratory nematodes in the soil can increase the severity of the disease. Soil-borne infections of *Verticillium* and its associated plant wilts are limited mainly by holistic measures at pre- and post-infection stages. Perhaps, precautionary pre-plant soil fumigation with chloropicrin will help to reduce the amount of *Verticillium* in the soil in known infested sites. Adequate protection against invading fungi is possible with propane flaming of microsclerotia that persists within the plant residues, planting of resistant rootstocks or cultivars, and restricting the wilt-susceptible intercrop or weed hosts. Although soil mulching with dense, black, poly spreads has been proven to overwhelm fungal structures in many stone fruit trees, newly planted rootstocks may topple in hot summer environments. Owing to the extended survival of inactive microsclerotia in the top layers of soil and the wider host choice of most plant species, combating *Verticillium* wilt is challenging. Furthermore, once the infection has reached vascular plant tissue, it is difficult to control, and even fungicidal applications seem ineffective. Once a tree has been infected with the *Verticillium* sp., it may be likely to enhance innate tolerance or defence modules by promoting crop environments, although frequent irrigation and nitrogen supplementation should be avoided (Berlanger and Powelson 2000). Table 8.1 shows the list of major diseases in plum.

TABLE 8.1
List of Major Plant Diseases in Plum

Disease	Pathogen	Cultivars	References
Fungal diseases			
Wilt	*Verticillium albo-atrum*	*P. domestica*	(Pollaci 1933)
Brown rot; blossom wilt; parda rot; twig and blossom blight	*Monilinia fructicola*; *M. laxa*; *M. fructigena*	*P. domestica*	(Michailides et al. 2007)
Powdery mildew	*Podosphaera tridactyla*	*P. salicina*	(Lee et al. 2012)

(*Continued*)

TABLE 8.1 *Continued*

Disease	Pathogen	Cultivars	References
Plum pocket; witches' broom; leaf curl	*Taphrina* spp.; *T. communis*; *T. pruni*; *T. insititiae*	*P. domestica*; *P. salicina*	(Horst 2013; Oh et al. 2020)
Black knot	*Dibotryon morbosum*	*Prunus americana*; *P. domestica*; *P. salicina*	(Biggs 1993)
Sooty blotch and flyspeck	*Zygophiala cryptogama*; *Z. wisconsinensis*; *Microcyclosporella mali*; *Stomiopeltis* spp.; *Pseudocercosporella* sp.	*P. domestica*	(Mirzwa-Mróz et al. 2011)
Rust	*Tranzschelia discolour*; *T. pruni-spinosae*	*P. domestica*	(Vidal et al. 2021)
Scab	*Cladosporium carpophilum*	*P. domestica*	(Pineau et al. 1991)
Anthracnose; bitter rot	*Collectotrichum gloeosporioides* sensu stricto; *C. acutatum*; *C. nymphaeae*; *C. foriniae*; *C. siamense*	*P. salicina*	(Lee et al. 2017; Børve and Vangdal 2004)
Pink mould rot	*Trichothecium roseum*	*P. domestica*	(Hasija and Agarwal 1978)
Root rot	*Armillaria mellea*; *Phymatotrichum omnivorum*	*Prunus* sp.	(Horst 2013)
Silver leaf rot	*Stereum purpureum*	*P. domestica*	(Ferdinandsen 1923)
Fruit rot	*Alternaria* sp.; *Botrytis cinerea*; *Cladosporium cladosporioides*; *Lambertella pruni*	*P. domestica*	(Ogawa and English 1991; Horst 2013; Grantina-Ievina and Stanke 2015)
Heart rot	*Fomes applanatus*; *F. fulvus*; *Lenzites saepiaria*; *Polyporus hirsutus*; *P. versicolor*	*P. domestica*	(Baxter 1925)
Mucor rot	*Mucor piriformis*	*P. domestica*	(Børve and Vangdal 2004)
Leaf blotch	*Phyllosticta congesta*	*P. salicina*	(Roberts 1921)
Thread blight	*Pellicularia koleroga*	*P. domestica*	(Horst 2013)
Leaf spot; shot hole; *Coryneum* blight	*Cercospora circumscissa*; *Coccomyces prunophorae*; *Coryneum carpophilum*; *Higginsia prunophorae*; *Phyllosticta circumscissa*; *Septoria pruni*; *Alternaria* sp.	*P. domestica*	(Pirnia et al. 2012; Grove and Maloy 1994; Horst 2013)

TABLE 8.1 *Continued*

Disease	Pathogen	Cultivars	References
Red spot	*Polystigma rubrum*	*P. domestica*	(Borovinova 2001)
Brown spot	*Alternaria alternata*	*P. domestica*; *P. salicina*	(Kim et al. 2005; Moosa et al. 2019)
Blossom blight; twig blight; grey mould	*Botrytis cinerea*; *B. prunorum* sp. nov.	*P. salicina*	(Børve and Vangdal 2004; Ferrada et al. 2016)
Twig blight	*Diplodia seriata*	*P. domestica*	(Reeder 2020)
Gummosis	*Botryosphaeria* sp.	*P. salicina*	(Ko et al. 2008)
Leucostoma (*Cytospora*/*Valsa*) canker	*Leucostoma cincta*; *Valsaria insitiva*; *Leucostoma persoonii*	*P. domestica*	(Adams et al. 2006)
Fusicoccum canker	*Fusicoccum amygdali*	*P. domestica*	(Diekmann and Putter 1996)
Charcoal canker	*Biscogniauxia rosacearum* sp. nov.	*P. domestica*	(Raimondo et al. 2016)
White root rot	*Rosellinia necatrix*	*P. domestica*	(Bouzari et al. 2021)
Eutypa die-back; gummosis; canker	*Eutypa lata*; *E. cremea*; *Eutypella citricola*; *E. microtheca*; *Cryptovalsa ampelina*	*P. salicina*	(Moyo et al. 2018)
Blue mould	*Penicillium* sp.	*P. domestica*	(Børve and Vangdal 2004)
Oomycetes disease			
Root rot	*Phytophthora cactorum*; *P. megasperma*	*P. domestica*; *P. salicina*	(Dang et al. 2004)
Bacterial diseases			
Bacterial crown gall	*Agrobacterium tumefaciens*	*P. domestica*	(Alburquerque et al. 2017)
Bacterial fire blight	*Erwinia amylovora*	*P. domestica*; *P. salicina*	(Jones and Benson 2001)
Bacterial canker	*Pseudomonas syringae* pv. *syringae*	*P. domestica*; *P. salicina*	(Popović et al. 2021)
Bacterial spot; shot hole; black spot	*Xanthomonas arboricola* pv. *pruni*	*P. salicina*; *P. japonica*; *P. simonii*	(Stefani 2010)
Fire blight	*Erwinia amylovora*	*P. domestica*	(Végh et al. 2012)
Leaf scorch/scald	*Xylella fastidiosa*	*P. domestica*	(Baldi and La Porta 2017)
Phytoplasmal diseases			
European stone fruit yellows	*Candidatus* Phytoplasma prunorum	*P. domestica*; *P. salicina*	(Delić et al. 2010)
X-disease	*Candidatus* Phytoplasma pruni	*P. domestica*; *P. salicina*	(Davis et al. 2013)
Witches' broom	*Candidatus* Phytoplasma asteris	*P. domestica*	(Zwolińska et al. 2019)

(*Continued*)

TABLE 8.1 *Continued*

Disease	Pathogen	Cultivars	References
Witches' broom	*Candidatus* Phytoplasma ziziphi	*P. salicina*	(Gao et al. 2020)
Late bloom	*Candidatus* Phytoplasma mali	*P. domestica*	(Ben Khalifa and Fakhfakh 2011; Mehle et al. 2006)
Witches' broom	*Candidatus* Phytoplasma solani	*P. domestica*	(Salem et al. 2020)
Viral diseases			
Yellow line pattern	*American plum line pattern virus* (APLPV)	*P. domestica*; *P. salicina*	(Choueiri et al. 2006; Pallas et al. 2012)
Chlorotic leaf spot	*Apple chlorotic leaf spot virus* (ACLSV)	*P. domestica*; *P. salicina*	(Rehman et al. 2017; Osman et al. 2017)
European plum line pattern	*Apple mosaic virus* (ApMV)	*P. domestica*	(Gospodaryk et al. 2013)
Stem grooving	*Apple stem grooving virus* (ASGV)	*P. domestica*	(Negi et al. 2010)
Chlorotic blotch	*Apricot latent virus* (ApLV)	*P. domestica*	(El Maghraby et al. 2006)
Leaf spot	*Apricot pseudo-chlorotic leaf spot virus* (APsCLSV)	*P. domestica*	(Safarova et al. 2012)
Ring mottle	*Cherry green ring mottle virus* (CGRMV)	*P. domestica*	(Wang et al. 2009)
Rasp leaf and die-back	*Cherry rasp leaf virus* (CRLV)	*P. domestica*	(Osman et al. 2017)
Ringspot and shot hole	*Cherry necrotic rusty mottle virus* (CNRMV)	*P. domestica*	(Zhou et al. 2013)
Resetting and chlorosis	*Cherry virus A* (CVA)	*P. domestica*	(Osman et al. 2017)
Bronzing	*Little cherry virus 1* (LChV-1)	*P. domestica*	(Tahzima et al. 2017)
Mosaic	*Peach mosaic virus* (PcMV)	*P. domestica*	(Pine and Cochran 1962)
Bark necrosis and stem pitting	*Plum bark necrosis stem pitting-associated virus* (PBNSPaV)	*P. domestica*; *P. salicina*	(Jo et al. 2016)
Sharka	*Plum pox virus* (PPV)	*P. domestica*; *P. salicina*	(Atanasoff 1932; Fiore et al. 2016)
Dwarfing	*Prune dwarf virus* (PDV)	*P. domestica*	(Fiore et al. 2016)
Chlorosis	*Prunus latent virus* (PrLV)	*P. salicina*	(Al Rwahnih et al. 2018)
Necrotic ringspot	*Prunus necrotic ringspot virus* (PNRSV)	*P. domestica*; *P. salicina*	(Fiore et al. 2016)
Chlorosis	*Prunus virus T* (PrVT)	*P. domestica*	(Marais et al. 2015)

TABLE 8.1 *Continued*

Disease	Pathogen	Cultivars	References
Mosaic	*Sowbane mosaic virus* (SoMV)	*P. domestica*	(Sutic and Juretic 1976)
Chlorosis	*Stocky prune virus* (StPV)	*P. domestica*	(Candresse et al. 1998)
Bushy stunt	*Tomato bushy stunt virus* (TBSV)	*P. domestica*	(Novak and Lanzova 1977)
Ringspot	*Tomato ringspot virus* (ToRSV)	*P. domestica*	(Fiore et al. 2016)
Viroid diseases			
Plum dapple fruit	*Hop stunt viroid* (HpSVD)	*P. domestica*; *P. salicina*	(Kaponi and Kyriakopoulou 2013; Cho et al. 2011)
Plum spotted fruit	*Peach latent mosaic viroid* (PLMVd)	*P. domestica*	(Giunchedi et al. 1998)
Marbling and corky flesh	*Plum viroid I* (PVd-I)	*P. salicina*	(Bester et al. 2020)

8.2.1.2 Brown Rot

Brown rot, also referred to as blossom blight or twig blight, is an economically and environmentally significant disease caused by various destructive necrotic fungi. *Monilinia fructicola* (G. Wint.) Honey (Syn: *Sclerotinia fructicola* [G. Winter] Rehm); *M. fructigena* (Aderhold & Ruhland) Honey (Syn: *S. fructigena* Aderhold & Ruhland); and *M. laxa* (Aderhold & Ruhland) Honey are common caustic pathogens of plum orchards worldwide, being especially severe in the wet season and warm, humid conditions (Michailides et al. 2007). Extreme loss notices from the fruit decay in plantation, in shipping, on the market, and even before consumers' hands (Obi et al. 2018). The first narrative of a brown rot pathogen on a moldering drupe was in 1796. Kirtland, in 1855, published a report of plum brown rot and its microscopic observation and stated that he had known about the disease for 30 years. Moreover, brown rot has been comprehensively studied with different species in the Mediterranean for more than 170 years; in the United States and Canada for more than 120 years; and later in Australia, Iran, New Zealand, China, Japan, Korea, Turkey, and Vietnam (Martini and Mari 2014). Brown rot produced by *M. fructicola* on *P. salicina* was recently reported in south-western Spain (Arroyo et al. 2012) and Turkey (Uysal-Morca and Kinay-Teksür 2019). These pathogens are very aggressive, with rapid colonizing and disseminating potential on fruit orchards in Asia, the Mediterranean, Oceania, and North and South America; it is also listed as a quarantine pathogen in Europe. *Monilinia* fungal infection remains on fruit epicarp as visible or invisible latent until the maturation of fruits (Tate and Corbin 1978). A quiescent infection of *Monilinia* will be detected earlier by dipping immature fruits in a paraquat herbicide that induces brown rot symptoms (Northover and Cerkauskas 1994). Actual brown rot symptoms produced by *M. fructicola* exhibit diversified symptoms by primary infections as blossom blights, twig blights, twig cankers in early or mid-season, and secondary infectious rots in the late season. Severely affected blossoms wilt, wither, and die, coated with greyish fungal propagules under favourable environmental conditions. Infections can extend to the twigs and appear as an oval, brownish spur canker. These cankers can ultimately break the twig, exude gummy substances, causing the branches to die. On infected fruit, symptoms initiate as minute, circular, brownish necrotic spots that surge promptly in size and eventually penetrate deeply, resulting in destruction or mummification and abortion of the entire fruit or in clusters that cling to the twigs. Clustered fruits have characteristics that are highly prone to rapid disease. Under wet and humid conditions, ash-grey, powdery tufts appear around the fruit, a typical diagnostic indication of this disease. *M. laxa*, also called European brown rot, is known to cause shoot tips to wither, blossoms to wilt and die and fruit to rot in orchards in California (Martini and Mari 2014). *M. fructigena* only produces a mild to severe fruit rot. The fungus can be defined with microscopic observation in comparison to published features. The proficient enactment of effective and holistic control strategies befits significant aspects in disease management (Michailides et al. 2007). Hygiene orchard management and best sanitation practices are more significant in preventing the disease in plums. The pathogenesis-related protein-5 (*PdPR5–1*) from the *P. domestica*, induced the innate gene *phenylalanine ammonia-lyase* (*PAL*), thereby increasing the content of camalexin phytoalexin at the earlier colonization of *M. fructicola* in the cloned plants; it also widens the development of resistant varieties or fortification of existing varieties in stone fruits (El-Kereamy et al. 2011). Mummified fruits should be removed and destroyed from the infected orchard, although blighted foliage and twigs should be trimmed thoroughly and removed from the plantings. Furthermore, because foliage interaction predisposes fruits to infection, timely thinning can minimize inoculum in plums (Michailides and Morgan 1997). Cut ends were immediately dressed with copper-based fungicides as a preventive measure. Prophylactic application of 4.5% carnauba wax considerably abridged occurrences of *M. fructicola* in plums (Gonçalves et al. 2010). *Aureobasidium pullulans* has been identified as a possible biological control agent for brown rot in numerous pome and stone fruits. More research into acetic acid and thymol vapour revealed that they were also effective in preventing brown rot (Liu et al. 2002). Registered fungicides, such as fluxapyroxad, iprodione, metconazole, triforine, thiophanatemethyl, trifloxystrobin, and fluopyram are regularly applied at the blooming time to reduce the risk of blighting.

8.2.1.3 Grey Mould

The genus *Botrytis* is among the "first defined genera of fungi", Pier Antonio Micheli reported, based on the appearance of a "bunch of grape berries" at the infection site in 1729. Grey mould rot symptoms caused by *Botrytis cinerea* Pers.:Fr. and brown rot symptomatic blossom blight caused by *M. laxa* are commonly confused because of their similar symptoms, even though grey mould is limited to flowers and does not spread to lignified limbs and branches like blossom blight. Grey mould is an aggressive necrotrophic post-harvest fungal decay, often called blossom blight; it is a well-known omnipresent disease in all plum-producing countries. Grey mould symptoms from *B. cinerea* have been reported on plums in the United States (Ogawa and English 1960) and South Africa (Fourie and Holz 1994). In addition, the new complex fungal species *B. prunorum* sp. nov., which causes blossom blight, is also reported on Japanese plums in central Chile (Ferrada et al. 2016). *Botrytis* sp. is most predominant on fruits picked late in the season and exposed to wet or humid conditions before harvesting. *Botrytis* blossom blight infections comprise pale brown necrosis on petals, beginning as discrete tanned mottles or V-shaped necrotic lesions at the petal boundaries, browning the stamen and pistils. It occurs when the temperature is chill and moist during the flowering period (Ferrada et al. 2016). The blossom blight phase did not directly cause post-harvest infection, but it may have been a source of infection. Fruit infections of *B. cinerea* on plums usually happen by cracks and wounds in the fruit, but they can also occur by contact with diseased adjacent fruit, developing a cluster of decayed fruits. Lesions typically develop on epicarp as light brown, fixed spots with a discrete periphery anywhere on the fruit. In severe cases, when the lesion has quickly spread and infected the whole ripe fruit, it becomes soft, brownish-black, and may even become leathery. Superficial greyish, dense, bunchy conidia and quiescent mycelia constantly ensued on the fruit surface at varying levels. Generally, immature drupes were moderately resistant, and the expression of the disease was mostly found on maturing fruit. To prevent the disease, avoid mechanical brushes and wounds in orchards and storage. The chemical botryticides cyprodinil, fludioxonil, fenhexamid, mepanipyrim, and pyrimethanil, are effective in treating disease and is the current mandate to implement effective anti-resistance management strategies (Rosslenbroich and Stuebler 2000).

8.2.1.4 Powdery Mildew

Plums and their hybrids are affected by several powdery mildew fungi, mainly *Podosphaera* spp., originating from infected roses or peaches, and near apricots and causing considerable yield loss up to 50%. *Podosphaera clandestina* var. *tridactyla* (Wallr.:Fr.) Lév. and *Sphaerotheca pannosa* (Wallr.:Fr.) Lév. are widely reported in early-season fruits of *P. domestica* from the United States (Gary and Wendy 2006) and in *P. salicina* from Northern Israel (Reuveni et al. 2006). *Podosphaera tridactyla* complex with cryptic speciation is widely reported in late-season plum foliage in Australia, Europe, Korea, India, Japan, Mexico, and New Zealand (Lee et al. 2012; García-Ruiz et al. 2019). Though there are two significant forms of pathogens with different symptoms, *P. tridactyla* colonized on *P. domestica* fruits exhibit circular extended white mycelial growth, followed by a necrotic or scabby exterior ten days after initial infection. The initial symptoms of *P. tridactyla* on *P. salicina* included white, transitory mycelia appearing as discrete patches on foliage (Lee et al. 2012). *S. pannosa* can infect young fruits, causing chlorotic spots and necrotic patches with browning and russeting that enlarges and leads to surface cracking on mature fruits (Reuveni et al. 2006). In comparison, *P. tridactyla* ascocarp and cleistothecia, which make spores, are used to find infections. As they mature, brown or black, globose or spherical-shaped cleistothecia with one to eight fasciculate appendages are formed on the infected shoots, leaves, and floors of orchards. Still, cleistothecia on plums are rare or absent (Grove 1995). *P. tridactyla* conidia produce simple germ tube and extend to nipple-shaped hyphal appressoria to draw the nutrients from the host during cool, moist nights and warm days. The proliferating conidiophores were colourless, erect, and long. Infective conidia were colourless, ovoid to ellipsoid, and with fibrosin bodies on the surface (García-Ruiz et al. 2019). The mildew survives as dormant mycelia or cleistothecia in the buds and is spread by airborne ascospores or anamorphic conidia. The powdery mildew susceptible plum cultivars/rootstocks, such as Kelsey, Gaviota, Black Beaut, and Wickson should be avoided in disease-prone areas. *S. panosa* fungal propagules on a host can be reported to control by the spraying of fungicides, sulphur and phosphates (Reuveni and Reuveni 1998), trifloxystrobin and kresoxym-methyl (Reuveni 2000), and antifungal biotics polyoxin B (Reuveni et al. 2000). *Podosphaera* can be effectively managed in foliage by applying fungicides trifloxystrobin, quinoxyfen, sulphur, mineral oil, and potassium bicarbonate at recommended doses.

8.2.1.5 Black Knot Disease

Black knot is a cancerous disease instigated by *Apiosporina morbosa* (Schw. Ex Fr.) Arx. (Syn: *Dibotryon morbosum* [Schweinitz] Theissen & Sydow, *Plowrightia morbosum* [Schweinitz] Saccardo) ensues on several edible, ornamental, and wild plums in temperate countries, which are particularly very serious in susceptible cultivars of plum in eastern North America, Canada, and Mexico (Snover and Arneson 2002). The disease was first reported in Massachusetts in 1811 and Pennsylvania in 1821 as a destructive form in plums. Thomas Taylor (1820–1910) was a US Department of Agriculture microscopist and first described the microscopic structures of black knot disease 1871 (Biggs 1993). It has been reported on over 24 *Prunus* spp. The disease is most often observed in poorly managed and abandoned trees; *P. americana*, *P. domestica*, and *P. salicina* are amongst the most vulnerable cultivated types. Black knot disease is rightly termed for the noticeable thick, black, irregular, tumour-like cankerous swellings or knotty growths that are expressed on diseased twigs. The diagnosable symptoms are expressed on the woody parts of trees, mainly on twigs, fruit spurs, branches, and occasionally on trunks and scaffold limbs. The initial indication of the black knot is usually a small, light-brown-to-black swelling on one side of the twigs that rupture as they enlarge, though they often girdle the entire twigs, mostly during the winter season. Within a year, the enlarged swellings vary in size (2.5–20 cm long and 8–25 mm thick), become olive green in surface colour with a soft, velvety texture that hardens over time. As the infection progresses, cylindrical or spindle-shaped knots endure swelling each year until girdled branches become a foot long and eventually die. It can cause substantial twig and branch mortality, severe stunting, and sometimes death of the trees (Ogawa 1995). The colonized fungus, *A. morbosa*, one year or older, may die, and the galls may become covered with pinkish white secondary fungi and riddled with insects. Infectious fungus overwinters as teleomorphic pseudothecia that are embedded on the black stroma on the surface of knots; the collective appearance of pseudothecia is referred to as ascostroma. The pseudothecia release the sexual clavate, bitunicate asci, and eight ascospores. Each ascospore resembles *Venturia inaequalis*, a closely related fungus, which produces two unequal-sized cells. The anamorphic stage is *Fusicladium* with copious olive-green spores on year-old knots; asexual conidial dispersal is limited in plums. Rain is required for ascospore discharge and is dispersed by rain splashing and wind current to new infection locations; the infection favours a warm, wet environment (McFadden-Smith et al. 2000). Temperatures between 16°C and 27°C are perfect for the spreading, germination, and colonization of new tree twigs (El Kayal et al. 2021). The ascospores penetrate unwounded, succulent green twigs and, occasionally, wounded tissues. Knots develop to appear slowly a few months after infection. One to two years are requisite for the knots to produce ascostroma (Snover and Arneson 2002). Management of black knots depends on cultural measures intended to limit the sources of inoculum and the damage they cause. Before ascospore discharge, all twigs and branches with fungal knots should be pruned out. Even after being removed from the tree, diseased twigs can continue to release ascospores. Thus, pruned debris should be burned or buried from the site immediately after the pruning takes place. Proper sanitation is crucial to controlling this disease in individual plantings. When planting new saplings over a large area, consider selecting resistant varieties. The American plum variety President has shown a high degree of resistance. Moderately resistant varieties in American and European types include the Bradshaw, Bluefree, Early Italian, Methley, Milton, Fellenberg, Formosa, Shiro, and Santa Rosa. Avoid planting with damson, Shropshire, and Stanley plums, which are very susceptible to black knot (Smart et al. 2019). Biological management offers protection of trees at an earlier stage of infection by applying *Trichothecium roseum*. Fungicides are efficient as a protectant against black knots, though they are unlikely to be successful if pruning and sanitation are neglected. Inoculum levels and weather conditions should be taken into account when scheduling fungicide applications. Spray with a Bordeaux mixture or liquid lime sulphur at the delayed dormant stage. Apply the sulphur as a protective application in three stages: lime sulphur with water 1:8 at bud break, lime sulphur with water 1:50 at full bloom, and the same dose two to three weeks later at stuck fall (Northover and McFadden-Smith 1995).

8.2.1.6 Leucostoma Canker

Leucostoma canker, also referred to as *Cytospora* canker, *Valsa* canker, or perennial canker affects over 85 suscepts of tree and shrub hosts. Molecular phylogeny and our follow-up of the 'one fungus one name' proposal, the name *Cytospora* (formerly called *Leucostoma*) is used in place of *Valsa* (Rossman et al. 2015). Since its detection more than 110 years ago in Canada and the northeastern United States (Biggs 2019), the disease has emerged as destructive in plum trees, especially in temperate countries. The pathogen has a wide range of hosts; besides plum, it also affects other stone fruits, pome fruits, and flowering plants and limits their yield potential and longevity (Adams et al. 2006). *Leucostoma*, or *Cytospora*, is considered an anamorphic stage of the teleomorphic genus *Valsa*. The disease is symptomatic with the perennial development of canker and dieback of twigs and branches caused by *L. cincta* (Fr. Ex.:Fr.) Höhn, clade in *P. domestica* reported from Italy (Romanazzi et al. 2012). It is a facultative wound parasite that invades wounded or dried tissues and colonizes weakened trees and overwinters as dark pycnidia on the infected bark. The fungal inoculum enters at pruned cuts, winter-injured buds, mechanical wounds, insect borings, sun-scaled injured bark crevices, and leaf scars, mostly on stressed trees. The infestation may occur as gradual tree decline through the drying of twigs with sunken cankers in the cortical tissue, xylem, phloem vessels, and pith, sometimes with amber-colour gummosis at the point of infection. Then, the fungus advances rapid death, resulting in girdling, depending on the tree's physiological condition and surrounding environmental factors. Underneath the infected sapwoods, the cankerous portion turns reddish brown by the growth of intra- and intercellular hyphae into the vascular region of trunks and crotches. Visibly, numerous tiny, brownish black, pimple-like, flask-shaped pycnidial fruiting bodies erupt from the infected barks; under humid conditions pycnidia exude orange pycnidiospores in gelatinous tendrils called cirrhi. Most pycnidiospores are dispersed by rain, boring insects, birds, and pruning practices (Sholberg and Kappel 2008). The disease is challenging to manage and requires the use of fungicides and the pruning-off of infected branches. Effectively, canker can be controlled by hygiene crop cultivations, such as only planting resistant varieties, planting sites with well-drained soil, and ensuring good drainage to reduce the chance of bark infection. The removal of highly cankerous trees from the orchard, avoiding overdoses of nitrogenous fertilizers and potassium deficits in the trees, mechanical injuries, periodical pruning, and dressing cut ends with copper-based fungicides along with borer control can also effecitvely manage the disease (Biggs and Grove 2005; Sholberg and Kappel 2008).

8.2.1.7 Plum Pocket

An ascomycetous genus, *Taphrina* Fries (1832), is the best-known dimorphic fungi, exhibiting hyphal and yeast-like stages. It is found solely responsible for most of the hyperplasia (overgrowth of cells) malformations unsightly expressed as a blister or curls in the foliage of *Prunus* sp. or, occasionally, like pockets in European plums (Horst 2013) and Japanese plums (Oh et al. 2020). The *Taphrina* species, comprising *Taphrina pruni* (Fuck.) Tul, *T. communis* (Sadeb.), and *T. mirabilis* (Atkinson) Giesenhagen, are responsible for the plum pocket in *T. domestica*, and *T. deformans* is responsible in *P. salicina*. Plum pocket disease affects fruit structure, which becomes abnormally enlarged, and the epicarp becomes misshapen with thick, distorted, soft, puffed-up, spongy flesh that is hollow without a stone and bladder-like, rendering it inedible. *T. pruni* infection can also cause distortions and swollen twigs, small branches, and young leaves of trees, and a witches' broom appearance produces numerous shortened shoots. In the fortitude of the anamorphic state, it is known that yeast forms grow in artificial culture producing pseudohyphae and ovoid-shaped, holoblastic, or enteroblastic budding cells. Infected fruits are harboured by the filamentous teleomorphic state, which reveals forming fragmented intercellular or subcuticular dikaryotic mycelium that give rise to naked asci in a subcuticular palisade layer. Eight ellipsoidal or fusiform-shaped ascospores are produced, and habitually budded cells inside the ascus develop as tiny blastospores (Fonseca and Rodrigues 2011). The fungus is difficult to control once established on the fruits. However, the preventive application of copper or sulfur fungicides could be an effective strategy combined with the pre-spraying pruning-off of infected branches. The fungus *Cladosporium delicatulum* was reported as a mycoparasite on *T. pruni* (Amanelah Baharvandi and Zafari 2015). Furthermore, clear evidence is required for developing it as an antagonistic agent for the management of plum pocket disease.

8.2.1.8 Sooty Blotch and Flyspeck

The convoluted tale of sooty blotch and flyspeck (SBFS) infection began 200 years ago. The fungi SBFS in the diverse complex, ectophytic plant-pathogen includes over

100 species accommodated in 30 saprophytic fungi genera that cause late-season darkly pigmented blemishes cosmetically surface-dwelling on the cuticle of pome and stone fruits, often on leaves and shoots, triggering economic fatalities in major growing regions worldwide (Gleason et al. 2019). More specifically on plum, five different putative SBFS fungal associates were described on *P. americana*, namely, *Zygophiala cryptogama* Batzer & Crous, *Zygophiala wisconsinensis* Batzer & Crous, *Pseudocercosporella* sp., and two strains of *Stomiopeltis* spp. in Iowa, in the United States (Latinović et al. 2007). Even though flyspeck, apparently caused by the *Schizothyrium pomi* teleomorph of *Zygophiala jamaicensis*, was reported on *P. salicina* in Japan (Nasu and Kunoh 1987), there was a new report of *Microcyclosporella mali* producing SBFS on *P. domestica* in Poland (Mirzwa-Mróz et al. 2011). A new cryptic species of *Z. emperorae* G. Y. Sun & Liu Gao, sp. nov., Z. *musae* G. Y. Sun & Liu Gao, sp. nov., and *Z. wisconsinensis* Batzer & Crous was recently reported on *P. salicina* in China (Gao et al. 2014). SBFS is a two-disease paradigm close to soot or flyspeck-like dots. The diagnostic symptoms of sooty blotch are superficial aggregations of olive green, dark mycelium with or without pin-headed, black structures implanted on the epicuticular layer and do not invade the peel. Indeed, flyspeck is a mass of minute, black sclerotium-like structures as shiny pycnothyria of varied sizes with no discernible mycelia. To establish sustainable and environmentally friendly disease management techniques, a precise diagnosis of the fungal cause is required. Cultural practices are the mainstay of SBFS control in plum, premeditated to condense the risk of epidemic, through planning orchards on better slopes and wind-exposed sites with resistant cultivars; producers should also prune out infected and unfertile branches, untie the bush canopy, speed up the drying process, and build the orchard simply so fungicides can penetrate.

8.2.1.9 Rust

Rust fungi cause severe pathophysiological symptoms and damages to crop plants. Rust in commercial orchards of plum caused by *Tranzschelia discolor* (Fuckel) (Syn: *T. pruni-spinosae* [Persoon] Dietel), mainly infecting *P. domestica* sub. sp. *domestica* in Brazil (Vidal et al. 2021), the United States (Lopez-Franco and Hennen 1990), the UK, and northern Europe (D'urban Jackson 2018). *T. discolor* sp. *domestica* is shortlisted as a harmful pathogen in Korea, and *T. discolor* sp. *persicae* is harmful in Canada and Korea (USDA PCIT 2020). It is a well-studied heteroecious, macrocyclic complex rust pathogen with five reproductive life stages that grow on two different host plants, though the alternate host plant is not essential for epidemic development on plum. Plum is the cultivated main host and produces urediniospores (stage II), teliospores (stage III), and basidiospores (stage IV). The secondary alternate host is *Anemone coronaria*, which produces spermatia (stage 0) and aeciospores (stage I). Urediniospores are repeating vegetative spores and can infect their respective *Prunus* sp. The diagnostic cyclic symptoms of rust take place in leaves, twigs, and fruits. Foliar rust lesions develop from the twig cankers as yellow speckling visible on the dorsal side of the foliage and light orange-brown uredinial sori shedding the urediniospores on the lower side. In severe stages, urediniospores develop as dark, rusty brown teliospores with brown pustules. Infected young leaves become crinkled and distorted, aged leaves become shriveled, necrotic, and shredded off.

The fruit reddens, pustules develop as brownish erumpent lesions with halos that become green-yellow tinged, which causes more widespread fruit decay and secondary invading fungi. Heavy infection results in twig canker development as masses of rusty-brown asexual urediniospores on surface tender woods, expressing as blisters and longitudinal splits in the bark. Twig cankerous inoculum is a source of infection for next season blooms. Rust pustules modify respiration rates, photosynthesis, and transpiration and cause a decline in vigour and the yield potential of plants. The inoculum produced by both hosts as airborne interchanging teliospores or basidiospores depends on host and wetness periods of 12–18 hours for infection. The disease is challenging to control and entails applying fungicides and the pruning off of cankered twigs to reduce the humidity in the canopy. Previous studies proved that calcium hydroxide, fenhexamid, tebuconazole, propiconazole, and myclobutanil reduced the disease pressure in plum (Ogawa and English 1991). Precautionary application of *Bacillus amyloliquefaciens* subsp. *plantarum* strain D747, *Gliocladium catenulatum* strain J1446 and *Bacillus subtilis* QST713 are recommended for effective management (D'urban Jackson 2018).

8.2.1.10 Anthracnose

Utmost, *Colletotrichum* sp. are necrotrophic, omnipresent foliar pathogens that cause anthracnose or fruit rot in a wider host range, including plum. The complex of fungus, *C. acutatum* J.H. Simmonds, *C. foriniae* (Marcelino & Gouli), *C. siamense* Prihastuti, L. Cai & K.D. Hyde, *C. gloeosporioides* sensu stricto (Penz.) Penz. & Sacc, and *C. nymphaeae* (Pass.) have been described as anthracnose causative of Japanese plum (*P. salicina*) in Korea (Lee et al. 2017; Chang et al. 2018; Hassan et al. 2019). Recently, anthracnose lesions were observed in leaves of Japanese plum trees in China. Initially, lesions were brown, pale, and sunken, and brown spots appeared nearer to the margin. The lesions then grew uneven and became distorted, wrinkled, and wilted (Cui et al. 2021). Affected fruits were shown sunken and slightly round with and brown necrotic lesions on the fruit surface (Hassan et al. 2019). Preventing *Colletotrichum* infections is a concern of complete orchard management and appropriate tree care: plant saplings on a wide space, maintain the hygiene of the vegetation and prune timely to facilitate airflow on the foliage, and dress wounds and prune cuts with wound dressers.

8.2.1.11 Shot Hole

Shot hole disease of plum trees, produced by *Wilsonomyces carpophilus* (Lév.) Adaskaveg, Ogawa & Butler (Syn: *Coryneum beijerinckii* Oud. and *Stigmina carpophila* [Lév.] M. B. Ellis) (Adaskaveg et al. 1990), is a significant problem on other *Prunus* species in temperate to semiarid regions (Ogawa 1995; Ahmadpour et al. 2009). The diagnostic symptoms are minute purple lesions extended to brownish spots on the leaves. Later, infected leaves abscise in a dry and warm environment and are peppered with holes. The infection of *W. carpophilus* is tiny, dark specks consisting of hyaline, septate mycelia, and anamorphic sporodochia; it also produces a mass of fusoid conidia over conidiophores. Conidia are holoblastic, rhexolytic, initially subhyaline to a golden brown, later olivaceous to dark black (Ahmadpour et al. 2009). To manage the disease, zinc sulphate is sprayed on leaves to quicken leaf fall and prevent the inoculum. Prune out lower twigs or unproductive foliage to avoid infections. Hygiene sanitation and directing water away from the leaves provide better prevention of disease.

8.2.1.12 Red Spot

Red spot disease is an important disease on *P. domestica* in years with a rainy spring, especially in poorly maintained orchards or older trees. It is also referred to as blackthorn dotty, caused by the distinctive ascomycete fungus *Polystigma rubrum* (Pers.) DC. in the Middle East and southern Europe. *P. rubrum* is a more widespread 'vulnerable' fungal pathogen of the early 20th century, listed in the conditional British Fungi Red Data List (Evans et al. 2006). A *P. rubrum* infection looks like bright red blisters on living leaves of *P. domestica* and related trees. Infected twigs look splashed with the circular, yellow-red elevated lesion, although manifestations may be confined to a few spots on a few leaves (Suzuki et al. 2008). A *P. rubrum* infection decreases the cellular turgescent, the deterioration of the chlorophyll leads to defoliation, and the fruits become small. The fungus produces the mycelium with intercellular growth, which forms coloured stroma. Young stromata of *P. rubrum* may be yellowish-brown, becoming orange to red upon maturity, reddish-brown without purple-coloured, "shot-holes," or necrotic patches on the leaf tissues. The infected portion produces roughly spherical conidiomata 150–250 μm dia, the ostiole showing on the upper surface of inconspicuous stromata. Conidiomata consist of distinctive hooked or curved conidia. The teleomorphic stage on fallen overwintered leaves exhibits black stromata (Roberts et al. 2018). The source of reducing the disease infestation is mainly by resistant genotypes and post-infectional control with fungicide mancozeb (Glišić et al. 2017).

8.2.2 Oomycete Diseases

8.2.2.1 Phytophthora Root and Crown Rot

Oomycetes (kingdom: Chromista) are earlier pseudo-fungi that evolved its pathogenicity to infect hosts independently of other eukaryotic pathogens and is likely to have distinct pathogenic processes. *Phytophthora* root and crown rot (occasionally by girdling the scion, called collar rot) are quite common oomycetes in most orchard soils and the most fatal diseases of pome and stone fruit trees worldwide. More than ten species of *Phytophthora* are known to distress commercial *Prunus* sp., affecting trunk and scaffold cankers, crown rot, root rot, and post-harvest fruit rot (Browne 2017). *Phytophthora cactorum* (Lebert & Cohn) J. Schöt. has been widely stated to cause havoc in plums, especially in European, Mediterranean, American, and south-east Asian countries (Grígel et al. 2019). The erratic symptoms depend on tree species, soil nutrients, and region and commonly present as sparse or poor growth, foliage discoloration, and sudden wilt. Internally, damaged root systems, vascular systems, or both can be found. Despite its diversified infection, *P. cactorum* was also reported to cause a foliar black spot on *P. salicina*, which seriously reduced crop yields (around 20%) in Vietnam. Diagnostic symptoms are typically whitish-grey water-soaked lesions on immature fruit, progressing into sunken black spots with brown margins. In circumstances of intensive infection, the entire fruit shrivels and trees are fruitless (Dang et al. 2004). In Turkey, commercial *P. salicina* orchards infected by *Phytophthora megasperma* Drechs exhibit diagnostic symptoms of foliar discoloration, dieback, and severe defoliation. The infected basal stem shows reddish-brown cankers (Kurbetli et al. 2017). Abad and Abad reported that *Phytophthora niederhauserii* caused little crown rot in plum hybrids (Browne 2017). Similarly, the oomycete *Globisporangium intermedium* (de Bary) was also associated with nursery infections in the Czech Republic (Grígel et al. 2019). The dispersal of the pathogen occurs when soils are wet, oospores developing white fungal hyphae, directly infecting plant collar or root portions. When grounds are waterlogged, anamorphic zoospores produce within sporangia, releases upon exhaustive burst, disperses in soil, and swims onto the new roots to attack. However, in dozens of plum hybrid rootstocks, a judiciously high range of resistance were detected to *Phytophthora* species and may have long-term consequences for clonal rootstock propagation. Most plum hybrids were greatly and reliably resistant to infections of *P. niederhauserii*, e.g., Mariana 2624 was vastly resistant to ground-level infections by all of the *Phytophthora* spp. (Browne 2017). Preventing the source of infection is important, as are timely monitoring and limiting the water and soil from an infected area to a healthy area. Copper oxychloride, mancozeb, metalaxyl, and fosetyl aluminium are progressively active when used along with hygiene practices.

8.2.3 Bacterial Diseases

8.2.3.1 Bacterial Canker

Bacterial canker is a severe problem in *P. domestica* plum trees worldwide; recently *P. salicina* was also found to be susceptible in Montenegro (Popović et al. 2021). Bacterial canker is instigated by two distinct pathovars, *Pseudomonas syringae* pv. *syringae* van Hall and *P. syringae* pv. *morsprunorum* (Wormald) Young, Dye & Wilkie (Wenneker et al. 2012), and is further diversified into two races. *Pseudomonas syringae sensu lato* is widely defined as a complex of strains, delineated into nine genomospecies with a broad host range, including herbaceous plants, woody plants, and some ornaments. The most noticeable indications are blighting of dormant buds resulting in small, elongated, flattened cankerous growths on twigs and branches that ooze dark, sticky gum; spur dieback; and often complete death. Devitalized buds and leaf lesions also can occur. Leaves on affected branches exhibit round, pale brown lesions on the leaves that develop into holes encircled by dry brown margins. In severe cases, affected trunks are discoloured, necrotic, and girdled by cankers; subsequently, bacteria invade woody tissue, succeeding colonization of blooms or latent buds, causing rapid death of infected trees. Canker oozes can easily be confused with gummosis, a disorder of plum fruits that frequently occurs after icy weather, whereas with bacterial canker, the gum oozes from necrotized tissue and has a sour acidic smell. *P. syringae* in inhabiting trees both endophytically and epiphytically shows a conspicuous role in retarding effective bacterial management (Kennelly et al. 2007). Avoid winter pruning in plum since the cut ends may facilitate the entry of bacterial cells. Eradicate cankered twigs by summer pruning with dis-infected tools at least 20–30 cm from the healthy portion. The pruned regions should be sealed with copper-based contact fungicides (Borkar and Yumlembam 2016). Copper resistance in bacteria has been found in stone fruit–growing countries under conditions favourable for disease, exposing chemical control measures at risk. Indeed, prioritize cultural control measures first, then add chemical control if needed. Endophytic infections could be managed by applying registered streptomycin or oxytetracycline antibiotics or antagonistic strains, *B. amyloliquefaciens* strains D747 or *B. subtilis* strain QST 713 at initial stages of infection.

8.2.3.2 Bacterial Spot

The *Xanthomonas arboricola* is a bacterial complex predominantly pathogenic with leaf spots, shot hole, bacteriosis, and blackspot of *Prunus* sp. affected by seven different pathovars with a high host specialization (Fischer-Le Saux et al. 2015), including *X. arboricola* pv. *pruni* (Xap) that attacks only *Prunus* spp. and their hybrids, particularly *P. domestica* and *P. salicina*. It was first described on *P. japonica* in Michigan (USA) by Smith in 1902 and was termed *Xanthomonas pruni* (Smith). Then, it was re-named as *X. arboricola* pv. *pruni* through DNA–DNA hybridization by Vauterin et al. (1995). The *X. arboricola* pathovars *corylina*, *juglandis*, and *pruni* are closely related and are the most virulent strains causing economic threats to stone fruit production. It has caused severe epidemic yield losses on European plum in Georgia (USA), reported around 25–75% production loss (Dunegan 1932), and Australia, with 10% production loss. Initial, *X. arboricola* epidemics of 30% on Japanese plum (cv. Calita, Golden Plum, Angeleno) in northern Italy (Stefani 2010) and in south-western Spain (Roselló et al. 2012) developed swiftly over the years; presently, the infected have dispersed in epidemic form throughout five continents. Undeniably, *X. arboricola* pv. *pruni* are designated as quarantine pathogens in many European countries (EFSA PLH Panel 2014). The remaining pathovars are opportunistic or saprophytic pathogens, not being in the authority of pandemics in a specific host range. Symptoms are witnessed on leaves, twigs, branches, and fruits. Typical symptoms are leaf spot and shot hole; defoliation; dieback; branch and trunk cankers; and sunken, necrotic fruit lesions (Scortichini et al. 2001), but these vary through the pretentious species. Significantly, lesions on leaves appear as small, angular, pale green to yellow spots that expand and become dark or purple necrotic lesions over time. These necrotized lesions can be dropped, providing a shot-holed, tattered look to the leaf, typically with a brown ring on retained leaf tissues. Often, twigs symptomatic with small, water-soaked, slightly darkened, superficial blisters may exhibit gumming, developing cankers that may result in the death of the plants. Lesions on the fruit surface appear as large, slimy, sunken, water-soaked spots with light green halos. The bacterium *X. arboricola* pv. *pruni* is Gram-negative, gamma rod, flagellated, motile, aerobic, non-sporulating, and produces pale yellow colonies on nutrient media (Garita-Cambronero et al. 2018). The parasitic infections mainly through epiphytic bacteria exist in woody cankers, infected buds, scars, and leaf debris in the soil. Sources of resistance are trustworthy for preventing epidemics in plum and, in practice, disease control depends on tree hygiene, cultural practices, periodical application with copper-based bactericides, sulphates, oxy-chlorides, and hydroxides, liable on the phenological phases A and B of the host plant (Stefani 2010). Combined sulphur and captan exposed specific efficacy in reducing the disease during post-flowering periods (McLaren et al. 2005). Major epidemics in young orchards are frequently endorsed to deprived cultural operations, though it must be practised to reduce inoculum levels. In affected plums, the pruning of twigs with lesions and cankers must be done during winter (Morales et al. 2017). Intuitively, marker-assisted-selection-based quantitative trait loci with resistance is being employed to improve new varieties with an augmented tolerance within the charter of the recent FruitBreedomics in the EU and RosBREED in the EU USA. Alternative strategies are being established with phages, pseudomonads, as antagonists, and resistance inducers, such as plant elicitor peptides and copper glucohumates.

8.2.4 Phytoplasma Diseases

8.2.4.1 European Stone Fruit Yellows Phytoplasma

Phytoplasmas are microorganisms related to hundreds of plant diseases in almost every country on the planet. Several phytoplasma groups impend *Prunus* cultivation in growing countries. The strains of the 16SrI-B; 16SrII-B,C; 16SrVI-A,D; 16SrIX-B,C,D; 16SrX-F; and 16SrXII-A are the prominent phytoplasma groups affecting *Prunus* species in the Middle East (Hemmati et al. 2021) and other parts of the world. *Candidatus* Phytoplasma mali was first reported in *P. domestica*, causing a plum decline in Tunisia (Ben Khalifa and Fakhfakh 2011; Mehle et al. 2006). Recently, *Candidatus* Phytoplasma solani has also been known to cause infections in *P. domestica* in Jordan (Salem et al. 2020), exhibiting leaf chlorosis and reddish pigmentation, restricted growth, and a witches' broom appearance. Among the diverse phytoplasmas, European stone fruit yellows phytoplasma (ESFY) is a prime epidemic disease in *P. domestica* in Europe (Bongiorno et al. 2020), plum leptonecrosis in the Japanese plum in Italy (Giunchedi et al. 1982), and Molières decline in France (Bernhard et al. 1977). Additionally, apple proliferation, apricot chlorotic leaf roll, peach decline, pear decline, and plum yellowing are the symptoms caused by Candidatus *Phytoplasma prunorum* (16SrX-F). ESFY is an epidemic disease, well known by its rapid and extensive movement when conditions are favourable for host plants, and insects can disseminate the disease. It is currently prevalent in Albania, Bosnia-Herzegovina, Greece, Italy, Austria, Hungary, Bulgaria, Romania, Slovenia, Belgium, the Czech Republic, England, France, Germany, Switzerland, Serbia, Montenegro, Spain, and Turkey. Even numerous Mediterranean plum orchards become unfertile eight to ten years after initial establishing due to a large proportion of ESFY phytoplasma infection, which is more usually accompanied by a significant mortality rate in plum. After seven years of growth, up to 100% of infection proportions have been observed on disposed *P. salicina* cultivars Ozark Premier and Shiro (Carraro et al. 1998). In mid-summer, distinctive signs consist of small and narrow foliage, upward leaf roll, and a brown-red appearance that becomes profuse and fragile. Infected buds are shorter and bear smaller, deformed leaves. Declined fruit yield is witnessed; fruits are reduced, ripen later, and may fall prematurely, then the entire plant can be affected and die. The psyllid *Cacopsylla pruni* transmits

Ca. *P. prunorum* between *Prunus* sp. (Carraro et al. 2001) in northern Italy and England.

8.2.5 Viral Diseases

8.2.5.1 Plum Pox Potyvirus

Plum pox virus (PPV) is the most solemn viral infection in *Prunus* sp. and belongs to the genus *Potyvirus*. PPV causes infections in a wider plant species with greater intraspecific diversity, which are transmitted predominantly by insects. The PPV infection was first reported in *P. domestica* cv. Kyustendil by plum growers in the Zemen village of Bulgaria in 1915–1918, although certain documents specify indications were noticed in Macedonia as early as 1910. It was described as a viral disease and given the name Sarka po slivite (sharka), or 'Pox of Plum', in 1932 (Atanasoff 1932). Between 1932 and 1960 the virus spread relentlessly across the Eurasian, African, and Mediterranean regions. Later, the progressive spread was recorded between 1970 and 2000 in Argentina, Belgium, Chile, China, Cyprus, Egypt, France, India, Spain, Syria, Italy, Portugal, and the United States (Levy et al. 2000; Sastry et al. 2019). PPV is most detrimental in *P. domestica*, *P. salicina*, and their hybrids. Symptoms may be detected on foliage, sprouts, bark, blossoms, fruits, and even endocarphic stones; conversely, they differ on host plants, cultivars, virus strain, age, physiology, and prevailing environmental conditions in the orchard. The characteristic symptoms in leaves include pale-green bruising, circular yellow spots, bands, blotch rings, oak-leave pattern, vein clearing, and premature fall off. Infected flowers may exhibit pinkish streak discoloration on the petals (Barba et al. 2011). Fruits with infected flesh have shallow yellow rings, line patterns, or arabesque depressions, as well as brown or reddish necrotic flesh that becomes uneven or malformed and gummy. The drupe of infected plums shows yellow rings. In certain cases, shoots split and die. European plums may also show early fruit fall of nearly 100% in the most susceptible cultivars of plums, whereas Japanese plums exhibit ring spots on fruits (García et al. 2014). Foliar symptoms of PPV in the plums are almost similar to another stone fruit virus, namely, American plum line pattern virus (APLPV). Mixed infection with Prune dwarf virus (PDV), *Prunus necrotic* ringspot virus (PNRSV), or Apple chlorotic leafspot virus (ACLSV) may advance the severity of PPV infections (Bamford and Zuckerman 2020). PPV has gained much importance as a plant-infecting ss(+) RNA virus. The virions are non-enveloped, flexuous, and approximately 750 × 15 nm. Genomic particles are composed of the linear genomic ss(+)RNA of 9741–9795 nt in length, with encapsidated coat protein (CP) subunits (Bamford and Zuckerman 2020), polarized with polyadenylated (poly-A tract) and covalently linked virus-encoded protein (VPg). It contains a single open reading frame (ORF) translated into a large polyprotein precursor of 355 kDa, 3125–3143 amino acids that is proteolytically catalysed by virus-encoded proteinases (HC-Pro, NIa-Pro, and P1) to produce functional protein products. Additionally, P3N-PIPO is synthesized by a polymerase slippage of the viral RNA polymerase in a P3-encoded GA6 motif, which probably acts as a movement protein. PPV possesses limited genome constituents to accomplish its life cycle in the selected host. PPV isolates that possess definite symptoms, host range, infectivity, genome, epidemiology, and vector transmissibility have been structured into ten monophyletic strains: PPV-An, PPV-C, PPV-CE, PPV-CR, PPV-EA, PPV-D, PPV-M, PPV-Rec (recombinant between strains D and M), PPV-T, and PPV-W (Sihelská et al. 2017). Among the PPV strains, M is concomitant with peach, D and Rec are concomitant with plum, both are most infective and widespread across the *Prunus* sp. (Ilardi and Tavazza 2015). PPV strains An, T, M, D, Rec, and W are frequently detected in *P. domestica* (Rodamilans et al. 2020). The virus coded HC-Pro facilities the transmission by aphids, *Aphis craccivora, A. gossypii, A. fabae, A. spiraecola, Brachycaudus helichrysi, B. cardui, B. persicae, Hyalopterus pruni, Myzus humuli, M. persicae, M. varians*, and *Phorodon humuli*, in a non-circulative, non-persistent manner from neighbouring infected reservoirs. PPV is transmissible by seed (especially PPV-M), virus-infected propagating materials, and grafting, which seriously contributes as the main source of infection under orchard conditions. Plant viral infections cannot be controlled directly by spraying chemicals on diseased plants, though insecticidal application over migratory or over-wintering aphids reduces the inoculum level. Resistance breeding of plum genotypes with a hypersensitive response to PPV infection may be sustainable for developing multigenic resistant plum (Hartmann 1998); the use of resistance rootstocks is recognized mostly within the *P. armeniaca, P. davidiana, P. domestica*, and *P. ferganensis*. Even though *P. domestica* is a PPV-resistant variety, the Jojo plum is characterized as an oligogenically organized avirulence (avr) factor by recognizing host plant R gene encode nucleotide-binding leucine-rich repeat proteins (Rodamilans et al. 2014). Deployment of resistance in wide cultivars provoked the high diversity of PPV strains, tolerant cultivars (i.e. Cacak Best, Kalipso, and Stanley) which may not express fruit infections but permits PPV replication and transmission in many countries. Genetic transformation and regeneration is a promising tool to develop transgenes under the lack of resistant sources. The US and France worked together to make C5, a clone resistant to PPV. C5 is a PPV *CP* gene construct in the plasmid pGA482GG/PPVCP-33 using *Agrobacterium tumefaciens* (Scorza et al. 1994) or, hpPPV-*CP* (Scorza et al. 2010), now renamed Honeysweet in the United States (Scorza et al. 2013), and the resistance can be conveyed through hybridization to other plums and its hybrids. The transgenic plum expresses the reduced level of PPV *CP*-mRNA, and it is processed effectively due to post-transcriptional gene silencing (PTGS) inducing intron-spliced hairpin RNAs (ihpRNAs) (Smith et al. 2000), which induce the resistance, which is systematically analogous to RNAi. C5 provides a unique source of resistance for ongoing and upcoming breeding experiments worldwide. In addition, knockdown of *eIF(iso)4E* protein (eukaryotic translation initiation factors or their isoforms) in susceptible plum cultivars by gene replacement (Wang et al. 2013). As a result, mining such recessive genes or core proteins will be future research aims for designing new interfering tactics against viral infections (Ilardi and Tavazza 2015).

8.2.5.2 American Line-Pattern Virus (APLPV)

The American line-pattern virus (APLPV) (genus: *ilarvirus*), is observed in *P. salicina* and *P. domestica*, reported wherever it has been grown widely, including

Canada, New Zealand, Poland, Greece, Norway, India, Mexico, Israel, and the United States (Choueiri et al. 2006). On *P. salicina*, the characteristic symptoms start with yellow vein banding and chlorosis in the leaf lamina with an occasional oak-leaf pattern, accompanied by dwarfing with the yellow appearance transitioning to a creamy white. Young petioles and stems exhibit chlorosis, although old branches show aerial rotting. The fruits are faintly mottled, even when green. Fruits found on severely affected trees show diffuse, chlorotic spots during ripening. The virions comprise four different quasi-isometric particles. The genome is ss(+)RNA with a tripartite genome: RNA1, RNA2, and RNA3 (Scott 2011). The disease is contagious by somatic propagation and grafting. It is also infectious by dodder *Cuscuta hyalina*. Apricot seedlings are used as resistant rootstock against plum line-pattern virus to check the virus transmission in scion varieties through rootstock.

8.2.5.3 Plum Bark Necrosis Stem Pitting-Associated Virus (PBNSPaV)

Plum bark necrosis stem pitting-associated virus (PBNSPaV) (genus: *Ampelovirus*) is an infection in *P. salicina* (Black Beaut) that was detected in Dinuba, California (Stouffer et al. 1969) and later noticed on *P. domestica* (Jo et al. 2016), with the Shiro and Friar cultivars being the most susceptible. The virus spreads in China, Egypt, France, Jordan, Italy, Korea, Serbia, Turkey, Morocco, and the United States. PBNSPaV-infected plum shows bark necrosis, destruction of scaffold twigs, gummosis, and stem pitting. The matured particles are flexuous, non-enveloped, and 1500 × 10–13 nm. The genome is a linear, monopartite, ss(+)RNA of 14,214 nt. The virus is spread mainly by mealybugs and/or grafting (Boscia et al. 2010).

8.3 INSECT PESTS

Insects are the most diverse group of living creatures on the planet. They can be found in all varied environments, even habitats in extreme weather. They are the most adaptable form of living organism, as their total numbers far exceed that of any other animal category. Insect pests have harmful effects on plants, farm animals, and humans (Williams 1947). In general, insect pests are divided into two categories based on the damage caused to plants: chewing insects and sucking insects. Chewing insects chew off plant portions and devour them, causing crop damage, whilst sucking insects pierce the epidermis and draw the sap from the plants. Many of the sap feeders act as vectors of plant viruses and cause devastation to the host by injecting their salivary toxins. Peaches, plums, and berries are among the most commonly grown small fruits in the world. They have many insect pests, which are reported on various *Prunus* spp. in all different geographical locations (Table 8.2). The plum fruit moth, plum sawfly, San Jose scale, aphids, twig borers, thrips, and mites are all well-known destructive plum tree pests. They cause severe economic damage for farmers by affecting the quality and marketability of the farmer's plum fruits. Identifying these pests and understanding their life cycles and damage can help you to choose the most effective control practices. Pests of peaches and plums are especially difficult to control because the fruit is susceptible to many kinds of pests over a long period, from petal fall through harvest.

TABLE 8.2
List of Major Insect Pests and Nematodes in Plum

Insect pests	Species	Cultivars	References
Insects			
Plum fruit moth	*Grapholita funebrana*	*P. domestica*; *P. salicina*	(Rizzo et al. 2019; Butturini et al. 2000)
Prune or peach twig borer	*Anarsia lineatella*	*P. domestica*; *P. salicina*	(Damos and Savopoulou-Soultani 2008)
Plum sawflies	*Hoplocampa minuta*; *H. flava*	*P. domestica*; *P. salicina*	(Njezic and Ehlers 2020)
San Jose scale	*Quadraspidiotus perniciosus*	*P. domestica*; *P. salicina*	(Gupta 2005)
Lecanium scales or European fruit lecanium	*Parthenolecanium corni*	*P. domestica*; *P. salicina*	(Ben-Dov 1993)
Leaf curl plum aphid	*Brachycaudus helichrysi*	*P. domestica*; *P. salicina*	(Arora et al. 2009)
Apple and thorn skeletonizer	*Choreutis pariana*	*P. americana*; *P. domestica*	(Heppner 2004)
Spotted cutworm	*Xestia c-nigrum*	*P. americana*; *P. domestica*	(Landolt et al. 2010)
American plum borer	*Euzophera semifuneralis*	*P. americana*	(Biddinger and Leslie 2014)
Branch and twig borer	*Melalgus* (=*Polycaon*) *confertus*	*P. domestica*	(Solomon 1995)
Mealy plum aphid	*Hyalopterus pruni*	*P. domestica*; *P. salicina*	(Monia et al. 2013)
Peachtree borer	*Synanthedon exitiosa*	*P. americana*; *P. salicina*	(Teixeira et al. 2010)
Grape mealybug	*Pseudococcus maritimus*	*P. americana*; *P. domestica*	(Waterworth and Millar 2012)
Shothole borer	*Scolytus rugulosus*	*P. domestica*	(Ozgen et al. 2012)
Western flower thrips	*Frankliniella occidentalis*	*P. americana*; *P. domestica*; *P. salicina*	(Allsopp 2010)
Plum curculio	*Conotrachelus nenuphar*	*P. americana*; *P. domestica*; *P. salicina*	(Lampasona et al. 2020)
Oriental fruit fly	*Bactrocera dorsalis*	*P. domestica*; *P. salicina*	(Vargas et al. 2010)
Mediterranean fruit fly	*Ceratitis capitata*	*P. domestica*; *P. salicina*	(Ali et al. 2016)
Mites			
Two-spotted spider mite	*Tetranychus urticae*	*P. domestica*; *P. salicina*	(Baldo et al. 2018)
Pacific spider mite	*Tetranychus pacificus*	*P. americana*; *P. salicina*	(Baldo et al. 2018)

TABLE 8.2 *Continued*

Insect pests	Species	Cultivars	References
Plum rust mite	*Aculus fockeui*	*P. americana*; *P. domestica*; *P. salicina*	(Abou-Awad et al. 2010)
Brown mite	*Bryobia rubrioculus*	*P. americana*; *P. domestica*	(Jamshidian et al. 2005)
Nematodes			
Ring nematode	*Mesocriconema (=Criconemella) xenoplax*; *Macroposthonia xenoplax*	*P. domestica*	(McKenry 2004; Sharma et al. 2003; Nyczepir et al. 2009; Nyczepir 2011)
Dagger nematode	*Xiphinema americanum*	*P. domestica*	(McKenry 2004; Nyczepir et al. 2009)
Root lesion nematode	*Pratylenchus vulnus*; *Pratylenchus prunii*	*P. domestica*; *P. salicina*	(McKenry 2004; Sharma et al. 2003)
Root-knot nematode	*Meloidogyne incognita*	*P. domestica*; *P. salicina*	(McKenry 2004; Nyczepir et al. 2009)
Lesion nematode	*Pratylenchus penetrans*	*P. salicina*	(Sharma et al. 2003)
Stunt nematode	*Tylenchorhynchus mashhoodi*	*P. salicina*	(Sharma et al. 2003)
Steiner's spiral nematode	*Helicotylenchus dihystera*	*P. salicina*	(Sharma et al. 2003)
Needle nematode	*Longidorus attenuates*; *Longidorus elongatus*	*P. domestica*; *P. salicina*	(EPPO 2009)
Dagger nematode	*Xiphinema diversicaudatum*	*P. domestica*; *P. salicina*	(EPPO 2009)

8.3.1 Insects

8.3.1.1 Plum Fruit Moth

The plum fruit moth (PFM), *Grapholita funebrana* Treitschke (Lepidoptera: Tortricidae) is a widespread threat to European plums and is also referred to as the plum fruit maggot or red plum maggot. It is an oligophagous pest on wide host plants, especially European plums, cherries, and peaches (Rizzo et al. 2019). One or two generations can be seen per year in many parts of Europe (Butturini et al. 2000), even three generations per year are also noticed in Italy (Rizzo and Lo Verde 2011). *G. funebrana* presently increases in European countries, the Middle East, and northern Asia, with plant damages extending 25–100%. Destruction is due to the voracious feeding of the larval instars in the fruits, which causes early colour change and premature fruit fall. Besides, infected plums show gummy holes tunnelled by neonate larvae leaving from the immature fruit, resulting in significant yield loss and abridged marketability. Nocturnal adult moths mate at 17–22°C. Eggs are laid singly or in small groups on the surface of the developing fruits of host plants, most often in the afternoon and evening. Eggs hatch in five to ten days, near the stem. Early instars stamp the entry hole with deposits of chewed fruit skin with silk exudate. Larvae complete their development in 15–17 days and leave a large exit hole in the fruit. Pupation takes place under the trunk or inside crevices. It can be managed in orchards by the use of wide-spectrum insecticides, and virtuous fruit protection can be attained by interchanging a low frequency of spinosad and mineral oil sprays (Rizzo et al. 2012; Rizzo et al. 2019).

8.3.1.2 Plum and Prune Twig Borer

The prune twig borer, *Anarsia lineatella* Zeller (Lepidoptera: Gelechiidae), is an important pest of apricots, peaches, plums, and prunes worldwide (Damos and Savopoulou-Soultani 2008). The origin of this pest may be Europe, but it was first reported in California in the 1880s. In northern Greece, *A. lineatella* has three to four generations per year, conditional to prevalent temperatures (Damos and Savopoulou-Soultani 2007). Prune twig borer larvae mostly feed on buds, tender twigs, and ripening plum fruits. Larvae nourish inside terminal sprouts of apricots, peaches, plums, and prunes, causing wilt, stunted growth, and reduced tree vigour. Ripened fruits are tunnelled with sticky sap protuberant by larval development inside the fruit. The dark brown-headed larvae are about 12.7 mm long and have distinctive alternate light and dark bands on their bodies. Caterpillars become more active at the pink bud stage, and newly hatched larvae feed on buds and young leaves. Larvae mature and pupate within two to three weeks. It pupates in crevices of the trunk bark and limbs. Light and dark grey mottled wings are present in small adult moths size between 7.6 mm and 12.7 mm long. A single mated female adult can lay 80–90 eggs, which are yellowish-white to orange in colour and are laid on the undersides of leaves, shoots, and fruits. The well-timed application of ecologically safe insecticides during the bloom period of the crop can help prevent infestation. *Bacillus thuringiensis*, spinosad, methoxyfenozide, and diflubenzuron are recommended for integrated pest management programmes. Additionally, carbamates, organochlorines, organophosphates, and synthetic pyrethroids can also be used for effective management of borer.

8.3.1.3 Plum Sawfly

Plum sawflies, namely, *Hoplocampa minuta* C. (black plum sawfly) and *H. flava* L. (yellow plum sawfly) (Hymenoptera: Tenthredinidae), are host pests of European plum (Njezic and Ehlers 2020). They are spread over several countries in Europe, Asia Minor, and the Middle East. They can damage anywhere from 36% 96% of fruit production (Andreev and Kutinkova 2010; Rozpara et al. 2010). The black plum sawfly is a monophagous species which only attacks the plum tree. The two species of plum sawflies are univoltine insects and habitat in the soil as diapausing larvae and pre-pupae in a cocoon at 5–20 cm soil depth. Adults emerge from the ground during the flowering period. Adult flies feed on nectar and pollen. The adult body is 4–5 mm long with a black, glossy colour. It has black antennae, yellow legs, and transparent wings, with a smoky base and black stigma. Females lay the eggs by inserting into the calyx or often in flowers by a saw-like ovipositor, which causes visible swelling in the epidermis. The larva is a false caterpillar, and at full development it reaches

6–10 mm and has a pale white body with a brown head. The hatched larvae penetrate the fruit when it starts to develop, making galleries to the kernel and then consuming it. Fully developed larvae fall onto the uppermost soil and burrow to build a cocoon to hibernate. Deep autumn and spring ploughing are recommended, as a large part of the cocoons overwintering in the soil get destroyed, as well as gathering and destroying the infested fruit, before the larvae emerge. Plum sawflies can be managed by a group of pyrethroids and neonicotinoids, although three species of entomopathogenic nematodes, *Steinernema feltiae*, *S. carpocapsae*, and *Heterorhabditis bacteriophora*, were used to effectively manage the larval stage on plants (Njezic and Ehlers 2020).

8.3.1.4 San Jose Scale

San Jose scale, *Quadraspidiotus perniciosus* (Comstock) (Hemiptera: Diaspididae), is a major insect pest of apples, cherries, peaches, pears, and plums. *Q. perniciosus* is native to China and was introduced to India in the first decade of the 20th century (Gupta 2005). Scale insects will not be noticed at initial infestation. The notable signs are when large populations often exude gummy substances and show lumpy bark and dieback. The presence of scale insects during the dormant season can be identified by dead leaves adhering to fruit spurs. Forbes scale (*Q. forbesi*) can be distinguished from the San Jose scale by a raised reddish pattern in the middle of the scale. Nymphs and adults feed on branches, leaves, and fruits by withdrawing plant fluids with their stylet-like mouthparts. San Jose scale feeding reduces plant vigour, growth, and yield. New infestations usually start near the base of new growth, but over time branches can be covered completely if infestations are extensive. San Jose scale infestations can often destroy twigs or the entire tree, depending on the age and size of the tree. It can feed on young and mature fruit, leaving indentations or blemishes surrounded by a red to the purple halo. This damage can reduce the quality and marketability of the fruit. Adult winged males mate with adult females that are wingless, legless, yellow, and 12.7 mm in length. After mating, each female scale can produce about 400 first-instar nymphs (crawlers) over a six-week period, with the nymphs developing under the edge of the scale. The first-instar nymphs migrate on the leaves and bark for inserting their stylet-like mouth into host tissues to draw fluids. During acquisition, the nymphs discharge a waxy cover over the body that protects them from hazardous environments and insecticides. Nymphs are present from spring through summer, although this depends on temperature and geographic location. During the summer, all life stages can be present on a tree simultaneously. It hibernates as an immobile immature (second-instar nymph or black cap stage) on the tree. Prune out branches which are heavily infested with the San Jose scale and remove them from the vicinity. Pheromone traps will attract the winged males, which helps to monitor San Jose scale activity. Select nursery stock free from scale infestation. Fumigate nursery stocks with hydrogen cyanide or methyl bromide gas. Parasitoids *Aphytis* (proclia group), *Aspidiotophagus* sp., *Encarsia perniciosi*, *Prospaltella perniciosi* and predators *Chilocorus circumdatus* and *Coccinella infernalis* are widely adopted biological control agents against *Q. perniciosus* (Gupta 2005). Application with contact or systemic insecticides like methyl demeton, fenitrothion, and phosalone during the summer will reduce an infestation. Spraying with diesel oil emulsions during the winter period is a combatively effective approach for managing the scales.

8.3.1.5 Lecanium/European Fruit Lecanium Scale

Lecanium scales, or European fruit lecanium, *Parthenolecanium corni* Bouché (Hemiptera: Coccidae), are about 5 mm in size, rounded and reddish-brown in colour and can be found on twigs, branches, and the adaxial part of leaves. Nymphs are pinkish-brown and oval in shape. Twigs can be killed by severe infestations and at these scale insects can produce sugary substances which makes a conducive environment for sooty mould fungus colonization. It is a highly polyphagous insect that attacks around 300 plant species globally (Ben-dov 1993). The immature scale hibernates on branches and twigs, continues feeding in the spring season, and then lays eggs beneath the scales. After four to six weeks, the scale migrates to the stems and twigs to feed, mate, and overwinter. This scale insect passes one generation per year. The larvae of coccinellid, chrysopidae, and other predatory insects act as cantankerous predators of scale. Insecticides are ever practicing approaches to combat scales worldwide, like acetamiprid, azadirachtin, carbaryl, gamma-cyhalothrin, lambda-cyhalothrin, imidacloprid, pyrethrins, buprofezin, pyriproxyfen, and diazinon, all of which can provide effective control (Camacho and Chong 2015).

8.3.1.6 Leaf Curl Plum Aphid

The leaf curl plum aphid, *Brachycaudus helichrysi* Kalt. (Homoptera: Aphididae), is often referred to as the peach leaf curl aphid and is frequently observed inside curled leaves. The aphid has been found associated with almonds and plums throughout India, though the most severe damage is found in peaches (Arora et al. 2009). The shiny aphids are brownish-green or brownish-yellow and overwinters as an egg close to the base of tree buds. It transmits PPV, which aggravates the incidence and provides a diseased appearance (Maison and Massonie 1982). The aphids infest all plant parts, viz., leaves, flower buds, and newly formed fruits, which makes fruits fall off prematurely or results in poor fruit set. *B. helichrysi* infestation results in leaf curl with a large amount of sugary secretion, retarded growth, and reduced sugar content of fruits. Release of the *Aphidius colemani* parasite can largely scavenge aphids. Important predators include *Chauliognathus pensylvanicus*, *Chrysoperla carnea*, *Chaitophorus* sp., *Micromus variegatus*, and *Syrphus opinator* and can be used to manage this pest (Arora et al. 2009).

8.3.1.7 Apple-and-Thorn Skeletonizer

The apple-and-thorn skeletonizer, *Choreutis pariana* Clemens (Lepidoptera: Choreutidae), is one of the most widespread species in Europe, the Pacific Northwest, and America with lesser incidence (Heppner 2004). The reddish-brown moth is about 12.7 mm in size and has an irregular pattern of light and dark coloured bands on its wings. The larvae have yellowish or greenish bodies about 12.7 mm long with black spots and brown heads. The pupae are yellow to brown and covered with a silky white cocoon. The larvae attack apples, crabapples, cherries, plums, and hawthorn and cause skeletonization and inward rolling of the leaves. *C. pariana* hibernates as an adult in the crevices of the tree. Eggs are laid in small clusters on the adaxial surface of the leaves. The emergent young larvae feed on the interveinal leaf tissues, then roll the leaves by tying the sides together. More than one larva is present in one roll and makes the leaves skeletonized. The mature larvae take three to four weeks to become pupae, and pupation takes place in the rolled leaf itself. The adults emerge out of the pupae after about two weeks. This pest passes at least two generations per year. Control measures mostly include cultural and physical methods in the orchards, rather than spraying chemical insecticides.

8.3.1.8 American Plum Borer

The American plum borer, *Euzophera semifuneralis* Walker (Lepidoptera: Pyralidae), is a cambium-feeding grey moth with wavy brown and black colorations on its forewings; its wingspan is about 19 mm. *E. semifuneralis* has been considered the most important pest of the Michigan plum (Biddinger and Leslie 2014). The new larvae are white with big dusky brown heads. Matured larvae are almost 2.54 cm long, and they are greyish white, pinkish purple, or dull green in colour. *E. semifuneralis* presence in plums is indicated by reddish-orange frass, webbing, and gum sacks. It hibernates as a mature larva in a white silky cocoon within a plum tree. It makes horizontal boreholes, which are most injurious to the scaffold crotches or graft unions of establishing orchards. To retard the infestation, prune out the infected woods from the orchards and apply systemic insecticides on the scaffold crotches using hand-held sprayers.

8.3.1.9 Mealy Plum Aphid

The mealy plum aphid, *Hyalopterus pruni* Geoffroy (Hemiptera: Aphidinae), is a notable sap-feeder of several stone fruits, such as almonds, nectarines, peaches, and plums (Halima et al. 2013). Two forms of aphids are noticed in plums; the wingless forms of adult aphids are pale green or whitish green in colour and have dark green stripes on their backs. Black thorax and crosswise bands are present on the abdomen, covered with white mealy wax. The nymphs and adults stay inside the curled plum leaves. *H. pruni* overwinters as an egg stage near the cores of the buds, and it hatches at the flowering time of trees. It used to complete 3–13 generations per year on plum trees. Wingless forms of aphids are persisting on plum trees during the summer season, and they will not lay eggs in the fall season. The progeny of the winged adults in the plum trees can only lay the overwintering eggs (Mdellel and Ben Halima 2012). The population dynamically proliferate on the underside of the leaves, causing curl and stunt. Aphids can devitalize the tree, limit growth, crack the fruit, and diminish the sugar content of the fruit in severe infestations.

8.3.1.10 Plum Curculio

The plum curculio, *Conotrachelus nenuphar* Herbst (Coleoptera: Curculionidae), is a serious pest of stone fruits, mostly in the United States and Canada. If the trees are left uncontrolled, this pest may result in up to 80–85% damaged fruits at harvest and early fruit abortion. It is an oligophagous pest, which attacks several Rosaceous plants comprising apples, cherries, quinces, peaches, pears, and plums (Lampasona et al. 2020). Additionally, in the limited geographic region, it can harbour in alternate hosts, such as the highbush blueberry and Muscadine grape. The adult is a typical snout beetle that is 6.25 mm long and dark brown in colour. The wing covers have four bumps and black, greyish white patches. Mature larvae are 7.5 mm long, legless, and white with brown heads. The half-moon scar on fruits is the most recognizable type of wound caused by this weevil. The female first makes a small circinoid flap in the fruit skin earlier to the oviposition. Larvae take 16 days to become pupae with four instars, and the larvae remain inside the fruit. In general, the premature dropping of fruits can be observed.

8.3.2 Mites

8.3.2.1 Plum Rust Mite

The adult plum rust mites, *Aculus fockeui* (Nalepa & Trouessart) (Eriophyidae), are four-legged and wedge-shaped. It is a widespread pest of the almond, nectarine, peach, and plum (Abou-Awad et al. 2010). The adult mites are in yellow, pinkish white, or purple in colour, and they can barely be seen by the naked eye (Baldo et al. 2018). Immature mites look similar to the adults in structure but are smaller and white in colour. The mites suckle on the leaves by drawing sap from the cells. Infected mature plum leaves expose dwarfed or curl upward. Small yellow spots can be seen on younger leaves, followed by shot hole. The adult female mite hibernates under bud scales. As the buds enlarge, the female mites move out of the buds and spread on green foliage and feed for several days.

8.3.2.2 Brown Mite

The brown mite, *Bryobia rubrioculus* (Scheuten) (Acari: Tetranychidae), is a worldwide mite pest of apples, pears, peaches, plums, and other deciduous fruit trees. It can be identified by its flattened body and front legs (Jamshidian et al. 2005). Nymphs are red in colour, while adults are brownish green. *B. rubrioculus* overwinters as an egg stage on twigs and branches. The young nymphs move leaves after hatching and feed on the leaves but do not produce webbing. *B. rubrioculus* prefers to feed only during early morning and late evening and travels out from the leaves at noon. Brown mites feed by drawing the sap from the shoot cells. Leaf damage caused by *B. rubrioculus* initiates as mottling and browning of the leaves. The mite damage results in reduced tree vitality and the development of malformed fruits.

8.3.3 Nematodes

Plant nematodes are generally unsegmented parasitic roundworms in soil and plant spheres. The rhizosphere of plum trees has been parasitized by a variety of nematode species. The nematodes that infest plum trees are obligate parasites that mostly inhabit moist soil or host roots. The major parasitic nematodes associated with plums are *Meloidogyne incognita* (root-knot nematode), *Mesocriconema xenoplax*, *Criconemoides xenoplax* (ring nematode), *Pratylenchus vulnus* (root lesion nematode), and *Xiphinema americanum* (dagger nematode), which causes detrimental loss (McKenry 2004; Nyczepir et al. 2009; Nyczepir 2011). In the same orchard, two or more species may coexist. In addition to plums, they also parasitize on other host plants. *Paratylenchus* sp. (pin nematodes) are the least infectious nematodes that are often found in plum orchards. Mature trees may withstand high populations of nematodes before showing signs of damage, but young trees that are transplanted in nematode-infested areas develop poorly. In a damp environment, nematodes are spread unevenly throughout the plum orchard, resulting in low-vigour plants. It is associated with fungal and bacterial infestations in orchards, which cause blighted buds, and cankers, as well as girdling and mortality of limbs and trees (Verdejo-Lucas and Talavera 2009). Integrated nematode management necessitates the deployment of a variety of methods over a long period of time. The cost-effective and environmentally friendly technique of controlling parasitic nematodes in stone fruits is to use resistant rootstocks. Root-knot nematode–resistant clones (Mariana and Myrobalan), which express Ma genes, entirely suppress nematode multiplication and provide a broad range of resistance with long-term stability in intraspecific hybrids. Also, the Bruce hybrid and wild types of plum have shown durable field resistance towards *M. incognita* and *P. vulnus* (Ciancio and Mukerji 2009).

8.4 DEFICIENCIES

8.4.1 Copper Deficiency

European plum (*P. domestica*) production is hampered by poor soil fertility caused by continual micronutrient deficiency. Low quantities of organic matter and calcareous materials in soil limit the biological availability of copper to plums (Rafiullah et al. 2020). Copper (Cu) is important because of its oxidation-reduction activities in the plant system. It is a catalyst in chlorophyll formation and plays a significant role in enzymes. Deficiency symptoms are rarely expressed since it is immobile (Rusjan 2012). Copper deficiency affects the photosynthesis and respiration processes in many plants. It can induce interveinal chlorosis by inhibiting photosystem II on matured leaves, which looks like a magnesium or manganese deficit. Severely deficit plants exhibit leaf scorches near the margins. Also, leaves may have a wavy margin and fall prematurely (Leece 1975). The interveinal pattern of chlorosis is often more evident in magnesium deficiency, and internodes at the tip may be lengthened. In contrast, they are probable to be reduced in copper-deficient plants. Copper deficiency can be reclaimed by applying CuO or $CuSO_4$ at frequent intervals for prolonged maintenance and avoiding over-liming acid soils since it may condense copper uptake. Foliar spraying of $CuSO_4$ might be a superior substitute to topsoil drenching.

8.4.2 Iron Chlorosis

Iron (Fe) is a di- or trivalent cation absorbed through the roots or foliar way. Iron deficiency–induced chlorosis, referred to as lime-induced chlorosis or jaundice in plants, is a universal problem in plum fruit production on calcareous and alkaline soils, specifically European and Mediterranean orchards (Marschner 1995). In perennial stone fruit, iron chlorosis is a more complex phenomenon than in annual fruit crops. The occurrence of iron chlorosis is mild to severe in plum orchards where soils are high in lime content, bicarbonate ions, calcium and magnesium carbonates, pH (8–8.5), and waterlogged conditions. Sometimes, iron may accumulate in leaves in an immobilized form due to bicarbonate-induced tissue damage (Morales et al. 1998). Superfluous lime in soil obstructs with absorption and use of iron by the plant. While the plant does not absorb a usual iron supply, leaves fail to form the adequate chlorophyll pigment (Yoshikawa et al. 1982). Iron is immobile. Thus, deficiency symptoms are first noticed on younger tissues. Young leaves exhibit interveinal chlorosis or chlorosis paradox, veins remain green, and severity induces scorch. A severe deficiency stunts tree growth and decreases fruit production. Iron deficiency causes declines in fresh weight and yield of fruits in the majority of plum cultivars. Thus, necrotic patches and defoliation might develop, beginning at the tip of the shoot. Excess lime can be detected by pouring droplets of 0.2 N hydrochloric acid onto the soil. It foams out, indicating abundant lime in the soil to induce chlorosis (Morales et al. 1998). Trunk injection of ferrous sulfate 1–2% solutions into three-year-old trees corrected iron deficiency in *Prunus* sp. (Yoshikawa et al. 1982). Generally, in plum, foliar spraying of micronutrients mixture (Zinc 0.5% + Iron 0.5% + Manganese 0.5% + Copper 0.2% + Boron 0.1%) is effectual to advance yield attributes and quality parameters of plum fruits in calcareous lands (Rafiullah et al. 2020).

8.5 DISORDERS

8.5.1 Sunburn

Sunburn is a widely noticed disorder caused by high irradiation and temperatures in Japanese plums (*P. salicina*), which appears as a russet brown to chlorotic yellow colour on the fruit peel. Plum trees open to extreme temperature in the lack of irradiance can grieve with two forms of heat injury, namely, pit-burn and gel disintegration (Kapp and Jooste 2006; De Kock 2015). The extreme level of temperature hastens the pace of respiration, reducing oxygen levels in the fruit. This results in anaerobic respiration with the generation experiencing more ethanol and internal heat suffocation (Bufler and Bangerth 1982). Fruit at the top canopy grows larger, matures faster, and is prone to sunburn. Insufficient light coverage, particularly in the lower canopy, results in reduced fruit that ripens later. Delaying summer pruning can make the fruit more prone to sunlight. Timely summer pruning reduces sunburn, increases fruit size and total soluble sugar level, and enhances the colour of the fruit. These highlight the need to maintain consistent light coverage and temperature for optimal fruit quality. As a result, proper light modification tactics to alleviate sunburn-inducing conditions, such as direct light, water stress, and delayed summer pruning while maintaining optimal fruit quality are essential (Makeredza et al. 2018).

8.6 INTEGRATED PEST AND DISEASE MANAGEMENT IN PLUMS

Pest and diseases severely limit plum production worldwide, and management strategies can constitute a large amount of production expenses in some areas. Plum rootstocks are at risk of being affected by diverse symptoms, such as crown rot (*Phytophthora* spp.), root rot (*Armillaria mellea*), bacterial canker (*P. syringae* pv. *syringae*), and crown gall (*A. tumefaciens*) (Gainza et al. 2015). Many insects are present in plum orchards, of which a few important pests may cause destruction to the extent that the introduction of resistance becomes crucial. Keep at least two pheromone traps per orchard to monitor and manage the pest population. Pheromone traps can be used in combination with organic spinosad, kaolin clay, horticultural oil, *Bacillus thuringiensis*, predators, and parasitoides. The integration of cultural, physical, and biological methods of pest management and pesticides can provide effective control of multiple pests at the same time. Deep autumn and spring ploughing is recommended, as it demolishes most of the cocoons or pathogen resting structures stagnating in the soil; gathering and destroying the infested fruit before the larvae emerge or fructification of fungal pathogens occurs is also recommended. Many natural enemies of insect pests and pathogens are called predators and parasites. Recognize these beneficial insects and microbes found in unsprayed orchards and reduce pest and disease damage in orchards. Eradicate wild or untended trees near the orchard to reduce the harbouring pest and pathogen population. Prune in the dormant season to maintain an open canopy to promote air circulation and improve plant protection. Coherent adaptation of avoidance, elimination, resistance, and remediation strategies guided by traits of specific host–insect–pathogen interactions using environmental ethics to create conditions favourable for host growth and development while adverse to biotic agents' reproduction and evolution are required for long-term pest and disease management in plums (He et al. 2016).

8.7 CONCLUSION

The pervasive biotic and abiotic problems of plum discussed in this chapter have a global distribution, while others are only detected locally or regionally on rare occasions. Many biotic and abiotic stressors can affect crop production if they are not appropriately handled. Plant quarantine regulated movement of plum rootstocks and orchard certification programmes can limit the spread of hazardous infections and pests while allowing the exchange of different genetic plant materials worldwide. Long-term pest and disease control solutions based on the exact diagnosis, deployment of resistance, and broad-spectrum pesticides are also being developed. To close, with the promise of genetic engineering, plum could be imbued with resistance genes, or a cluster of genes, similar to other engineered crops resistant to biotic and abiotic stresses. Consequently, while plant security in plum orchards may become complicated, the trials will almost certainly be met with various traditional and modern defence techniques.

8.8 REFERENCES

Abou-Awad, B.A., Al-Azzazyb, M.M., and El-Sawi, S.A. 2010. The Life—History of the Peach Silver Mite, *Aculus fockeui* (Acari: Eriophyidae) in Egypt. *Archives of Phytopathology and Plant Protection* 43: 384–389.

Adams, G.C., Roux, J., and Wingfield, M.J. 2006. *Cytospora* Species (Ascomycota, Diaporthales, Valsaceae): Introduced and Native Pathogens of Trees in South Africa. *Australasian Plant Pathology* 35: 521–548.

Adaskaveg, J.E., Ogawa, J.M., and Butler, E.E. 1990. Morphology and Ontogeny of Conidia in *Wilsonomyces carpophilus*, Gen. Nov., and Comb. Nov., Causal Pathogen of Shot Hole Disease of *Prunus* Species. *Mycotaxon* 37: 275–290.

Ahmadpour, A., Ghosta, Y., Javan-Nikkhah, M., Fatahi, R., and Ghazanfari, K. 2009. Isolation and Pathogenicity Tests of Iranian Cultures of the Shot Hole Pathogen of *Prunus* Species, *Wilsonomyces carpophilus*. *Australasian Plant Disease Notes* 4: 133–134.
Ahmed, T., Sadia, H., Khalid, A., Batool, S., and Janjua, A. 2010. Report: Prunes and Liver Function: A Clinical Trial. *Pakistan Journal of Pharmaceutical Sciences* 23: 463–466.
Alburquerque, N., Faize, L., and Burgos, L. 2017. Silencing of *Agrobacterium tumefaciens* Oncogenes ipt and iaaM Induces Resistance to Crown Gall Disease in Plum but Not in Apricot. *Pest Management Science* 73: 2163–2173.
Ali, A.Y., Ahmad, A.M., Amar, J.A., Darwish, R.Y., Izzo, A.M., and Al-Ahmad, S.A. 2016. Field Parasitism Levels of *Ceratitis capitata* Larvae (Diptera: Tephritidae) by *Aganaspis daci* on Different Host Fruit Species in the Coastal Region of Tartous, Syria. *Biocontrol Science and Technology* 26: 1617–1625.
Allsopp, E. 2010. Investigation into the Apparent Failure of Chemical Control for Management of Western Flower Thrips, *Frankliniella occidentalis* (Pergande), on Plums in the Western Cape Province of South Africa. *Crop Protection* 29: 824–831.
Al Rwahnih, M., Alabi, O.J., Westrick, N.M., and Golino, D. 2018. *Prunus geminivirus A*: A Novel Grablovirus Infecting *Prunus* spp. *Plant Disease* 102: 1246–1253.
Amanelah Baharvandi, H., and Zafari, D. 2015. Identification of *Cladosporium delicatulum* as a Mycoparasite of *Taphrina pruni*. *Archives of Phytopathology and Plant Protection* 48: 688–697.
Andreev, R., and Kutinkova, H. 2010. Possibility of Reducing Chemical Treatments Aimed at Control of Plum Insect Pests: IX International Symposium on Plum and Prune Genetics, Breeding and Pomology. *ISHS Acta Horticulturae* 874: 215–220.
Arora, R.K., Gupta, R.K., and Bali, K. 2009. Population Dynamics of the Leaf Curl Aphid, *Brachycaudushelichrysi* (Kalt.) and Its Natural Enemies on Subtropical Peach, *Prunus persica* cv. Flordasun. *Journal of Entomology and Nematology* 1: 36–42.
Arroyo, F.T., Camacho, M., and Daza, A. 2012. First Report of Fruit Rot on Plum Caused by *Monilinia fructicola* at Alcalá del Río (Seville), Southwestern Spain. *Plant Disease* 96: 590.
Atanasoff, D. 1932–1933. *Plum Pox: A New Virus Disease*, 4969. Year Book Faculty Agricultural University.
Atkinson, N.J., and Urwin, P.E. 2012. The Interaction of Plant Biotic and Abiotic Stresses: From Genes to the Field. *Journal of Experimental Botany* 63: 3523–3543.
Baldi, P., and La Porta, N. 2017. *Xylella Fastidiosa*: Host Range and Advance in Molecular Identification Techniques. *Frontiers in Plant Sciences* 8: 944.
Baldo, F.B., de Carvalho Mineiro, J.L., and Raga, A. 2018. Diversity and Population Dynamics of Mites in Peach and Plum Trees (Rosaceae) in the Southwest State of São Paulo, Brazil. *International Journal of Acarology* 44: 129–137.
Bamford, D.H., and Zuckerman, M. 2020. *Encyclopedia of Virology*. Academic Press.
Barba, M., Hadidi, A., Candresse, T., and Cambra, M. 2011. Plum pox virus. *Virus and Virus-Like Diseases of Pome and Stone Fruits* 185–198.
Baxter, D.V. 1925. The Biology and Pathology of Some of the hardwood Heart-Rotting Fungi-Part II. *American Journal of Botany* 9: 553–576.
Ben Khalifa, M., and Fakhfakh, H. 2011. Note Detection of 16S rDNA of '*Candidatus* Phytoplasma mali' in Plum Decline in Tunisia. *Canadian Journal of Plant Pathology* 33: 332–336.
Ben-Dov, Y. 1993. *A Systematic Catalogue of the Mealybugs of the World (Insecta: Homoptera: Coccoidea: Pseudococcidae and Putoide) with Data on Geographical Distribution, Host Plants, Biology and Economic Importance*, 536. Intercept.
Berlanger, I., and Powelson, M.L. 2000. *Verticillium* wilt. *The Plant Health Instructor*. DOI:10.1094/PHI-I-2000-0801-01.
Bernhard, R., Marenaud, C., Sechet, J., Fos, A., and Moutous, G. 1977. A Complex Disease of Certain *Prunus*: "Molieres decline". *Comptes Rendus des Seances de l'Academie d'Agriculture de France* 3: 178–188.

Bester, R., Malan, S.S., and Maree, H.J. 2020. A Plum Marbling Conundrum: Identification of a New Viroid Associated with Marbling and Corky Flesh in Japanese Plums. *Phytopathology* 110: 1476–1482.

Biddinger, D.J., and Leslie, T.W. 2014. Observations on the Biological Control Agents of the American Plum Borer (Lepidoptera: Pyralidae) in Michigan Cherry and Plum Orchards. *The Great Lakes Entomologist* 47: 51–65.

Biggs, A.R. 1993. *Cytology, Histology and Histochemistry of Fruit Tree Diseases*. CRC Press.

Biggs, A.R. 2019. Pathological Anatomy and Histochemistry of *Leucostoma* Canker on Stone Fruits and Other Selected Fungal Cankers of Deciduous Fruit Trees. In *Handbook of Cytology, Histology, and Histochemistry of Fruit Tree Diseases*. CRC Press.

Biggs, A.R., and Grove, G.G. 2005. *Leucostoma* Canker of Stone Fruits. *The Plant Health Instructor*. DOI:10.1094/PHI-I-2005-1220-01.

Bongiorno, V., Alessio, F., Curzel, V., Nome, C., Fernández, F.D., and Conci, L.R. 2020. '*Ca*. Phytoplasma pruni'and '*Ca*. Phytoplasma meliae' Are Affecting Plum in Argentina. *Australasian Plant Pathology* 15: 1–5.

Borkar, S.G., and Yumlembam, R.A. 2016. *Bacterial Diseases of Crop Plants*. CRC Press.

Borovinova, M. 2001. Susceptibility of Plum Cultivars to Red Leaf Spot *Polystigma rubrum* (Persoon) De Candolle. *VII International Symposium on Plum and Prune Genetics, Breeding and Pomology* 577: 255–258.

Børve, J., and Vangdal, E. 2004. Fungal Pathogens Causing Fruit Decay on Plum (*Prunus domestica* L.) in Norway. *VIII International Symposium on Plum and Prune Genetics, Breeding and Pomology* 734: 367–369.

Boscia, D., Myrta, A., and Uyemoto, J.K. 2010. Plum Bark Necrosis Stem Pitting Associated Virus. In *Virus and Virus-Like Diseases of Pome and Stone Fruits*, edited by A. Hadidi, M. Barba, T. Candresse, and W. Jelkmann, 177–183. APS Press.

Bouzari, N., Khabbaz Jolfaee, H., Ahmadzadeh, S., Sadeghi Garmaroodi, H., and Hosseini, S.S. 2021. Exploitation of Plum Genetic Diversity to Identify Soil-Borne Fungi Resistance Rootstocks. *International Journal of Fruit Science* 21: 681–692.

Browne, G.T. 2017. Resistance to *Phytophthora* Species Among Rootstocks for Cultivated *Prunus* Species. *HortScience* 52: 1471–1476.

Bufler, G., and Bangerth, F. 1982. Pyruvate Decarboxylase in "Golden Delicious" Apples; Kinetics and Relation to Acetone and Ethanol Production in Different Storage Atmospheres. *Scientia Horticulturae* 16: 137–146.

Butturini, A., Tiso, R., and Molinari, F. 2000. Phenological Forecasting Model for *Cydiafunebrana*. *EPPO Bulletin* 30: 131–136.

Camacho, E.R., and Chong, J.H. 2015. General Biology and Current Management Approaches of Soft Scale Pests (Hemiptera: Coccidae). *Journal of Integrated Pest Management* 6 (1): 17.

Candresse, T., Delbos, R.P., Le Gall, O., Dunez, J., and Desvignes, J.C. 1998. Characterization of Stocky Prune Virus, a New Nepovirus Detected in French Prunes. *Acta Horticulturae* 472: 175–182.

Carraro, L., Loi, N., and Ermacora, P. 2001. Transmission Characteristics of the European Stone Fruit Yellows Phytoplasma and Its Vector *Cacopsylla pruni*. *European Journal of Plant Pathology* 107: 695–700.

Carraro, L., Loi, N., Ermacora, P., and Osler, R. 1998. High Tolerance of European Plum Varieties to Plum Leptonecrosis. *European Journal of Plant Pathology* 104: 141–145.

Chang, T., Hassan, O., and Lee, Y.S. 2018. First Report of Anthracnose of Japanese Plum (*Prunus salicina*) Caused by *Colletotrichum nymphaeae* in Korea. *Plant Disease* 102: 1461.

Cho, I.S., Chung, B.N., Cho, J.D., Choi, S.K., Choi, G.S., and Kim, J.S. 2011. Hop Stunt Viroid (HSVd) Sequence Variants from Dapple Fruits of Plum (*Prunus salicina* L.) in Korea. *Research in Plant Disease* 17: 358.

Choueiri, E., Myrta, A., Herranz, M.C., Hobeika, C., Digiaro, M., and Pallas, V. 2006. First Report of *American Plum Line Pattern Virus* in Lebanon. *Journal of Plant Pathology* 88: 225–229.

Ciancio, A., and Mukerji, K.G. 2009. *Integrated Management of Fruit Crops and Forest Nematodes*, Vol. 4. Springer Science & Business Media.

Cui, Y., Peng, A., Song, X., Chen X., and Ling, J. 2021. First Report of Japanese Plum Anthracnose Caused by *Colletotrichum fioriniae* in China. *Journal of Plant Pathology*. https://doi.org/10.1007/s42161-021-00827-z.

Damos, P., and Savopoulou-Soultani, M. 2007. Flight Patterns of *Anarsia lineatella* (Lepidoptera: Gelechiidae) in Relation to Degree-Days Heat Accumulation in Northern Greece. *Communications in Agricultural and Applied Biological Sciences* 72: 465–468.

Damos, P., and Savopoulou-Soultani, M. 2008. Temperature Dependent Bionomics and Modeling of *Anarsia lineatella* (Lepidoptera: Gelechiidae) in the Laboratory. *Journal of Economic Entomology* 101: 1557–1567.

Dang, V.T.T., Ngo, V.V., and Drenth, A. 2004. *Phytophthora* Diseases in Vietnam. In *Diversity and Management of Phytophthora in Southeast Asia*, 83–89. Australian Centre for International Agricultural Research (ACIAR).

Davis, R.E., Zhao, Y., Dally, E.L., Lee, M., Jomantiene, R., and Douglas, S.M. 2013. '*Candidatus* Phytoplasma pruni', a Novel Taxon Associated with X-Disease of Stone Fruits, *Prunus* spp.: Multilocus Characterization Based on 16S rRNA, secY, and Ribosomal Protein Genes. *International Journal of Systematic and Evolutionary Microbiology* 63: 766–776.

De Kock, A. 2015. *Quality Management of Plums Following Heat Waves During the Harvesting Season*. Fresh Notes 115, Experico.

Deketelaere, S., Tyvaert, L., França, S.C., and Höfte, M. 2017. Desirable Traits of a Good Biocontrol Agent Against *Verticillium* Wilt. *Frontiers in Microbiology* 8: 1186.

Delić, D., Mehle, N., Lolić, B., Ravnikar, M., and Đurić, G. 2010. European Stone Fruit Yellows Phytoplasma in Japanese Plum and Myrobalan Plum in Bosnia and Herzegovina. *Julius-Kühn-Archiv* 427: 415.

Diekmann, M., and Putter, C.A.J. 1996. FAO/IPGRI Technical Guidelines for the Safe Movement of Germplasm. *Stone Fruits* 16. https://hdl.handle.net/10568/104433.

Dunegan, J.C. 1932. The Bacterial Spot of Peach and Other Stone Fruits. *US Department of Agriculture Technology Bulletin* 273: 53.

D'urban Jackson, R. 2018. *A Review of Key Control Measures for Plum Rust in the UK and Overseas*. RSK ADAS Ltd. Project Report.

EFSA, PLH Panel (EFSA Panel on Plant Health). 2014. Scientific Opinion on Pest Categorisation of *Xanthomonas arboricola* pv. *pruni* (Smith, 1903). *EFSA Journal* 12: 1–25.

El Kayal, W., Chamas, Z., El-Sharkawy, I., and Subramanian, J. 2021. Comparative Anatomical Responses of Tolerant and Susceptible European Plum Varieties to Black Knot Disease. *Plant Disease*. https://doi.org/10.1094/PDIS-07-20-1626-RE.

El-Kereamy, A., El-Sharkawy, I., Ramamoorthy, R., Taheri, A., Errampalli, D., Kumar, P., and Jayasankar, S. 2011. Prunus Domestica Pathogenesis-related Protein-5 Activates the Defense Response Pathway and Enhances the Resistance to Fungal Infection. *PLoS One* 6: e17973.

El Maghraby, I., Sanchez-Navarro, J., Matic, S., Myrta, A., and Pallas, V. 2006. First Report of Two Filamentous Particles from Stone Fruit Trees in Egypt. *Journal of Plant Pathology* 88: S69.

EPPO. 2009. PM 4/35 (1) Soil Test for Virus-Vector Nematodes in the Framework of EPPO Standard PM 4 Schemes for the Production of Healthy Plants for Planting of Fruit Crops, Grapevine, *Populus* and *Salix*. *Bulletin OEPP / EPPO Bulletin* 39: 284–288.

Evans, S., Henrici, A., and Ing, B. 2006. *The Red Data List of Threatened British Fungi: Preliminary Assessment*. British Mycological Society. www.britmycolsoc.org.uk/mycology/conservation/red-data-list/ (accessed May 12, 2021).

Ferdinandsen, C. 1923. Silver Leaf: A Disease Produced by the Purple Bark Fungus (*Stereum, purpureum* Fr.). *Medd. Foren. til Svampekundsk. Fremme* 22–32.

Ferrada, E.E., Latorre, B.A., Zoffoli, J.P., and Castillo, A. 2016. Identification and Characterization of *Botrytis* Blossom Blight of Japanese Plums Caused by *Botrytis cinerea* and *B. prunorum* sp. nov. in Chile. *Phytopathology* 106: 155–165.

Fiore, N., Zamorano, A., Pino, A.M. et al. 2016. Survey of Stone Fruit Viruses and Viroids in Chile. *Journal of Plant Pathology* 98: 631–635.

Fischer-Le Saux, M., Bonneau, S., Essakhi, S., Manceau, C., and Jacques, M.A.A. 2015. Aggressive Emerging Pathovars of *Xanthomonas arboricola* Represent Widespread Epidemic Clones Distinct from Poorly Pathogenic Strains, as Revealed by Multilocus Sequence Typing. *Applied and Environmental Microbiology* 81: 4651–4668.

Fonseca, Á., and Rodrigues, M.G. 2011. *Taphrina* fries (1832). In *The Yeasts*, edited by C. Kurtzman, J. W. Fell, and T. Boekhout, 823–858. Elsevier.

Fourie, J.F., and Holz, G. 1994. Infection of Plum and Nectarine Flowers by *Botrytis cinerea*. *Plant Pathology* 43: 309–315.

Fries, E. 1832. Taphrina Fries. *Systema Mycologicum* 3: 520.

Gainza, F., Opazo, I., Guajardo, V. et al. 2015. Rootstock Breeding in *Prunus* Species: Ongoing Efforts and New Challenges. *Chilean Journal of Agricultural Research* 75: 6–16.

Gao, L., Zhang, M., Zhao, W. et al. 2014. Molecular and Morphological Analysis Reveals Five New Species of Zygophiala Associated with Flyspeck Signs on Plant Hosts from China. *PloS One* 9: e110717.

Gao, R., Yang, S.K., Yan, H.H., Wang, J., Wang, H.Y., and Lu, X.B. 2020. First Report of '*Candidatus* Phytoplasma ziziphi' Subgroup 16SrV-B Associated with *Prunus salicina* Witches'-Broom in China. *Plant Disease* 104: 564–564.

García, J.A., Glasa, M., Cambra, M., and Candresse, T. 2014. Plum Pox Virus and Sharka: A Model Potyvirus and a Major Disease. *Molecular Plant Pathology* 15: 226–241.

García-Ruiz, M., Galicia-Buendia, T., Leyva-Mir, N.G. et al. 2019. First Report of Powdery Mildew of European Plum Caused by *Podosphaera Tridactyla* in Mexico. *Plant Disease* 103: 587.

Garita-Cambronero, J., Palacio-Bielsa, A., and Cubero, J. 2018. *Xanthomonas arboricola* pv. *pruni*, Causal Agent of Bacterial Spot of Stone Fruits and Almond: Its Genomic and Phenotypic Characteristics in the *X. arboricola* Species Context. *Molecular Plant Pathology* 19: 2053–2065.

Gary, P., and Wendy, M.S. 2006. Powdery Mildew and Fruit Quality. In *Crops: Growth, Quality and Biotechnology Chapter: IV. Control of Pests, Diseases and Disorders of Crops Publisher*, edited by R. Dris. WFL Publisher.

Gavrilenko, L.A., and Kropis, E.P. 1978. Resistance of Plum to Verticillium Dahliae Wilt and to Red Spot Disease. *Selektsiia i tekhnologiia vyraschcivaniia plodovykh kul'tur* 1978: 184-190.

Giunchedi, L., Gentit, P., Nemchinov, L., Poggi-Pollini, C., and Hadidi, A. 1998. Plum Spotted Fruit: A Disease Associated with Peach Latent Mosaic Viroid. *Acta Horticulturae* 472: 571–579.

Giunchedi, L., Poggi-Pollini, C., and Credi, R. 1982. Susceptibility of Stone Fruit Trees to the Japanese Plum Tree Decline Causal Agent. *Acta Horticulturae* 130: 285–290.

Gleason, M.L., Zhang, R., Batzer, J.C., and Sun, G. 2019. Stealth Pathogens: The Sooty Blotch and Flyspeck Fungal Complex. *Annual Review of Phytopathology* 57: 135–164.

Glišić, I.S., Paunović, S.A., Milatović, D., Jevremović, D., and Milošević, N. 2017. Evaluation of Promising Plum (*Prunus domestica* L.) Genotypes for Resistance to Causal Agents of the Most Important Diseases. *II International Symposium on Fruit Culture Along Silk Road Countries* 1308: 325–332.

Gonçalves, F.P., Martins, M.C., Junior, G.J.S., Lourenço, S.A., and Amorim, L. 2010. Postharvest Control of Brown Rot and *Rhizopus* Rot in Plums and Nectarines Using Carnauba Wax. *Postharvest Biology and Technology* 58: 211–217.

Gospodaryk, A., Morocko-Bicevska, I., Pupola, N., and Kale, A. 2013. Occurrence of Stone Fruit Viruses in Plum Orchards in Latvia. *Proceedings of the Latvian Academy of Sciences. Section B* 67: 116–123.

Grantina-Ievina, L.E.L.D.E., and Stanke, L.A.S.M.A. 2015. Incidence and Severity of Leaf and Fruit Diseases of Plums in Latvia. *Communications in Agricultural and Applied Biological Sciences* 80: 421–433.

Grígel, J., Černý, K., Mrazkova, M. et al. 2019. Phytophthora Root and Collar Rots in Fruit Orchards in the Czech Republic. *Phytopathologia Mediterranea* 58: 261–275.

Grove, G.G. 1995. Powdery Mildew. In *Compendium of Stone Fruit Diseases*, edited by J.M. Ogawa et al., 12–14. American Phytopathological Society.

Grove, G.G., and Maloy, O.C. 1994. *Coryneum Blight of Stone Fruits*. Washington State University Extension.

Gupta, P.R. 2005. Biological Control of San Jose Scale in India—An Overview. *Acta Horticulturae* 696: 427–432.

Halima, M.K.B., Bouagga, S., and Lachab, N. 2013. Preferential host instars of Aphidius colemani Vierreck (Hymenotera, Braconidae) to mealy plum aphid Hyalopterus pruni Geoffroy. *American-Eurasian Journal of Sustainable Agriculture* 7: 27–31.

Hartmann, W. 1998. Breeding of Plums and Prunes Resistant to *Plum pox virus*. *Acta Virologica* 42: 230–232.

Hasija, S.K., and Agarwal, H.C. 1978. Nutritional Physiology of *Trichothecium roseum*. *Mycologia* 70: 47–60.

Hassan, O., Lee, Y.S., and Chang, T. 2019. *Colletotrichum* Species Associated with Japanese Plum (*Prunus salicina*) Anthracnose in South Korea. *Scientific Reports* 9: 1–12.

He, D.C., Zhan, J.S., and Xie, L.H. 2016. Problems, Challenges and Future of Plant Disease Management: From an Ecological Point of View. *Journal of Integrative Agriculture* 15: 705–715.

Hemmati, C., Nikooei, M., Al-Subhi, A.M., and Al-Sadi, A.M. 2021. History and Current Status of Phytoplasma Diseases in the Middle East. *Biology* 10: 226.

Heppner, J.B. 2004. Metalmark Moths (Lepidoptera: Choreutidae). In *Encyclopedia of Entomology*, edited by J.E. Capinera, vol. 2, 1386–1387. Kluwer Academic Publ.

Hiemstra, J.A. 1998. *A Compendium of Verticillium Wilts in Tree Species*. CPRO.

Horst, R.K. 2013. *Field Manual of Diseases on Fruits and Vegetables*. Springer.

Ilardi, V., and Tavazza, M. 2015. Biotechnological Strategies and Tools for Plum Pox Virus Resistance: Trans-, Intra-, Cis-Genesis, and Beyond. *Frontiers in Plant Science* 6: 379.

Jamshidian, M.K., Hatami, B., and Saboori, A. 2005. Biology of Brown Mite, *Bryobia Rubrioculus* (Acari: Tetranychidae) in Baraghan Region of karaj (Iran). *Acarologia* 4: 287–289.

Jayasankar, S., Dowling, C., and Selvaraj, D. 2016. Plums and Related Fruits. In *Encyclopedia of Food and Health*, edited by B. Caballero, P. Finglas, and F. Toldrá, 401–405. Elsevier.

Jo, Y.H., Chu, H.S., Cho, J.K., Lian, S., Choi, H.S., and Cho, W.K. 2016. First Report of *Plum Bark Necrosis Stem Pitting-Associated Virus* Infecting Plum Trees in Korea. *Plant Disease* 100: 2541.

Jones, R.K., and Benson, D.M. 2001. *Diseases of Woody Ornamentals and Trees in Nurseries*. American Phytopathological Society.

Kaponi, M.S., and Kyriakopoulou, P.E. 2013. First Report of Hop Stunt Viroid Infecting Japanese Plum, Cherry Plum, and Peach in Greece. *Plant Disease* 97: 1662–1662.

Kapp, A., and Jooste, M. 2006. Report on the Possible Reasons for the Development of Gel Breakdown and Internal Browning in Laetitia Plums in the 2005 Season. *South African Fruit Journal* 5: 15–19.

Kennelly, M.M., Cazorla, F.M., de Vicente, A., Ramos, C., and Sundin, G.W. 2007. *Pseudomonas syringae* Diseases of Fruit Trees: Progress Toward Understanding and Control. *Plant Disease* 91: 4–17.

Kim, Y., Lee, H.B., and Yu, S.H. 2005. First Report of Leaf Spot on Japanese Plum Caused by an *Alternaria* sp. in Korea. *Plant Disease* 89: 343–343.

Ko, Y., Yao, K.S., Chen, C.Y., Liu, C.W., Maruthasalam, S., and Lin, C.H. 2008. First Report of Gummosis Disease of Plum (*Prunus salicina*) Caused by a *Botryosphaeria* sp. in Taiwan. *Plant Disease* 92: 483–483.

Kurbetli, İ., Sülü, G., Aydoğdu, M., and Polat, I. 2017. First Report of Crown and Root Rot of Plum Caused by *Phytophthora megasperma* in Turkey. *Plant Disease* 101: 260.

Lampasona, T.P., Rodriguez-Saona, C., Leskey, T.C., and Nielsen, A.L. 2020. A Review of the Biology, Ecology, and Management of Plum Curculio (Coleoptera: Curculionidae). *Journal of Integrated Pest Management* 11: 22.

Landolt, P.J., Gue dot, C., and Zack, R.S. 2010. Spotted Cutworm, *Xestia c-nigrum* (L.) (Lepidoptera: Noctuidae) Responses to Sex Pheromone and Blacklight. *Journal of Applied Entomology* 135: 593–600.

Latinović, J., Batzer, J.C., Duttweiler, K.B., Gleason, M.L., and Sun, G. 2007. First Report of Five Sooty Blotch and Flyspeck Fungi on *Prunus Americana* in the United States. *Plant Disease* 91: 1685–1685.

Lee, S.C., Han, K.S., Cho, S.E., Park, J.H., and Shin, H.D. 2012. Occurrence of Powdery Mildew of Japanese Plum Caused by *Podosphaera tridactyla* in Korea. *Research in Plant Disease* 18: 49–53.

Lee, Y.S., Ha, D.H., Lee, T.Y., Park, M.J., Chung, J.B., and Jeong, B.R. 2017. Isolation and Characterization of *Colletotrichum* Isolates Causing Anthracnose of Japanese Plum Fruit. *Hanguk Hwangyong Nonghakhoe Chi* 36: 299–305.

Leece, D.R. 1975. Diagnostic Leaf Analysis for Stone Fruit. 4. Plum. *Australian Journal of Experimental Agriculture* 15: 112–117.

Levy, L., Damsteegt, V., Scorza, R., and Kolber, M. 2000. Plum Pox Potyvirus Disease of Stone Fruits. *APSnet Features*. DOI:10.1094/APSnetFeature-2000-0300.

Liu, W.T., Chu, C.L., and Zhou, T. 2002. Thymol and Acetic Acid Vapors Reduce Postharvest Brown Rot Apricots and Plums. *Hort Science* 37: 151–156.

Lopez-Franco, R.M., and Hennen, J.F. 1990. The Genus *Tranzschelia* (Uredinales) in the Americas. *Systematic Botany* 15: 560–591.

Maison, P., and Massonie, G. 1982. First Observation on the Specificity of the Resistance of Peach to Aphid Transmission of Plum Pox Virus. *Agronomie* 2: 681–683.

Makeredza, B., Jooste, M., Lötze, E., Schmeisser, M., and Steyn, W.J. 2018. Canopy Factors Influencing Sunburn and Fruit Quality of Japanese Plum (*Prunus salicina* Lindl.). *Acta Horticulturae* 1228: 121–128.

Marais, A., Faure, C., Mustafayev, E., Barone, M., Alioto, D., and Candresse, T. 2015. Characterization by Deep Sequencing of *Prunus virus T*, a Novel Tepovirus Infecting *Prunus* Species. *Phytopathology* 105: 135–140.

Marschner, H. 1995. Functions of Mineral Nutrients: Micronutrients. In *Mineral Nutrition of Higher Plants*, edited by C. Marschner, 313–324. Academic Press.

Martini, C., and Mari, M. 2014. Monilinia fructicola, Monilinia laxa (Monilinia rot, brown rot). In *Postharvest Decay: Control Strategies*, edited by S. Bautista-Baños, 233–265. Elsevier.

McFadden-Smith, W., Northover, J., and Sears, W. 2000. Dynamics of Ascospore Release by *Apiosporina morbosa* from Sour Cherry Black Knots. *Plant Disease* 84: 45–48.

McKenry, M.V. 2004. Three Nematode Genera and the Damage They Cause for Plum Producers. *Journal of Nematology* 36: 333.

Mclaren, G.F., Vanneste, J.L., and Marshall, R.R. 2005. Sulphur as an Alternative to Copper for the Control of Bacterial Blast on Nectarine Fruit. *New Zealand Plant Protection* 58: 96–100.

Mdellel, L., and Ben Halima Kamel, M. 2012. Aphids on Almond and Peach: Biology and Lifecycle in Different Area of Tunisia. *Redia* 95: 3–8.

Mehle, N., Brzin, J., Boben, J., Hren, M., Frank, J., Petrovic, N., Gruden, K., Dreo, T., Zezlina, I., Seljak, G., and Ravnikar, M. 2006. First Report of '*Candidatus* Phytoplasma mali' in *Prunus avium*, *P. armeniaca*, and *P. domestica*. *New Disease Reports* 14: 42.

Michailides, T.J., Luo, Y., Ma, Z., and Morgan, D.P. 2007. Brown Rot of Dried Plum in California: New Insights on an Old Disease. *APSnet Features*. DOI:10.1094/APSnetFeature-2007-0307.

Michailides, T.J., and Morgan, D.P. 1997. Influence of Fruit-to-Fruit Contact on the Susceptibility of French Prune to Infection by *Monilinia fructicola*. *Plant Disease* 81: 1416–1424.

Mirzwa-Mróz, E., Wińska-Krysiak, M., Marcinkowska, J., and Gleason, M.L. 2011. First Report of *Microcyclosporella mali* Causing Sooty Blotch and Flyspeck Disease on Plum in Poland. *Plant Disease* 95: 493–493.

Mittler, R. 2006. Abiotic Stress, the Field Environment and Stress Combination. *Trends in Plant Sciences* 11: 15–19.

Monia, B.H.K., Lassaad, M., Hatem, K., and Sana, Z. 2013. Natural Enemies of Hyalopterus Pruni Species Complex in Tunisia. *Tunisian Journal of Plant Protection* 8: 119–126.

Moosa, A., Ahmad, T., Khan, S.A., Gleason, M.L., Farzand, A., Safdar, H., and Ali, M.A. 2019. First Report of *Alternaria alternata* Causing Postharvest Brown Spot of Plums (*Prunus domestica*) in Pakistan. *Plant Disease* 103: 1767–1767.

Morales, F., Grasa, R., Abadía, A., and Abadía, J. 1998. Iron Chlorosis Paradox in Fruit Trees. *Journal of Plant Nutrition* 21: 815–825.

Morales, G., Llorente, I., Montesinos, E., and Moragrega, C. 2017. A Model for Predicting *Xanthomonas arboricola* pv. *pruni* Growth as a Function of Temperature. *PLoS One* 12:0177583.

Moyo, P., Damm, U., Mostert, L., and Halleen, F. 2018. *Eutypa, Eutypella*, and *Cryptovalsa* Species (Diatrypaceae) Associated with *Prunus* Species in South Africa. *Plant Disease* 102: 1402–1409.

Nasu, H., and Kunoh, H. 1987. Scanning Electron Microscopy of Flyspeck of Apple, Pear, Japanese Persimmon, Chinese Quince, and Pawpaw. *Plant Disease* 71: 361–364.

Negi, A., Rana, T., Kumar, Y., Ram, R., Hallan, V., and Zaidi, A.A. 2010. Analysis of the Coat Protein Gene of Indian Strain of Apple Stem Grooving Virus. *Journal of Plant Biochemistry and Biotechnology* 19: 91–94.

Njezic, B., and Ehlers, R. 2020. Entomopathogenic Nematodes Control Plum Sawflies (*Hoplocampa minuta* and *H. flava*). *Journal of Applied Entomology* 144: 491–499.

Northover, J., and Cerkauskas, R.F. 1994. Detection and Significance of Symptomless Latent Infections of *Monilinia fructicola* in Plums. *Canadian Journal of Plant Pathology* 16: 30–34.

Northover, J., and McFadden-Smith, W. 1995. Control and Epidemiology of *Apiosporina morbosa* of Sour Cherry and Plum. *Canadian Journal of Plant Pathology* 17: 57–68.

Novak, J.B., and Lanzova, J. 1977. Identification of Tomato Bushy Stunt Virus in Cherry and Plum Trees Showing Fruit Pitting Symptoms. *Biologia Plantarum* 19: 234–237.

Nyczepir, A.P. 2011. Host Suitability of an Endophyte-Friendly Tall Fescue Grass to *Mesocriconema xenoplax* and *Pratylenchus vulnus*. *Nematropica* 41: 45–51.

Nyczepir, A.P., Nagel, A.K., and Schnabel, G. 2009. Host Status of Three Transgenic Plum Lines to *Mesocriconema xenoplax*. *HortScience* 44: 1932–1935.

Obi, V.I., Barriuso, J.J., and Gogorcena, Y. 2018. Peach Brown Rot: Still in Search of an Ideal Management Option. *Agriculture* 8: 125.

Ogawa, J.M. 1995. *Compendium of Stone Fruit Diseases*. American Phytopathological Society.

Ogawa, J.M., and English, H. 1960. Blossom Blight and Green Fruit Rot of Almond, Apricot and Plum Caused by *Botrytis Cinerea: Plant Disease Reporter* 44: 265–268.

Ogawa, J.M., and English, H. 1991. *Diseases of Temperate Zone Tree Fruit and Nut Crops*. University of California Publications.

Oh, N.K., Hassan, O., and Chang, T. 2020. First Report on Plum Pocket Caused by *Taphrina deformans* in South Korea. *Mycobiology* 48: 522–527.

Osman, F., Al Rwahnih, M., and Rowhani, A. 2017. Real-Time RT-qPCR Detection of Cherry Rasp Leaf Virus, Cherry Green Ring Mottle Virus, Cherry Necrotic Rust Mottle Virus, Cherry Virus A and Apple Chlorotic Leaf Spot Virus in Stone Fruits. *Journal of Plant Pathology* 99: 279–285.

Ozgen, I., Sarıkaya, O., and Çiçek, H. 2012. Damage of *Scolytus rugulosus* (Müller, 1818) (Coleoptera: Curculionidae, Scolytinae) in the Apricot Fruits. *Munis Entomology & Zoology Journal* 7: 1185–1187.

Pallas, V., Aparicio, F., Herranz, M.C., Amari, K., Sanchez-Pina, M.A., Myrta, A., and Sanchez-Navarro, J.A. 2012. Ilarviruses of *Prunus* spp.: A Continued Concern for Fruit Trees. *Phytopathology* 102: 1108–1120.

Pandey, P., Irulappan, V., Bagavathiannan, M.V., and Senthil-Kumar, M. 2017. Impact of Combined Abiotic and Biotic Stresses on Plant Growth and Avenues for Crop Improvement by Exploiting Physio-Morphological Traits. *Frontiers in Plant Sciences* 8: 537.

Parthasarathy, S. 2021. *Diseases of Fruit Crops and Their Management*, vol. II. Narendra Publishing House.

Pine, T.S., and Cochran, L.C. 1962. Peach Mosaic Virus in Horticultural Plum Varieties. *Plant Disease Reporter* 46: 495–497.

Pineau, R., Raymondaud, H., and Schiavon, M. 1991. Élaboration d'un modèle de prévision des risques d'infection du mirabellier (*Prunus domestica* L. var. *insititia*) par l'agent de la tavelure (*Cladosporium carpophilum* Thümen). *Agronomie* 11: 561–570.

Pirnia, M., Zare, R., Zamanizadeh, H.R., Khodaparast, A., and Javadi Estahbanati, A.R. 2012. Contribution to the Identification of the Genus *Passalora* in Iran. *Applied Entomology and Phytopathology* 80: 61–68.

Pollacci, G. 1933. *Rassegna sull'attività del R. Laboratório Crittogamico di Pavia (Osservatorio fitopatologico per le provincie di Cremona, Parma, Pavia e Piacenza) durante l'anno 1932*, series IV, vol. 4, 283–287. Atti Ist. Bot: 'Giovanni Briosi' e Lab. Crittogam University.

Popović, T., Menković, J., Prokić, A., Zlatković, N., and Obradović, A. 2021. Isolation and Characterization of *Pseudomonas syringae* Isolates Affecting Stone Fruits and Almond in Montenegro. *Journal of Plant Disease and Protection* 128: 391–405.

Rafiullah, R., Tariq, M., Khan, F. et al. 2020. Effect of Micronutrients Foliar Supplementation on the Production and Eminence of Plum (*Prunus domestica* L.). *Quality Assurance and Safety of Crops and Foods* 12: 32–40.

Raimondo, M.L., Lops, F., and Carlucci, A. 2016. Charcoal Canker of Pear, Plum, and Quince Trees Caused by *Biscogniauxia rosacearum* sp. nov. in Southern Italy. *Plant Disease* 100: 1813–1822.

Reeder, R. 2020. *Diplodia seriata (Grapevine Trunk Disease): Invasive Species Compendium.* CABI.

Rehman, S., Ahmad, J., Sediqi, H., et al. 2017. First Report of Apple Chlorotic Leafspot Virusin Motherstock Nurseries of Stone Fruits in Afghanistan. *Plant Disease* 101: 261.

Reuveni, M. 2000. Efficacy of Trifloxystrobin (Flint), a New Strobilurin Fungicide, in Controlling Powdery Mildews on Apple, Mango and Nectarine, and Rust on Prune Trees. *Crop Protection* 19: 335–341.

Reuveni, M., Cohen, H., Zahavi, T., and Venezian, A. 2000. Polar-a Potent Polyoxin B Compound for Controlling Powdery Mildews in Apple and Nectarine Trees, and Grapevines. *Crop Protection* 19: 393–399.

Reuveni, M., Cohen, M., and Itach, N. 2006. Occurrence of Powdery Mildew (*Sphaerotheca pannosa*) in Japanese Plum in Northern Israel and Its Control. *Crop Protection* 25: 318–323.

Reuveni, M., and Reuveni, R. 1998. Foliar Applications of Monopotassium Phosphate Fertilizer Inhibit Powdery Mildew Development in Nectarine Trees. *Canadian Journal of Plant Pathology* 20: 253–258.

Rizzo, R., Farina, V., Saiano, F., Lombardo, A., Ragusa, E., and Lo Verde, G. 2019. Do *Grapholita funebrana* Infestation Rely on Specific Plum Fruit Features?. *Insects* 10: 444.

Rizzo, R., and Lo Verde, G. 2011. Prim is tudisulla biologia e sulcontrollo di *Cydia funebrana* (Treitschke) in susineti biologicisiciliani. In *Progetto per lo Sviluppodell' Agricoltura Biologica in Sicilia-Atti del Convegno*, 239–248. Regione Siciliana.

Rizzo, R., Lo Verde, G., and Lombardo, A. 2012. Effectiveness of Spinosad and Mineral Oil for Control of *Grapholita funebrana* Treitschke in Organic Plum Orchards. *New Medit* 11: 70–72.

Roberts, H.R., Pidcock, S.E., Redhead, S.C., Richards, E., O'Shaughnessy, K., Douglas, B., and Griffith, G.W. 2018. Factors Affecting the Local Distribution of *Polystigma rubrum* Stromata on *Prunus spinosa: Plant Ecology and Evolution* 151: 278–283.

Roberts, J.W. 1921. *Plum Blotch, a Disease of the Japanese Plum Caused by Phyllosticta Congesta Heald and Wolf.* USDA Publications.

Rodamilans, B., San León, D., Mühlberger, L., Candresse, T., Neumüller, M., Oliveros, J.C., and García, J.A. 2014. Transcriptomic analysis of *Prunus domestica* undergoing hypersensitive response to plum pox virus infection. *PloS One* 9: e100477.

Rodamilans, B., Valli, A., and García, J.A. 2020. Molecular Plant-Plum Pox Virus Interactions. *Molecular Plant-Microbe Interactions* 33: 6–17.

Romanazzi, G., Mancini, V., and Murolo, S. 2012. First Report of *Leucostoma cinctum* on Sweet Cherry and European Plum in Italy. *Phytopathologia Mediterranea* 51: 365–368.

Roselló, M., Santiago, R., Palacio-Bielsa, A., García-Figueres, F., Montón, C., Cambra, M.A., and López, M.M. 2012. Current Status of Bacterial Spot of Stone Fruits and Almond Caused by *Xanthomonas arboricola* pv. *pruni* in Spain. *Journal of Plant Pathology* 94: 1–15.

Rosslenbroich, H.J., and Stuebler, D. 2000. *Botrytis cinerea*-History of Chemical Control and Novel Fungicides for Its Management. *Crop Protection* 19: 557–561.

Rossman, A.Y., Adams, G.C., Cannon, P.F. et al. 2015. Recommendations of Generic Names in Diaporthales Competing for Protection or Use. *IMA Fungus* 6: 145–154.

Rozpara, E., Badowska-Czubik, T., and Kowalska, J. 2010. Problems of the Plum and Cherry Plants Protection in Ecological Orchard. *Journal of Agricultural Engineering Research* 55: 73–75.

Rusjan, D. 2012. Copper in Horticulture. In *Fungicides for Plant and Animal Diseases*, edited by D. Dhanasekaran, N. Thajuddin, and A. Panneerselvam, 257–278. IntechOpen. DOI:10.5772/26964.

Safarova, D., Navratil, M., Faure, C., Candresse, T., and Marais, A. 2012. First Report of Apricot Pseudo-Chlorotic Leaf Spot Virus Infecting Plum (*Prunus domestica*) in the Czech Republic. *Plant Disease* 96: 461.

Salem, N.M., Tahzima, R., Odeh, S., Abdeen, A.O., Massart, S., Goedefroit, T., and De Jonghe, K. 2020. First Report of '*Candidatus* Phytoplasma Solani' Infecting Plum (*Prunus domestica*) in Jordan. *Plant Disease* 104: 563–563.

Sastry, K.S., Mandal, B., Hammond, J., Scott, S.W., and Briddon, R.W. 2019. *Encyclopedia of Plant Viruses and Viroids*. Springer.

Scortichini, M., Marchesi, U., and Di Prospero, P. 2001. Genetic Diversity of *Xanthomonas arboricola* pv. *juglandis* (synonyms: *X. campestris* pv. *juglandis*; *X. juglandis* pv. *juglandis*) Strains from Different Geographical Areas Shown by Repetitive Polymerase Chain Reaction Genomic Fingerprinting. *Journal of Plant Pathology* 149: 325–332.

Scorza, R., Callahan, A., Dardick, C. et al. 2013. Genetic Engineering of *Plum Pox Virus* Resistance: 'HoneySweet' Plum from Concept to Product. *Plant Cell, Tissue and Organ Culture* 115: 1–12.

Scorza, R., Georgi, L. et al. 2010. Hairpin *Plum Pox Virus* Coat Protein (hpPPV-CP) Structure in 'HoneySweet' C5 Plum Provides PPV Resistance When Genetically Engineered into Plum (*Prunus domestica*) Seedlings. *Julius Kühn Archiv* 427: 141–146.

Scorza, R., Ravelonandro, M., Callahan, A.M. et al. 1994. Transgenic Plum (*Prunus domestica* L.) Express the *Plum Pox Virus* Coat Protein Gene. *Plant Cell Reporter* 14: 18–22.

Scott, S.W. 2011. Bromoviridae and Allies. In *Encyclopedia of Life Sciences (ELS)*. Wiley. https://doi.org/ 10.1002/9780470015902.

Shamrao, B.S. 2020. Production Technology of Peach, Plum and Apricot in India. In *Prunus*, edited by A. Kudin. IntechOpen. https://doi.org/10.5772/intechopen.92884.

Sharma, G.C., Thakur, B.S., and Kashyap, A.S. 2003. Impact of NPK on the Nematode Populations and Yield of Plum (*Prunus salicina*). *VII International Symposium on Temperate Zone Fruits in the Tropics and Subtropics*, Part Two 696: 433–436.

Sholberg, A.P., and Kappel, F. 2008. Integrated Management of Stone Fruit Diseases. In *Integrated Management of Diseases Caused by Fungi, Phytoplasma and Bacteria*, edited by A. Ciancio and K.G. Mukerji, 3–25. Springer.

Sihelská, N., Miroslav, G., and Šubr, Z.W. 2017. Host Preference of the Major Strains of Plum Pox Virus—Opinions Based on Regional and World-Wide Sequence Data. *Journal of Integrative Agriculture* 16: 510–515.

Smart, A., Watt, B., and Novak, A. 2019. *University of Maine Cooperative Extension: Insect Pests, Ticks and Plant Diseases Pest Management Fact Sheet #5091 Black Knot of Plum and Cherry*. https://extension.umaine.edu/ipm/ipddl/publications/5091e/ (accessed May 12, 2021).

Smith, H.C. 1965. The Morphology of *Verticillium albo-atrum, V. dahliae*, and *V. tricorpus*. *The New Zealand Journal of Science and Technology* 8: 450–478.

Smith, N.A., Singh, S.P., Wang, M.B., Stoutjesdijk, P.A., Green, A.G., and Waterhouse, P.M. 2000. Gene Expression: Total Silencing by Intron-Spliced Hairpin RNAs. *Nature* 407: 319–320.

Snover, K.L., and Arneson, P.A. 2002. Black Knot. *The Plant Health Instructor*. DOI:10.1094/PHI-I-2002-092-01.

Solomon, J.D. 1995. Guide to Insect Borers in North American Broadleaf Trees and Shrubs: United States Department of Agricluture. *Forest Service Agriculture Handbook AH*-706: 372.

Stefani, E. 2010. Economic Significance and Control of Bacterial Spot/Canker of Stone Fruits Caused by *Xanthomonas arboricola* pv. *pruni*. *Journal of Plant Pathology* S99–S103.

Stouffer, R.F., Lewis, F.H., and Soulen, D.M. 1969. Stem Pitting in Commercial Cherry and Plum Orchards in Pennsylvania. *Plant Disease Reporter* 53: 434–438.

Sutic, D., and Juretic, N. 1976. Occurrence of Sowbane Mosaic Virus in Plum Tree. *Mitt Bundesanst Land Forstwirtsch* 170: 43–46.

Suzuki, Y., Tanaka, K., Hatakeyama, S., and Harada, Y. 2008. *Polystigma fulvum*, a Red Leaf Blotch Pathogen on Leaves of *Prunus* spp., Has the *Polystigmina pallescens* Anamorph/Andromorph. *Mycoscience* 49: 395–398.

Tahzima, R., Foucart, Y., Peusens, G., Belien, T., Massart, S., and De Jonghe, K. 2017. First Report of Little Cherry Virus 1 Affecting European Plum (*Prunus domestica*) in Belgium. *Plant Disease* 101: 1557.

Tate, K.G., and Corbin, J.B. 1978. Quiescent Fruit Infections of Peach, Apricot, and Plum in New Zealand Caused by the Brown Rot Fungus *Sclerotinia fructicola*. *New Zealand Journal of Crop and Horticultural Science* 6: 319–325.

Teixeira, L.A.F., Grieshop, M.J., and Gut, L.J. 2010. Effect of Pheromone Dispenser Density on Timing and Duration of Approaches by Peachtree Borer. *Journal of Chemical Ecology* 36: 1148–1154.

USDA Phytosanitary Certificate Issuance and Tracking System. 2020. Phytosanitary Export Database (PExD) Harmful Organisms Database Report. *Tranzschelia Discolor*. https://pcit.aphis.usda.gov/PExD/faces/ViewPExD.jsp (accessed July 15, 2021).

Uysal-Morca, A., and Kinay-Teksür, P. 2019. Brown Rot Caused by *Monilinia fructicola* on Japanese Plums in Turkey. *Journal of Plant Pathology* 1.

Vargas, R.I., Piñero, J.C., Mau, R.F.L. et al. 2010. Area-Wide Suppression of the Mediterranean Fruit Fly, *Ceratitis capitata*, and the Oriental Fruit Fly, *Bactrocera dorsalis*, in Kamuela, Hawaii. *Journal of Insect Science* 10: 135.

Vauterin, L., Hoste, B., Kersters, K., and Swings, J. 1995. Reclassification of *Xanthomonas*. *International Journal of Systemic Bacteriology* 45: 472–489.

Végh, A., Némethy, Z., Hajagos, L., and Palkovics, L. 2012. First Report of *Erwinia amylovora* Causing Fire Blight on Plum (*Prunus domestica*) in Hungary. *Plant Disease* 96: 759–759.

Verdejo-Lucas, S., and Talavera, M. 2009. Integrated Management of Nematodes Parasitic on *Prunus* spp. In *Integrated Management of Fruit Crops and Forest Nematodes*, edited by A. Ciancio and K.G. Mukerji, 177–193. Springer.

Vidal, G.S., Hahn, M.H., Pereira, W.V., Pinho, D.B., May-De-Mio, L.L., and Duarte, H.D.S.S. 2021. A Molecular Approach Reveals *Tranzschelia discolor* as the Causal Agent of Rust on Plum and Peach in Brazil. *Plant Disease*. DOI:10.1094/PDIS-11-20-2379-PDN.

Wang, L.P., Hong, N., Wang, G.P., Micheluti, R., and Zhang, B.L. 2009. First Report of *Cherry Green Ring Mottle Virus* in Plum (*Prunus domestica*) in North America. *Plant Disease* 93: 1073.

Wang, X., Kohalmi, S.E., Svircev, A., Wang, A., Sanfaçon, H., and Tian, L. 2013. Silencing of the Host Factor *eIF(iso)4E* Gene Confers *Plum Pox Virus* Resistance in Plum. *PLoS One* 8: e50627.

Waterworth, R.A., and Millar, J.G. 2012. Reproductive Biology of *Pseudococcus maritimus* (Hemiptera: Pseudococcidae). *Journal of Economic Entomology* 105: 949–956.

Wenneker, M., Janse, J.D., De Bruine, A., Vink, P., and Pham, K. 2012. Bacterial Canker of Plum Caused by *Pseudomonas syringae* Pathovars, as a Serious Threat for Plum Production in the Netherlands. *Journal of Plant Pathology* 94: 1–11.

Williams, C.B. 1947. The Field of Research in Preventive Entomology. *Annals of Applied Biology* 34: 175–185.

Yoshikawa, F., Reil, W., and Stromberg, L. 1982. Trunk Injection Corrects Iron Deficiency in Plum Trees. *California Agriculture* 36: 13.

Zhou, J.F., Wang, G.P., Qu, L.N. et al. 2013. First Report of Cherry Necrotic Rusty Mottle Virus on Stone Fruit Trees in China. *Plant Disease* 97: 290.

Zwolińska, A., Borodynko-Filas, N., Nowaczyk, D., and Hasiów-Jaroszewska, B. 2019. First Report of *Prunus domestica* as the Host of a Phytoplasma Belonging to Group 16SrI, Subgroup B/L. *Plant Disease* 103: 145–146.

9 Emerging Packaging and Storage Technologies of Plum

Kashif Ameer[1], *Muhammad Umair Arshad*[2], *Guihun Jiang*[3], *Mian Anjum Murtaza*[1], *Muhammad Nadeem*[1], *Muhammad Asif Khan*[4], *Ghulam Mueen-ud-Din*[1], *Shahid Mahmood*[1]

[1] Institute of Food Science and Nutrition, University of Sargodha, Sargodha, 40100, Pakistan

[2] Department of Food Science, Government College University Faisalabad, Pakistan

[3] School of Public Health, Jilin Medical University, Jilin, 132013, China

[4] University of Agriculture, Faisalabad, (Sub Campus Burewala) Pakistan

CONTENTS

DOI: 10.1201/9781003205449-9

9.1 INTRODUCTION

Prunus is a genus which includes several tree species which are also termed as *Prunus* classified under the Rosaceae family, and fruits of the plants from these species are characterized as drupaceous fruits. In general, temperate zones around the world are famous for the cultivation of plums. Several types of hybrids and varieties have been developed to suit the requirements related to specific soil and regional conditions. On average, almost 100 species of plum have been reported in literature, and out of these 100 species, North America cultivates almost 30 of them (Thakur et al. 2018). Prehistoric cultivation of plum has been reported in history, and its cultivation span is longer than any other type of fruit except apples. Plums are included in the stone fruit category and belong to the *Prunus* genus (Chocano et al. 2016). Plum is produced globally, at an estimated 12.6 million tons per year, with China as the largest producer (56% of global plum production), followed by Romania and Serbia. However, on the basis of per-hectare production, Chile ranks the highest, with an estimated production of 17.5 metric tons per hectare, which is almost quadruple the world average; China has reported yield of almost 2.9 metric tons/hectare (FAO 2021; Sultana et al. 2020). At maturity, plum fruits may exhibit a glaucous appearance owing to their wax coating, also known as wax bloom. When plums are subjected to drying, they are called prunes and are characterized by their wrinkled and dark appearance. Plum fruits range from 2 to 7 cm in diameter, and shape variation ranges from oval to globose. Plum fruits are drupes, characterized by their fleshiness surrounding an individual hard seed. Owing to their highly perishable nature, plum fruits exhibit very short seasonality (Siddiq 2006; Kumar et al. 2018). Their short shelf life poses major challenges regarding marketability and the availability of fresh chili plums. Plums can be categorized into three broad groups in which the first two groups comprise principle cultivars of plums cultivated at a commercial scale. The first group is the European plum (*Prunus domestica*), and its popular varieties include the German prune, the French prune, the Reine Claude, and the Stanley. The coloration of European plums ranges from purple to black. European plums are considered the best cultivars for consuming fresh and for canning. The second group of plum cultivars consists of Japanese types (*Prunus salicina*), and famous cultivars in this group include the Elephant Heart, the Burbank, the Ozark Premier, the Shiro, and the Methley (Siddiq 2006; Kumar et al. 2018;

Sultana et al. 2020). These plum cultivars may vary in coloration from yellow to crimson. The damson plum (*Prunus insititia*) is included in third group, and these are characterized by the production of tart fruit. In this regard, famous cultivars are the French damson and the Shropshire. In the United States, the majority of the species belong to the first two groups. In 1870, *P. salicina* was first introduced to the United States. Wild plum species (*P. americana*) has also been reported to exist along thickets and streams in areas such as New York and the Rocky Mountains. This plum is small in size and sweet in taste, and during harvesting this plum has a purple appearance. Native Americans consume these plums in various forms, such as fresh, dried, or cooked, and it is a staple dietary item (Siddiq 2006; Sultana et al. 2020).

Researchers are always interested in the development of packaging technologies intended for fresh produce from the agricultural sector. After the successful application of packaging technologies, it is possible to extend the shelf life and improve the safety of fresh produce (Corbo et al. 2010). Enhanced functionality of innovative packaging technologies is one of the main objectives in response to an increased demand of fresh produce, including plum, by consumers to fulfill several purposes, such as 1) nutritional requirements and 2) the preservation and maintenance of fresh characteristics of plum during distribution and extended periods of postharvest storage (Vakkalanka et al. 2012; Zhang et al. 2019). Global changes in international trade regulations, as well as market competitiveness, fresh produce logistics, and regulatory requirements, have fostered researchers' interest in developing emerging technologies to reduce postharvest losses in conjunction with improved safety and quality properties of fresh produce, like plum (Sousa-Gallagher et al. 2016). In this regard, these decisive factors have caused the emergence of several technological innovations, like active and intelligent packaging modalities, antimicrobial coatings and antimicrobial packaging, packaging systems equipped with chlorine dioxide, 1-MCP packaging and delivery systems, and synergistic application of active packaging and modified atmospheric packaging (MAP) (Krishna et al. 2021;. Zhang et al. 2019).

9.2 QUALITY PROPERTIES AND MATURITY INDICES OF PLUM

Volatile flavor components and the shelf life of plum fruits are significantly affected by the maturity stage of fruits, as harvest maturity exerts direct influence on plum fruit flavor components, physiological deterioration, susceptibility to invasion and physical injuries, ripening ability, market life, and degree of moisture loss (Sohail et al. 2018). Usually, the factors that act as determinants of plum maturity include modifications in fruit size and shape and skin-color changes. Apart from susceptibility to physical damage, a fully developed surface cuticle is absent in immature plums, and this consequently raises the degree of susceptibility to moisture loss from fruit surfaces. Usually, the highest amounts of organic acids and the lowest total soluble solids (TSS) are found in immature plums (Öztürk and Ağlar 2019; Suhag et al. 2020).

9.3 MODIFIED ATMOSPHERE PACKAGING (MAP) AS MODERN PACKAGING MODALITY

MAP has been widely recognized as one of the modern methods for shelf-life extension of fresh produce (minimally processed), including fruits and vegetables (Aglar 2018). As indicative from the terminology of the method, fruits and vegetables are usually packaged using polymeric films to store under MAP conditions, and commodity-specific modified environmental conditions are maintained, such as elevated carbon dioxide levels and lower oxygen levels (Aglar et al. 2017).

The concentrations and types of gaseous substances could be different in environments outside and inside the food packaging, therefore equilibrium of cumulative pressure in the inner package environment and outside of the package is usually maintained (Wang et al. 2021). For achieving this equilibrium, several factors must be taken into consideration, such as the mass of the produce, respiration rate, surface area, barrier properties and thickness of polymer film, migration rate of gases through the film, package atmospheric composition and initial volume of the packaging container, and environmental constraints, such as temperature, pressure, and relative humidity (Ozturk et al. 2019). In case of emerging packaging technologies, film permeability, seal bonding, and optical and physical properties are usually maneuvered using blending, coextrusion, and metallocene polymerization (Manganaris and Crisosto 2020). MAP storage for shelf-life extension of plum fruit showing parameters influencing long-term storage is shown in Figure 9.1. There are two popular ways to create MAP conditions, as given next.

9.3.1 Active MAP

There are two reported methods to create active MAP conditions: 1) first package evaluation is performed and then the desired gas mixture is flushed at the desired rate into the package, and 2) existing air inside the package is continuously replaced by the desired gaseous mixture through a lance (Peano et al. 2017). Oxygen, carbon dioxide, and ethylene scavengers are employed in the form of adsorbers or absorbers, which helps promote the creation of the environment required for shelf-life extension and the preservation of plum quality (Khan et al. 2018).

9.3.2 Passive MAP

Appropriate films are usually selected, and their respiration rate is matched to those of plum respiration rates for the generation of MAP conditions, which in response evolve carbon dioxide and consume oxygen owing to increased respiratory activity (Coskun et al. 2017). Film selection should be careful keeping in mind the adequate gas permeability in such a way that the oxygen entrance rate should be such that it must be consumed by the plum fruit (Mahajan et al. 2014). In a similar manner, venting of carbon dioxide from the package should occur at such a rate that equilibrates the offset of carbon dioxide production as result of plum fruit respiratory activity (Mohammed et al. 2019). The establishment of MAP should be carried out under ideal conditions so that elevated carbon dioxide and lower levels of oxygen are not

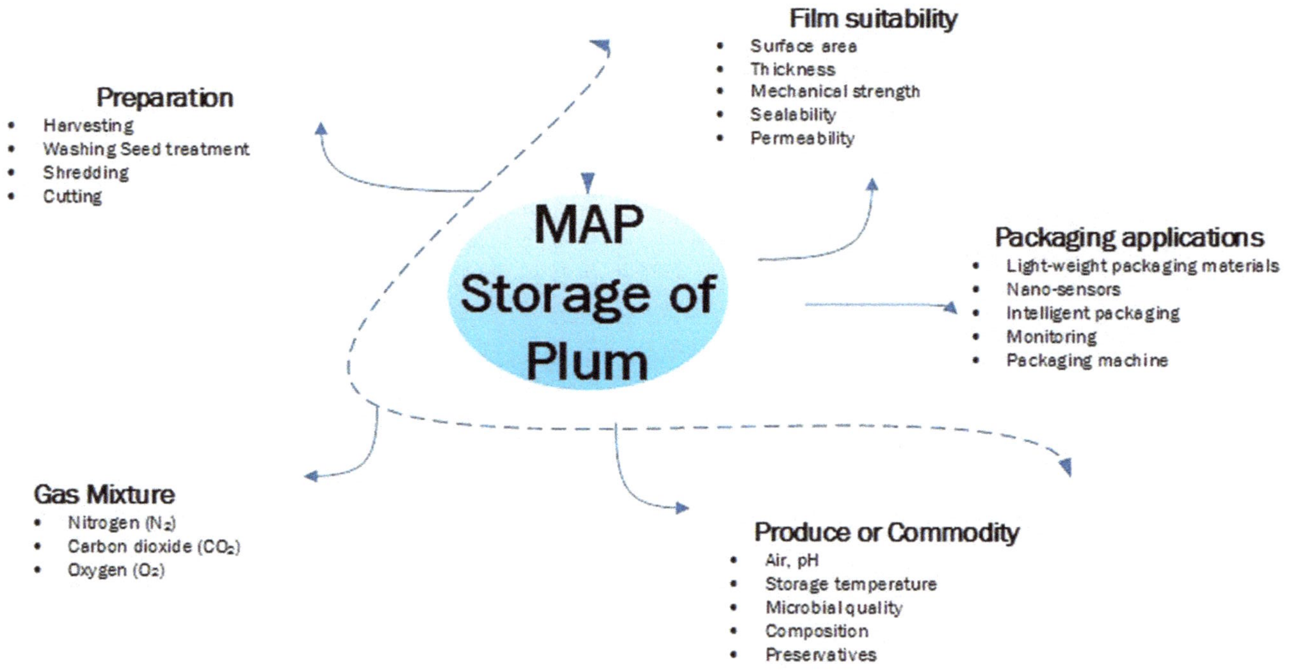

FIGURE 9.1 MAP storage for shelf-life extension of plum fruit showing parameters influencing long-term storage.

injurious to the fruit (Adepoju 2009). Hence, an intricate balance is of crucial significance between respiratory activity and gas permeability from the target atmosphere inside the package (Alia-Tejacal et al. 2012).

9.3.3 Low-Oxygen MAP

For fresh produce, low-oxygen MAP is the most commonly employed storage modality whereby the reduced oxygen concentration level is maintained at range of 1–10% compared to usual oxygen composition of about 21% in air under normal atmospheric conditions (Baños et al. 2003). Lower oxygen levels inside the package helps in the suppression of respiration rates. As far as carbon dioxide is concerned, concentration is elevated to a level above 0.03%. Respiratory activity is also suppressed due to the elevated carbon dioxide levels (Bridgemohan and Isaac 2017). This combined elevation of carbon dioxide and reduction in oxygen levels helps in achieving declining ethylene production, which in turn allows delayed ripening, prevention of texture softening, freshness maintenance, and shelf-life extension (Mohammed et al. 2019). In the case of a wide range of fresh produce, MAP gaseous concentration in the range of 1–10% is employed. For plums, carbon dioxide concentrations were reported to be in the range of 5–6.5 kPa, and oxygen concentrations ranged from 13 to 15 kPa irrespective of the cultivar type (Díaz-Mula et al. 2011; Hiwasa-Tanase and Ezura 2014).

9.3.4 High-Oxygen MAP

The application of high-oxygen MAP has also been employed in recent years to improve the storage stability of fresh produce, including plum. In high-oxygen MAP, usually higher concentrations of oxygen are employed, reaching beyond 40%, and oxygen concentration range from 70% to 95% (Isabelle et al. 2010). MAP under high-oxygen conditions is usually found to be beneficial and quite effective in the inhibition of enzyme-induced discoloration, enzymatic browning, prevention of microbial growth, and anaerobic fermentation reactions (Osuna García et al. 2011). It is also evident from the published literature that at elevated oxygen levels of 80–90%, the microbial growth was not inhibited; however, microbial growth reduced to lower levels in some tested microorganisms at low-temperature storage and led to an increased tendency in the lag phase of microbial growth (Sameh et al. 2018). Moreover, high oxygen levels lead to increased generation of reactive oxygen species, resulting in reduced cell viability and high damage to the cellular components. The synergistic application of high oxygen levels with elevated carbon dioxide levels exhibited an inhibitory effect on microbial growth in plum fruit during storage under high-oxygen MAP conditions (Sampaio et al. 2007). Storage under high oxygen levels may cause reduction or stimulation of the respiration rates depending on the oxygen concentration, maturity and ripeness levels, storage temperature and time, as well as ethylene concentration and carbon dioxide (Tiburski et al. 2011). Respiratory activity is directly related to the shelf life of fresh produce commodities like plum fruit. Until now, high-oxygen MAP could not be scaled up to the commercial scale because of several potential factors, including 1) lack of a profound understanding of the mechanisms pertaining to enzymatic browning and microbial growth, 2) extension of the influence of MAP on quality properties of plum fruit as well as nutritional quality, and 3) safety concerns regarding packaging produce under high-oxygen MAP conditions (Vargas et al. 2017; Zielinski et al. 2014). More than 25% oxygen concentration might be explosive, hence precautionary measures have to be taken. However, it should be noted that high-oxygen MAP effectiveness may differ with respect to commodity and packaging film types, as well as storage temperature (Vieira et al. 2011). Further research is needed to explore the effectiveness of high-oxygen MAP on the shelf-life extension of packaged plum fruits.

9.4 INFLUENCE OF POSTHARVEST ENVIRONMENTAL CONSTRAINTS ON QUALITY ATTRIBUTES OF PLUM

The shelf life of plums is highly influenced by variations in fruit temperature and relative humidity in the environment (Khan and Singh 2008). Field heat negatively impacts the fruit harvesting time, hence room cooling at 9–10 °C temperature and relative humidity of 80–95% or hydro cooling may lead to expedited harvesting of plum fruits (Liu et al. 2002). Refrigerated rooms are typically employed for the storage of chili plums, and good ventilation across the room can be ensured by creating air spaces between pallets and room walls (Navarro-Tarazaga et al. 2011). Vehicles used for the transit of fruit must be subjected to cooling prior to loading the fruit onto the carriage. Delays must be avoided between intervals of harvesting and loading. Therefore, cold supply chain and temperature management is of crucial importance for the optimization of quality and shelf life of plum fruits (Pérez-Marín et al. 2010).

9.4.1 Physical Damage Effect on Quality

Plums exhibit thin skin or peels on their outer surfaces and are usually consumed with the peel intact (Lin et al. 2018). Abrasion of fruits may cause physical damages owing to overpacking or to protrusion or punctures occurring during harvesting or from the harvester's fingernails or even vibrations during the loading or transport of fruits (Puerta-Gomez and Cisneros-Zevallos 2011). Fruit quality characteristics, such as appearance, space, size, and water loss, may be compromised due to physical injuries, and microbial contamination, stimulated ethylene production, and respiratory activity may accelerate deteriorative changes (Mer et al. 2015).

9.4.2 Transpiration Activity

The stage of maturity at harvest, storage duration, and storage temperature usually act as the most profound determinant factors responsible for fresh fruit weight losses (Chen and Zhu 2011). Usually, fresh fruit losses occur more in cases of immature fruits, regardless of storage temperature and storage interval, compared to mature fruits (Wu et al. 2019). This difference could be ascribed to cuticle thickness pertaining to the epidermal epicarp, which acts as the protective barrier against loss of moisture, with corresponding increases in fruit maturity stages. Graham et al. (2001) has also endorsed that fruit storage at temperatures ranging from 20°C to 21°C and 30°C to 31°C resulted in a high degree of moisture loss from fruit peels compared to fruits subjected to storage at a lower temperature range of 4–10°C. This implied that respiratory activity exhibited an increasing tendency at high storage temperatures compared to that was observed at low-temperature storage. However, regardless of the storage temperature and maturity or harvesting stage, chili plums are reported to exhibit moisture losses and declining tendency in fresh fruit weight with corresponding rises in storage time intervals (Valero et al. 2013). Moreover, rougher skin texture and a shriveled surface was exhibited by the fruits with high moisture loss. Consequently, such plum fruits were less juicy with thin layers of edible epicarp (Altaf et al. 2018).

9.4.3 Atmosphere

When sealed packages are subjected to the MAP conditions, it may incur significant benefits after packaging plum fruits with low-density polyethylene (LDPE) or high-density polyethylene (HDPE) bags (Martínez-Esplá et al. 2019). Such exposure of plum fruits caused a significant decline in moisture losses from peels owing to refrigerated storage under higher relative humidity and increased saturation of moisture within sealed plum fruit packages (Khan et al. 2013). Moreover, chili plums also exhibited declining trends in fruit weight losses and chilling injury symptoms accompanied by an enhancement in the overall appearance of plum fruits, which caused increased shelf stability of plum fruits with high marketability for longer time periods (Opara 2013). Evidently, quality characteristics of fruits exhibited an improvement in fruit flavor and juiciness with thicker epicarps (Nunes et al. 2019).

9.5 PHYSIOLOGICAL DISORDERS OF PLUM FRUIT DURING STORAGE

A high degree of sensitivity is reported for plums under refrigerated storage conditions. Among physiological disorders, chilling injuries have been reported as the most influential, which may cause alteration of plasma membrane function and associated enzymes, along with a disturbance of generally cellular metabolic activities (Argenta et al. 2003). Chilling injury may also result in skin pitting of fruits at harvest stages when fruits exist in their immature green forms at 4–9°C (Crisosto and Kader 2000).

9.5.1 Pathological Damages to Plum Fruits

Graham et al. (2001) has already reported that the storage of plum fruits at 20–31°C caused a decline in the shelf life of chili plums after eight days owing to fruit decay. The principal factor responsible for fruit decay was attributable to fungal-induced stem rot, and when fruits were stored at 15°C, it caused excessive softening of skin peels, with decay up to 65% (Aslam et al. 2019; Valero and Serrano 2010). Fruit storage at 12.5°C helped to achieved color maintenance of fruits without any discoloration and evidence of decay, with total TSS of 11% and acceptable fruit firmness (Li et al. 2021). After 14 days in storage, fruits were subjected to moderate temperatures (30–31°C) for 4 days, and results showed that fruit uniformity was maintained with acceptable organoleptic scores (Graham et al. 2001).

9.6 POSTHARVEST HANDLING PRACTICES

9.6.1 Unit Operation during Harvesting

Ripening of plum fruits is stimulated by physiological maturation, hence fruit harvesting is essential at the proper maturity stage to avoid postharvest losses during marketability and storage (Brar et al. 2020). When harvesting is carried out at an unripe stage, it may lead to deteriorative qualitative changes with respect to fruit firmness and color, which may possibly hamper the acceptability and marketability of plum fruits (Fanning et al. 2014). However, fruit harvesting at the half-ripened stage usually leads to toleration of longer storage periods. Plum fruits usually occur in the form of clusters on plants, whereas maturity of the fruits is subjected to variability depending on the degree of maturity and competition of neighboring plants for nutrients (Chang et al. 2019). Manual fruit harvesting is usually carried out by selecting fruits that have reached maturity. Fruit abscission occurs as result of the weakening of the abscission layer, and such fruit is easily dislodged by birds and strong winds (Ghaani et al. 2016). Damages caused by compression, microscopic punctures, and abrasion may also occur during fruit harvesting and hence may serve as secondary infection sites. Packing of harvested fruits should be carried out in light-colored containers with shallow and ventilated container features for achieving optimized quality and reduction of postharvest losses incurred by physical fruit damages. Fruits should be placed in a shady area during delays between the transportation of fresh produce to the packing facility for minimizing heat stress (Hashemi 2018; Shahzad et al. 2015).

9.6.2 Practices to Follow in Packaging

Plum fruits are climacteric in nature, hence pre-cooling of plum fruits is necessary for the removal of filed heat, both by cooling at room temperature and hydro cooling (Bal 2018). Moreover, it is also mandatory to wash the fruits with chlorine water (100–150 ppm) and then rinse and air dry them. Fruit placement is performed on packing lines of conveyor belts, which are well equipped with pads for minimizing bruising (Teruel et al. 2004). Uniformity in fruit maturity and size is achieved by sorting and grading prior to packaging to accommodate market requirements. During packing-house operations, sanitation and a hygienic environment are maintained throughout the facility (Serdyuk et al. 2016). Plums are eaten raw with associated unpeeled skin; therefore, sufficient decontamination is necessary for the elimination of any possible occurrence of food-borne pathogens (Duan et al. 2020).

9.6.3 Ripening Control and Senescence

Potentially, ripening, harvesting time, and possible shelf stability of plum fruit can be predicted, and good predictive factors in this regard include reduction of flesh firmness and transformation of skin color from green to light yellow (Farcuh et al. 2019). Both flesh firmness and skin coloration are indicative of a fruit's maturity. Mature green fruits at the proper harvesting stage do not need external application of ethylene gas. In general, fruits intended for harvest at the breaker stage need ethylene gas for ripening in a uniform manner; however, this ethylene application will not expedite the ripening process (Martínez-Esplá et al. 2018). Fruit shriveling is prevented during ripening through the application of adequate air circulation at a relative humidity range of 90–95%. This also ensures attainment of uniform fruit temperature. Fruits obtained at physiological maturity and stored at 28–30°C will experience ripening in three to four days (Mohammed et al. 2019). When fruits are stored at 9–10°C, they will need about two weeks to achieve full ripening. On the other hand, fruits stored at a chilling temperature (4–5°C), will result in shelf-life extension up to 15 days; however, this storage temperature may cause chilling injury and fruit decay (Mohammed et al. 2019; Vargas-Simón 2018). 1-MCP may exert its significant influence on shelf-life extension of plums, as was reported specifically in the case of Mexican plums by Vargas-Simón (2018). Plums were subjected to exposure of 1-MCP at applied levels ranging from 100 nL/L to 300 nL/L at an ambient temperature for 12 hours on fruits stored in experimental chambers (0.512 m^3) as compared to non-treated control samples (Mohammed et al. 2019). Post-MCP application, the fruits were subjected to storage under marketing condition of 22°C temperature and 70% relative humidity for a total storage period of nine days. Applied 1-MCP levels in the range of 100–300 nL/L resulted in shelf-life extension with maintained quality properties of yellow Mexican plum fruits up to three days compared to the control fruits (Graham et al. 2001). Moreover, both ripened and ¾-ripe fruits approximately exhibited shelf-life extension of nine and seven days, respectively. Exposure to 1-MCP resulted in a decline in respiratory activity as well as plum fruit weight loss in ripe fruits, whereas ¾-ripe fruits did not exhibit any decline in these parameters (Mohammed et al. 2019). It was also evident from the results that 1-MCP application caused a delay in external color development accompanied by maintained fruit firmness without negatively impacting TSS contents (Zhang et al. 2020).

9.6.4 Plum Fruit Shipping and Storage Conditions

Harvesting of chili plum fruits is carried out when fully mature fruits exhibit blemish-free, light yellow skin coloration, and then plum fruits are treated with hot water for 20 minutes at a temperature of 20°C for export purposes (Mohammed et al. 2019). Before fruits are shipped, they are stored in shallow, ventilated one-ply cardboard cartons at 12.5°C and relative humidity of 90–95% for up to seven or eight days (Manganaris and Crisosto 2020). Storage at 20–22°C temperature would allow fruit firmness maintenance for an extra four to six days (Fawole et al. 2020). When fruits are placed on display at retail centers, they should be packaged in such a way that LDPE bags should be sealed and comprise smaller portion sizes to achieve 90–95% marketability of fruits (Martínez-Esplá et al. 2019).

9.6.5 Fresh-Cut Processing of Plum Fruits

Fresh-cut processing has only been documented on level of cottage in which pit removal is carried out through use of knife and eaten by the intended consumers to fulfill calorie requirements (Khan et al. 2013). LDPE bags were also reported to carry out the packaging of sealed packages, with each package having 10–15 fruits.

Such fresh-cut products are popular ready-to-eat food items utilized by intended consumers at sporting events, cafeterias, bazaars, and other types of large public gatherings (Opara 2013). Moreover, some other variations of fresh-cut products are made by the vendors, including slicing plum fruits in a longitudinal manner on either side of the fruit and usually serves as ready-to-eat (RTE) item along with other fresh-cut fruits kept in wrapping or stretch-wrapped styrofoam (Krishna et al. 2021). Fresh-cut chili plums are also used as stuffing for fish filling in curried or baked fish (Shahzad et al. 2015).

9.6.6 Miscellaneous Processing Practices

In value-added products, chili plums processing and utilization has been reported for a diverse range of applications (Sammy 1994; Mohammed et al. 2019). At the ripened and mature green stages, chili plums are selected to transform through value-added in high-sugar value-added products, including squashes, cordials, candies, preserves, jellies, jams, and fruit cheeses (Fanning et al. 2014). Sammy (1994) has reported other value-added products, such as sauces, canned in brine, and salted pickles (Valero et al. 2002). Yogurt and wine can also be prepared from ripe fruits, as well as dehydrated or dried products, sliced or whole, and such as value-added products are manufactured throughout the Caribbean region (Wei et al. 2020). In the case of high-sugar products, almost 60–70% sugar concentration is utilized owing to its osmotic effects that help prevent microbial spoilage. For manufacturing high-salt products, granulated pure un-iodized salt (1–3%) was utilized for the immersion of fruits for flavor enhancement in conjunction with vinegar comprising 4–6% acetic acid for preservation and flavor-improvement purposes (Khan et al. 2018). Moreover, salt having a concentration of 2.5–8% was utilized to promote the growth of lactic acid bacteria as a selective growth-promoting agent for preparing fermented chili plum pickles (Peano et al. 2009). In the case of pickles prepared by the fermentation of chili plums, mature green fruits were sanitized by being dipped in a chlorine solution (50 ppm) and then fermented for two to three weeks in a brine solution comprising salt and vinegar concentrations of 25–50 g and 50 mL/L of water, respectively (Sandhya 2010). The fresh brine solution was utilized for acidification and flavoring by submerging fermented plums. Finally, pasteurization was carried out at a temperature of 85°C for 15–20 minutes or was utilized in a synergistic manner with that of a chemical preservation method involving potassium sorbate (0.1% w/w), sodium benzoate (0.1% w/w), and potassium metabisulfite (0.03% w/w) (Mahajan et al. 2014). For the production of dehydrated and dried products, ripe and mature fruits were subjected to blanching for time intervals of 5–15 seconds in boiling water with sodium hydroxide (10–20 g/L) added for skin roughening and the acceleration of drying kinetics (Panahirad et al. 2020). To prevent browning and discoloration, chili plum fruits were subjected to a sodium metabisulphite (0.5%) solution (5 g/L). After that, plum fruits were soaked in a sugar solution made up of two parts water and one part sugar (w/w) for 12–18 hours for moisture removal from the plum fruits through osmosis and were subjected to drying until they reached a final moisture content in the range of 12–14% (Alam et al. 2021).

9.7 AGRO-INDUSTRIAL WASTE UTILIZATION

In published literature, respiratory and climacteric pattern was already confirmed by Vargas-Simon (2018), who stressed the medicinal significance and nutraceutical values of chili plum fruits (*Spondias purpurea*). Plum fruits are particularly famous for their antispasmodic and diuretic properties and are also commonly employed as an antihistamine. Tree bark extract of plum is commonly used as a natural remedy for treating an upset stomach as well as dysentery (Kasole et al. 2019). This bark extract also exhibits potent larvicidal activity, and wastewater can be treated because the seeds, seminal coat, and endocarp have significant amounts of cellulose, which is beneficial for removing flocs of phosphate ions (Panahirad et al. 2020).

9.8 OXALIC ACID PREHARVEST TREATMENT

Oxalic acid is one of the functional compounds that plays role in the physiological functions of cellular matrices of plant tissues. Moreover, oxalic acid has been reported to play a profound role in inducing systemic resistance against food-borne pathogens and harmful microbes, such as viral, bacterial, and fungal diseases, by causing a rise in defense-related secondary metabolites and enzymes (Martínez-Esplá et al. 2019). Moreover, oxalic acid has been reported to cause a decline in postharvest chilling injuries in pears, peaches, and plums, owing to a delay in the ripening process and climacteric activities as a result of the inhibition of the biosynthetic production of ethylene (Wu et al. 2011). Furthermore, oxalic acid has been found to be effective in delaying ripening and extending the shelf life of plum fruits. Application of oxalic acid decreased the production of ethylene and enhanced the inherent concentrations of antioxidant enzymes, such as SOD, APX, CAT, and POD, compared to untreated samples. These antioxidant enzymes play a vital role in the generation of reactive oxygen species and scavenging free radicals produced during stages of fruit ripening and senescence (Serrano et al. 2016). Oxalic acid pretreatment enhanced the plum tissues' ability to remove the ROS during the maturing process and hence led to a delay in the onset of postharvest senescence and the ripening processes (Martinez-Espla et al. 2014). Moreover, oxalic acid treatment caused improvement in fruit-quality parameters and led to significant rises in bioactive compound contents having health-promoting effects at harvest with the extended prolonged storage of treated plum fruits compared to untreated samples (Wu et al. 2011).

9.9 CONCLUSIONS

Plum fruits are perishable commodities, and owing to their highly perishable nature these fruits exhibit very short seasonality. The short shelf life of plum fruits poses major challenges regarding the marketability and availability of fresh chili plums. Plum fruits have a low flesh-to-seed ratio, which necessitates the selection of fruits at the mature green stage. Plums are typically eaten fresh but are also used for other value-added plum-based food products. Cold storage, or atmospheric control-based temperature management, allows for a reduction of chilling injuries and damage to plum fruits. This also ensures acceptable fruit attributes to the intended consumers with proper ripening. Fruit fly infestation problems can be tackled through the use of hot water or moist air treatments. Further research is needed to elucidate the significance of enzymes on fruit softening to address issues pertaining to changes in fruit texture during extended storage. Moreover, the use of suitable packaging materials also reduces physical damage to ensure the best quality and shelf life of plum fruits.

9.10 REFERENCES

Adepoju, O.T. 2009. Proximate Composition and Micronutrient Potentials of Three Locally Available Wild Fruits in Nigeria. *African Journal of Agricultural Research* 4 (9): 887–892.

Aglar, E. 2018. Effects of Harpin and Modified Atmosphere Packaging (MAP) on Quality Traits and Bioactive Compounds of Sweet Cherry Fruits Throughout Cold Storage and Shelf Life. *Acta Scientiarum Polonorum-Hortorum Cultus* 17 (4): 61–71.

Aglar, E., Ozturk, B., Guler, S.K., Karakaya, O., Uzun, S., and Saracoglu, O. 2017. Effect of Modified Atmosphere Packaging and 'Parka' Treatments on Fruit Quality Characteristics of Sweet Cherry Fruits (*Prunus avium* L.'0900 Ziraat') During Cold Storage and Shelf Life. *Scientia Horticulturae* 222: 162–168.

Alam, A.U., Rathi, P., Beshai, H., Sarabha, G.K., and Deen, M.J. 2021. Fruit Quality Monitoring with Smart Packaging. *Sensors* 21 (4): 1509.

Alia-Tejacal, I., Astudillo-Maldonado, Y.I., Núñez-Colín, C.A., Valdez-Aguilar, L.A., Bautista-Baños, S., García-Vázquez, E., Rivera-Cabrera, F. et al. 2012. Caracterización de frutos de ciruela mexicana (*Spondias purpurea* L.) del sur de México. *Revista Fitotecnia Mexicana* 35 (SPE5): 21–26.

Altaf, U., Kanojia, V., and Rouf, A. 2018. Novel Packaging Technology for Food Industry. *Journal of Pharmacognosy and Phytochemistry* 7 (1): 1618–1625.

Argenta, L.C., Krammes, J.G., Megguer, C.A., Amarante, C.V.T., and Mattheis, J. 2003. Ripening and Quality of 'Laetitia' Plums Following Harvest and Cold Storage as Affected by Inhibition of Ethylene Action. *Pesquisa Agropecuária Brasileira* 38 (10): 1139–1148.

Aslam, A., Zahoor, T., Khan, M.R., Khaliq, A., Nadeem, M., Sagheer, A., Sajid, M.W. et al. 2019. Studying the Influence of Packaging Materials and Storage on the Physiochemical and Microbial Characteristics of Black Plum (*Syzygium cumini*) Jam. *Journal of Food Processing and Preservation* 43 (5): e13941.

Bal, E. 2018. Postharvest Application of Chitosan and Low Temperature Storage Affect Respiration Rate and Quality of Plum Fruits. *Journal of Agriculture Science & Technology* 15: 1219–1230.

Baños, S.B., Pérez, J.C.D., Necha, L.L.B., and Luna, L.B. 2003. Postharvest Study of Red-Mombin (*Spondias purpurea*) Fruit During Storage. *Revista Iberoamericana de Tecnología Postcosecha* 5 (2): 82–85.

Brar, H.S., Kaur, P., Subramanian, J., Nair, G.R., and Singh, A. 2020. Effect of Chemical Pretreatment on Drying Kinetics and Physio-Chemical Characteristics of Yellow European Plums. *International Journal of Fruit Science* 20 (2): S252–S279.

Bridgemohan, P., and Isaac, W.A.P. 2017. Postharvest Handling of Indigenous and Underutilised Fruits in Trinidad and Tobago. *Postharvest Handling* 165.

Chang, X., Lu, Y., Li, Q., Lin, Z., Qiu, J., Peng, C., Guo, X. et al. 2019. The Combination of Hot Air and Chitosan Treatments on Phytochemical Changes During Postharvest Storage of 'Sanhua' Plum Fruits. *Foods* 8 (8): 338.

Chen, Z., and Zhu, C. 2011. Combined Effects of Aqueous Chlorine Dioxide and Ultrasonic Treatments on Postharvest Storage Quality of Plum Fruit (*Prunus salicina* L.). *Postharvest Biology and Technology* 61 (2–3): 117–123.

Chocano, C., García, C., González, D., de Aguilar, J.M., and Hernández, T. 2016. Organic Plum Cultivation in the Mediterranean Region: The Medium-Term Effect of Five Different Organic Soil Management Practices on Crop Production and Microbiological Soil Quality. *Agriculture, Ecosystems & Environment* 221: 60–70.

Corbo, M.R., Speranza, B., Campaniello, D., D'amato, D., and Sinigaglia, M. 2010. Fresh-Cut Fruits Preservation: Current Status and Emerging Technologies. *Current Research, Technology and Education Topics in Applied Microbiology and Microbial Biotechnology* 2: 1143–1154.

Coskun, M.G., Omeroglu, P.Y., and Copur, O.U. 2017. Increasing Shelf Life of Fruits and Vegetables with Combined System of Modified Atmosphere Packaging and Edible Films Coating. *Eurasian Journal of Food Science and Technology* 1 (2): 47–53.

Crisosto, C.H., and Kader, A.A. 2000. *Plum and Fresh Prune Postharvest Quality Maintenance Guidelines*, 1–8. Department of Plant Sciences, University of California.

Díaz-Mula, H.M., Martínez-Romero, D., Castillo, S., Serrano, M., and Valero, D. 2011. Modified Atmosphere Packaging of Yellow and Purple Plum Cultivars. 1. Effect on Organoleptic Quality. *Postharvest Biology and Technology* 61 (2–3): 103–109.

Duan, Y., Wang, G.B., Fawole, O.A., Verboven, P., Zhang, X.R., Wu, D., Chen, K. et al. 2020. Postharvest Precooling of Fruit and Vegetables: A Review. *Trends in Food Science and Technology* 100: 278–291.

Fanning, K.J., Topp, B., Russell, D., Stanley, R., and Netzel, M. 2014. Japanese Plums (*Prunus salicina* Lindl.) and Phytochemicals—Breeding, Horticultural Practice, Postharvest Storage, Processing and Bioactivity. *Journal of the Science of Food and Agriculture* 94 (11): 2137–2147.

FAO. 2021. *Plum (and Sloe) Production In 2019; Crops/Regions/World/Production Quantity by picklists.* UN Food and Agriculture Organization, Statistics Division (accessed April 2, 2021).

Farcuh, M., Toubiana, D., Sade, N., Rivero, R.M., Doron-Faigenboim, A., Nambara, E., Blumwald, E. et al. 2019. Hormone Balance in a Climacteric Plum Fruit and Its Non-Climacteric Bud Mutant During Ripening. *Plant Science* 280: 51–65.

Fawole, O.A., Riva, S.C., and Opara, U.L. 2020. Efficacy of Edible Coatings in Alleviating Shrivel and Maintaining Quality of Japanese Plum (*Prunus salicina* Lindl.) During Export and Shelf Life Conditions. *Agronomy* 10 (7): 1023.

Ghaani, M., Cozzolino, C.A., Castelli, G., and Farris, S. 2016. An Overview of the Intelligent Packaging Technologies in the Food Sector. *Trends in Food Science and Technology* 51: 1–11.

Graham, O., Wickham, L.D., and Mohammed, M. 2001. Respiration and Ethylene Production Rates of Chili Plums (*Spondias purpurea* L.) During storage. *Proceedings of the 37th Annual Meeting of Caribbean Food Crops Society* 37: 243–251.

Hashemi, S.M.B. 2018. Effect of Pulsed Ultrasound Treatment Compared to Continuous Mode on Microbiological and Quality of Mirabelle Plum During Postharvest Storage. *International Journal of Food Science and Technology* 53 (3): 564–570.

Hiwasa-Tanase, K., and Ezura, H. 2014. Climacteric and Non-Climacteric Ripening. *Fruit Ripening, Physiology, Signalling and Genomics* 1–14.

Isabelle, M., Lee, B.L., Lim, M.T., Koh, W.P., Huang, D., and Ong, C.N. 2010. Antioxidant Activity and Profiles of Common Fruits in Singapore. *Food Chemistry* 123 (1): 77–84.

Kasole, R., Martin, H.D., and Kimiywe, J. 2019. Traditional Medicine and Its Role in the Management of Diabetes Mellitus: "Patients' and Herbalists' Perspectives". *Evidence-Based Complementary and Alternative Medicine* 1–12, Article ID 2835691. doi:10.1155/2019/2835691.

Khan, A.S., and Singh, Z. 2008. 1-Methylcyclopropene Application and Modified Atmosphere Packaging Affect Ethylene Biosynthesis, Fruit Softening, and Quality of 'Tegan Blue' Japanese Plum During Cold Storage. *Journal of the American Society for Horticultural Science* 133 (2): 290–299.

Khan, A.S., Singh, Z., and Ali, S. 2018. Postharvest Biology and Technology of Plum. In *Postharvest Biology and Technology of Temperate Fruits*, 101–145. Springer.

Khan, M.S., Zeb, A., Rahatullah, K., Ihsanullah, N.A., and Ahmed, S. 2013. Storage Life Extension of Plum Fruit with Different Colored Packaging and Storage Temperatures. *Journal of Environmental Science, Toxicology and Food Technology* 7 (3): 86–93.

Krishna, K.R., Smruthi, J., and Manivannan, S. 2021. Packaging and Storage of Stone Fruits. In *Production Technology of Stone Fruits*, 273–305. Springer.

Kumar, M., Sharma, D.D., Singh, N., and Shylla, B. 2018. Evaluation of Newly Introduced Plum (*Prunus salicina* Lindl.) Cultivars Under Mid-Hills of Himachal Pradesh. *International Journal of Chemical Studies* 6: 2925–2930.

Li, B., Li, S., Yuan, H., Ma, B., Jing, Y., Zhang, Z., Zeng, D. et al. 2021. Effects of Different Packaging Materials on Storage Quality of Crisp Plum. *E3S Web of Conferences* 251: 02041.

Lin, Y., Lin, H., Lin, M., Li, H., Yuan, F., Xiao, J. et al. 2018. Effects of Paper Containing 1-MCP Postharvest Treatment on the Disassembly of Cell Wall Polysaccharides and Softening in Younai Plum Fruit During Storage. *Food Chemistry* 264: 1–8.

Liu, W.T., Chu, C.L., and Zhou, T. 2002. Thymol and Acetic Acid Vapors Reduce Postharvest Brown Rot of Apricots and Plums. *HortScience* 37 (1): 151–156.

Mahajan, P.V., Caleb, O.J., Singh, Z., Watkins, C.B., and Geyer, M. 2014. Postharvest Treatments of Fresh Produce. *Philosophical Transactions of the Royal Society A: Mathematical, Physical and Engineering Sciences* 372 (2017): 20130309.

Manganaris, G.A., and Crisosto, C.H. 2020. Stone Fruits: Peaches, Nectarines, Plums, Apricots. In *Controlled and Modified Atmospheres for Fresh and Fresh-Cut Produce*, 311–322. Academic Press.

Martínez-Esplá, A., Serrano, M., Martínez-Romero, D., Valero, D., and Zapata, P.J. 2019. Oxalic Acid Preharvest Treatment Increases Antioxidant Systems and Improves Plum Quality at Harvest and During Postharvest Storage. *Journal of the Science of Food and Agriculture* 99 (1): 235–243.

Martinez-Espla, A., Zapata, P.J., Valero, D., García-Viguera, C., Castillo, S., and Serrano, M. 2014. Preharvest Application of Oxalic Acid Increased Fruit Size, Bioactive Compounds, and Antioxidant Capacity in Sweet Cherry Cultivars (*Prunus avium* L.). *Journal of Agricultural and Food Chemistry* 62 (15): 3432–3437.

Martínez-Esplá, A., Zapata, P.J., Valero, D., Martínez-Romero, D., Díaz-Mula, H.M., and Serrano, M. 2018. Preharvest Treatments with Salicylates Enhance Nutrient and Antioxidant Compounds in Plum at Harvest and After Storage. *Journal of the Science of Food and Agriculture* 98 (7): 2742–2750.

Mer, M.S., Attri, B.L., Narayan, R., and Kumar, A. 2015. Postharvest Studies Beyond Fresh Market Eating Quality: Phytochemical Changes in Peach Fruit During Ripening and Advanced Senescence. *Annals of Horticulture* 8 (1): 87–91.

Mohammed, M., Bridgemohan, P., Graham, O., Wickham, L., Bridgemohan, R.S., and Mohammed, Z. 2019. Postharvest Physiology, Biochemistry and Quality Management of Chili Plum (*Spondias purpurea var. Lutea*): A Review. *Journal of Food Research* 8 (3): 1–15.

Navarro-Tarazaga, M.L., Massa, A., and Pérez-Gago, M.B. 2011. Effect of Beeswax Content on Hydroxypropyl Methylcellulose-Based Edible Film Properties and Postharvest Quality of Coated Plums (Cv. Angeleno). *LWT-Food Science and Technology* 44 (10): 2328–2334.

Nunes, F.R., Steffens, C.A., Heinzen, A.S., Soethe, C., Moreira, M.A., and Amarante, C.V.T.D. 2019. Ethanol Vapor Treatment of 'Laetitia' Plums Stored Under Modified Atmosphere. *Revista Brasileira de Fruticultura* 41 (5). doi:10.1590/0100-29452019163.

Opara, U.L. 2013. A Review on the Role of Packaging in Securing Food System: Adding Value to Food Products and Reducing Losses and Waste. *African Journal of Agricultural Research* 8 (22): 2621–2630.

Osuna García, J.A., Barraza, P., Hilda, M., Vázquez Valdivia, V., and Gómez Jaimez, R. 2011. Application of 1-Methylcyclopropene (1-MCP) on Mexican Plum (*Spondias purpurea* L.). *Revista Fitotecnia Mexicana* 34 (3): 197–204.

Ozturk, A., Yildiz, K., Ozturk, B., Karakaya, O., Gun, S., Uzun, S., and Gundogdu, M. 2019. Maintaining Postharvest Quality of Medlar (*Mespilus germanica*) Fruit Using Modified Atmosphere Packaging and Methyl Jasmonate. *LWT* 111: 117–124.

Öztürk, B., and Ağlar, E. 2019. Effects of Modified Atmosphere Packaging (MAP) and Aloe Vera Treatments on Quality Characteristics of Cornelian Cherry Fruits During Cold Storage. *Akademik Ziraat Dergisi* 8 (1): 1–8.

Panahirad, S., Naghshiband-Hassani, R., Bergin, S., Katam, R., and Mahna, N. 2020. Improvement of Postharvest Quality of Plum (*Prunus domestica* L.) Using Polysaccharide-Based Edible Coatings. *Plants* 9 (9): 1148.

Peano, C., Girgenti, V., Sottile, F., and Giuggioli, N.R. 2009. Improvement of Plum Storage with Modified Atmosphere Packaging. *X International Controlled and Modified Atmosphere Research Conference* 876: 183–188.

Peano, C., Giuggioli, N.R., Girgenti, V., Palma, A., D'Aquino, S., and Sottile, F. 2017. Effect of Palletized MAP Storage on the Quality and Nutritional Compounds of the Japanese Plum cv. Angeleno (*Prunus salicina* Lindl.). *Journal of Food Processing and Preservation* 41 (2): e12786.

Pérez-Marín, D., Paz, P., Guerrero, J.E., Garrido-Varo, A., and Sánchez, M.T. 2010. Miniature Handheld NIR Sensor for the On-Site Non-Destructive Assessment of Post-Harvest Quality and Refrigerated Storage Behavior in Plums. *Journal of Food Engineering* 99 (3): 294–302.

Puerta-Gomez, A.F., and Cisneros-Zevallos, L. 2011. Postharvest Studies Beyond Fresh Market Eating Quality: Phytochemical Antioxidant Changes in Peach and Plum Fruit During Ripening and Advanced Senescence. *Postharvest Biology and Technology* 60 (3): 220–224.

Sameh, S., Al-Sayed, E., Labib, R.M., and Singab, A.N. 2018. Genus Spondias: A Phytochemical and Pharmacological Review. *Evidence-Based Complementary and Alternative Medicine* 1–13. doi:10.1155/2018/5382904.

Sampaio, S.A., Bora, P.S., Holschuh, H.J., and Silva, S.D.M. 2007. Postharvest Respiratory Activity and Changes in Some Chemical Constituents During Maturation of Yellow Mombin (*Spondias mombin*) Fruit. *Food Science and Technology* 27 (3): 511–515.

Sandhya, M. 2010. Modified Atmosphere Packaging of Fresh Produce: Current Status and Future Needs. *LWT-Food Science and Technology* 43 (3): 381–392.

Serdyuk, M., Stepanenko, D., Baiberova, S., Gaprindashvili, N., and Kulik, A. 2016. Substantiaton of Selecting the Method of Pre-Cooling of Fruits. *Eastern-European Journal of Enterprise Technologies* 4 (11): 62–68.

Serrano, M., Marínez-Esplá, A., Giménez, M.J., Valero, D., Zapata, P.J., Guillén, F., and Castillo, S. 2016. *Preharvest Application of Oxalic Acid Improves Antioxidant Systems in Plums*. VIII International Postharvest Symposium: Enhancing Supply Chain and Consumer Benefits-Ethical and Technological Issues 1194, 19–24.

Shahzad, M., Tahir, A., Jehan, N., and Luqman, M. 2015. Impact of Different Packaging Technologies on Post-Harvest Losses of Stone Fruits in Swat Pakistan. *Pakistan Journal of Agricultural Research* 28 (1): 53–63.

Siddiq, M. 2006. Plums and Prunes. In *Handbook of Fruits and Fruit Processing*, 553. Blackwell Publishing.

Sohail, M., Sun, D.W., and Zhu, Z. 2018. Recent Developments in Intelligent Packaging for Enhancing Food Quality and Safety. *Critical Reviews in Food Science and Nutrition* 58 (15): 2650–2662.

Sousa-Gallagher, M.J., Tank, A., and Sousa, R. 2016. Emerging Technologies to Extend the Shelf Life and Stability of Fruits and Vegetables. In *The Stability and Shelf Life of Food*, 399–430. Woodhead Publishing.

Suhag, R., Kumar, N., Petkoska, A.T., and Upadhyay, A. 2020. Film Formation and Deposition Methods of Edible Coating on Food Products: A Review. *Food Research International* 136: 109582.

Sultana, N., Haseeb-ur-Rehman, S.T.M., Haroon, Z., Fatima, D., and Fakhra, H. 2020. Prunus Domestica: A Review. *Asian Journal of Pharmacognosy* 4 (3): 21–29.

Thakur, R., Pristijono, P., Golding, J.B., Stathopoulos, C.E., Scarlett, C.J., Bowyer, M., Vuong, Q.V. et al. 2018. Development and Application of Rice Starch Based Edible Coating to Improve the Postharvest Storage Potential and Quality of Plum Fruit (*Prunus salicina*). *Scientia Horticulturae* 237: 59–66.

Teruel, B., Kieckbusch, T., and Cortez, L. 2004. Cooling Parameters for Fruits and Vegetables of Different Sizes in a Hydrocooling System. *Scientia Agricola* 61 (6): 655–658.

Tiburski, J.H., Rosenthal, A., Deliza, R., de Oliveira Godoy, R.L., and Pacheco, S. 2011. Nutritional Properties of Yellow Mombin (*Spondias mombin* L.) Pulp. *Food Research International* 44 (7): 2326–2331.

Vakkalanka, M.S., D'Souza, T., Ray, S., Yam, K.L., and Mir, N. 2012. Emerging Packaging Technologies for Fresh Produce. In *Emerging Food Packaging Technologies*, 109–133. Woodhead Publishing.

Valero, D., Díaz-Mula, H.M., Zapata, P.J., Guillén, F., Martínez-Romero, D., Castillo, S., and Serrano, M. 2013. Effects of Alginate Edible Coating on Preserving Fruit Quality in Four Plum Cultivars During Postharvest Storage. *Postharvest Biology and Technology* 77: 1–6.

Valero, D., Pérez-Vicente, A., Martínez-Romero, D., Castillo, S., Guillen, F., and Serrano, M. 2002. Plum Storability Improved After Calcium and Heat Postharvest Treatments: Role of Polyamines. *Journal of Food Science* 67 (7): 2571–2575.

Valero, D., and Serrano, M. 2010. *Postharvest Biology and Technology for Preserving Fruit Quality*. CRC Press.

Vargas, A.S., Juárez-López, P., López-Martínez, V., Flores, L.J.P., Sánchez, D.G., and Alia-Tejacal, I. 2017. Botany and Physiology Antioxidant Activity and Physicochemical Parameters in 'Cuernavaqueña' Mexican Plum (*Spondias purpurea* L.) at Different Ripening Stages. *Revista Brasileira de Fruticultura* 39 (4). Doi:10.1590/0100-29452017787.

Vargas-Simón, G. 2018. Ciruela/Mexican Plum—*Spondias purpurea* L. *Exotic Fruits* 141–152.

Vieira, L.M., Sousa, M.S.B., Mancini-Filho, J., and Lima, A.D. 2011. Total Phenolics and Antioxidant Capacity "in vitro" of Tropical Fruit Pulps. *Revista Brasileira de Fruticultura* 33 (3): 888–897.

Wang, L., Hong, K., Xu, R., Zhao, Z., and Cao, J. 2021. The Alleviation of Cold-Stimulated Flesh Reddening in 'Friar' Plum Fruit by the Elevated CO2 with Polyvinyl Chloride (PVC) Packaging. *Scientia Horticulturae* 281: 109997.

Wei, H., Seidi, F., Zhang, T., Jin, Y., and Xiao, H. 2020. Ethylene Scavengers for the Preservation of Fruits and Vegetables: A Review. *Food Chemistry* 337: 127750.

Wu, F., Zhang, D., Zhang, H., Jiang, G., Su, X., Qu, H., Duan, X. et al. 2011. Physiological and Biochemical Response of Harvested Plum Fruit to Oxalic Acid During Ripening or Shelf-Life. *Food Research International* 44 (5): 1299–1305.

Wu, W., Gao, H., Chen, H., Fang, X., Han, Q., and Zhong, Q. 2019. Combined Effects of Aqueous Chlorine Dioxide and Ultrasonic Treatments on Shelf-Life and Nutritional Quality of Bok Choy (*Brassica chinensis*). *LWT* 101: 757–763.

Zhang, J., Ma, Y., Dong, C., Terry, L.A., Watkins, C.B., Yu, Z., and Cheng, Z.M.M. 2020. Meta-Analysis of the Effects of 1-Methylcyclopropene (1-MCP) Treatment on Climacteric Fruit Ripening. *Horticulture Research* 7 (1): 1–16.

Zhang, X., Liu, Y., Yong, H., Qin, Y., Liu, J., and Liu, J. 2019. Development of Multifunctional Food Packaging Films Based on Chitosan, TiO2 Nanoparticles and Anthocyanin-Rich Black Plum Peel Extract. *Food Hydrocolloids* 94: 80–92.

Zielinski, A.A.F., Ávila, S., Ito, V., Nogueira, A., Wosiacki, G., and Haminiuk, C.W.I. 2014. The Association Between Chromaticity, Phenolics, Carotenoids, and In Vitro Antioxidant Activity of Frozen Fruit Pulp in Brazil: An Application of Chemometrics. *Journal of Food Science* 79 (4): C510.

10 Innovative Plum-Processing Technologies

Nabia Ijaz[1], *Bakhtawar Shafique*[2], *Syeda Mahvish Zahra*[2,3], *Shafeeqa Irfan*[2], *Rabia Kanwal*[4], *Saadia Zainab*[5], *Muhammad Modassar Ali Nawaz Ranjha*[2] *and Salam A. Ibrahim*[6]

[1] University Institute of Food Science and Technology, University of Lahore, 54000-Lahore, Pakistan

[2] Institute of Food Science and Nutrition, University of Sargodha, 40100-Sargodha, Pakistan

[3] Department of Environmental Design, Health and Nutritional Sciences, Allama Iqbal Open University, 44310-Islamabad, Pakistan

[4] School of Food Science and Engineering, South China University of Technology, 510641-Guangzhou, China

[5] College of Food Science and Technology, Henan University of Technology, 450001-Zhengzhou, China

[6] Food Microbiology and Biotechnology Laboratory, North Carolina Agricultural and Technical State University, Greensboro, 27411-NC, United States

CONTENTS

DOI: 10.1201/9781003205449-10

10.1 INTRODUCTION

Fruit products are exceptionally transient, and around 20–40% of the organic products are kept in safe storage from the hour of harvest until they arrive at the customers. Leafy foods are vital as far as nourishment and are known for their medical advantages. The plum tree (*Prunus domestica* L.) belongs to the *Rosaceae* family of deciduous trees. Plums come in over 2,000 assortments, with the most common species appearing in Europe and Australia. Only few studies have been conducted to examine the overall phenolic content of plum leaves and how it relates to antioxidant activity (Elsayed et al. 2020).

Plums (*Prunus domestica* L.) are one of the most significant fruit crops from the *Rosaceae* family. They are local to Europe and Asia and incline toward full sun and wet soils. When plum trees reach three to four years of age, they begin to produce fruit, maturing from August to September. Plums have both a sticky-sweet and acidic taste. Important contents of plums are malic acid, sugar, gelatin, especially amygdalin and prunasin. The dried organic products known as prunes are well appreciated for laxative action. Typically, plums have a shelf life of three to five days (Sunmathi et al. 2019). During its peak season, a generous amount of plum fruit goes to waste because of an excess of products in local markets, bringing about hefty postharvest losses. The postharvest loss falls somewhere in the range of 35–40%, or about $800 millions per year. Decreasing these issues can improve the economy and could encourage more consumers to purchase plum fruits (Sharmin et al. 2015) *Prunus* cultivars are developed all over the world for their characteristic items, which are considered a critical source of bioactive compounds such as anthocyanins, carotenoids, ascorbic corrosive, vitamin E, flavonoids, proanthocyanidins and tannins. Plums are one of the prevalent fruit products in Romania, accessible in a variety of forms, such as fresh, dried, jams, purees, and juices. Plums can be sed in combination with other fruits to have a combination of medicinal properties. Because of their enhanced phenolic synthesis, plums are particularly inclined to enzymatic browning in this specific circumstance. Right now, within the food industry, maintaining a strategic action to avoid enzymatic browning is one of the major concerns of the food industry. The existence of oxidative molecules, especially polyphenol oxidase and peroxidase, catalyzes this cycle. Polyphenol oxidase (POD) catalyzes the redox responses of a wide extent of phenolic and non-phenolic substrates. Within the presence of hydrogen peroxide as an electron acceptor, the unit catalyzes the oxidation of a couple of combinations, such as phenols, aromatic amines, ascorbic acid and indole (Chakraborty et al. 2015).

The inhibition of POD is basic for constraining the losses caused by enzymatic browning. Warm processing, which is additionally utilized to ensure product quality within the food industry, is one of the most broadly used ways for inactivating oxidative compounds. Because of its strong thermostability at sanitization temperatures, which can lead to organoleptic and nutritive characteristics to disintegration, the chemical poses an extreme challenge for handlers (Enachi et al. 2018).

In any case, understanding the deactivation limits is basic for streamlining a treatment. In this way, any unused mechanical development offers a surprising opportunity to control oxidative protein-catalyzed reactions. Because of its capacity to immobilize vegetative cells of microorganisms and catalysts included with enzymatic browning at temperatures around room temperature, high-pressure treatment is gaining popularity within the food industry. As a result, this interesting and discretionary advancement gives the opportunity to restrict the impact on flavor, shading and healthful characteristics for fabulous culinary outcomes source (Terefe et al. 2014).

This is often the primary time that POD from plums has been sifted using different chromatographic rules, as well as the primary time that the perfect confinements for the cleaned enzyme's warm treatment, high-pressure treatment, and combined high-pressure with warm treatment have been decided. The information gathered is of coherent and viable utilization in understanding societal changes related to drugs connected with mechanical planning and the security of attainable and current treatment forms for verdant foods, subsequently guaranteeing the quality of the products (Rodrigues and Fernandes 2012).

10.2 PLUM PRODUCTS AND STORAGE

Plum is a prominent commercial stone fruit that may be found in a variety of geological settings all over the world. The total annual production now exceeds 10 million tons (Karacha et al. 2014). The most important plum-growing areas are in the Emilia Romagna and Campania districts of Europe, where more than half of the total produce is amassed (Nencetti et al. 2008).

Depending on when the fruit is collected, plum fruits typically mature in about 8–10 days. Plum has a very short shelf life, and due to limited storage and utilization time, it is limited to far-flung commercial sectors (Khan et al. 2008). The expansion of ethylene production, mellowing, postharvest microorganisms, and post-reap infections result in a reduction in the time frame and character of plum usefulness (Manganaris et al. 2008). As a result, plums are highly fleeting, and their potential period is limited due to physiological difficulties (Chen and Zhu 2011).

Buyers' constant desire for innovative, high-quality natural goods with fewer or no synthetic ingredients have resulted in inventive processing techniques. Because of their prospects for organic product preparation with significant property preservation, these innovations may lead to the introduction of new items on the market, an increase in the severity of natural products, the elimination of the use of some added substances, and a reduction in energy costs (Evrendilek et al. 2012). New or enhanced innovations in fruit handling have been implemented to reduce disasters and increase the tangible features of manufactured fruit items (Rodrigues and Fernandes 2012).

10.2.1 Controlled Atmospheric Pressure

Controlled atmosphere (CA) is a capacity framework that considers greater fruit quality assurance as a result of a more defined decrease in fruit digestion. The capability of plums in controlled climates could be a point that has yet to be completely investigated, and there is constrained data on this stockpiling structure for Laetitia plums. In any case, a few plum and peach cultivars have appeared to have progressed in quality during due to a decrease in chilling injuries and the assurance of physicochemical characteristics.

After expulsion from the chamber and four days of utilization, plum fruit stored under controlled atmospheric conditions exhibited lower ethylene formation rates than regular atmosphere (RA). According to some researchers, low O_2 or potentially high CO_2 causes decreased ethylene production in a controlled setting. Reduced oxidation of 1-carboxylic-1-aminocyclopropane (ACC) by low O_2 or limitation of ethylene activity in commencing auto-catalysis by high CO_2 results from the suppressed ethylene production rate in a regulated environment. Incomplete pressure factors of 1-3 kPa O_2 and 8-12 kPa CO_2 are the optimal CA conditions for plums. Fruit stored within regulated climatic circumstances exhibited a lower skin shading pace (lower upsides of red-shading file) on expulsion from the chamber, as well as a less remarkable red tone upon expulsion from the chamber and after four days of realistic usefulness. The less severe red hue of the fruit should be distinguished with lower biosynthesis and ethylene activity in a regulated environment, as noted for tissue immovability and interface (Steffens et al. 2014).

10.2.2 Modified Atmospheric Pressure

Stockpiling in a modified environment is an approach for preserving the natural character of organic goods. Organic product digestion is influenced by a modified environment combined with low temperatures, which reduces ethylene production, weight loss, hardness, and the conservation of minerals and natural acids. Flow-pack bundling allows for modified atmospheric pressure stockpiling throughout the inventory network on various ephemeral natural items, such as plum, to improve the time frame of practical uses. According to current practices and recommendations, materials with suitable mixes of permeability and porousness should be packaged. The use and enhancement of novel eco-friendly materials, such as biodegradable films, might be particularly beneficial (Peano et al. 2009).

The modified atmospheric pressure approach for plums proved useful in postponing the maturing cycle of plum cultivars, as well as the drawbacks of immovability, and so increasing the time span of utilization (three to four weeks more depending on the evaluated boundary). Those plums bundled with film M [made out of polyester

(12 m) polypropylene (60 m)] had a better effect than those bundled with film H [made out of polyester (12 m) polypropylene (50 m)], most likely due to the greater prevention of ethylene generation caused by the use of film M (Díaz-Mula et al. 2011).

Using different barometrical bundling or box liners to cope with decreased Friar plum weight loss while maintaining the fruit tissue appearance for cold storage period up to 45 days is a good technique. However, because of improved tissue clarity, gel disintegration and "off flavor" progression, modified atmospheric bundling is not recommended for cold storage periods more than 45 days in this cultivar. To ensure Friar quality, box liners with no or low CO_2 or O_2 control are recommended for cold storage or delivery periods greater than 45 days. Great postharvest procedures, such as rot control, proper chilling, and temperature management during the postharvest life, must be used and executed in all circumstances if success is to be made. However, the use of modified atmospheric bundling box liners is recommended to extend the market life of Friar plums to 45 days in cold storage. In this approach, fruit cold-stored for prolonged periods of time may have adverse chilling injury (CI) effects (such as tissue transparency, gel disintegration, tissue dying, rot, and "off flavor") due to the use of box liners without gas-management capabilities (Cantin et al. 2008).

The use of modified atmosphere bundling has resulted in a significant reduction in the accumulation of anthocyanin and phenolic content in Monk plums stored at 0°C. Furthermore, the collection of phenolic content in strips and tissues at 2°C has been hampered. In any event, modern air bundling has completely reversed these carotenoid concentration advancements in all varieties (Fanning et al. 2014).

10.3 OZONE PRETREATMENT

The word "ozone" comes from the Greek word *ozein*, which suggests "spoiled." The gas is an imperative component of the climate in both the upper and lower stratosphere. Whereas stratospheric ozone acts as a defensive shield against possibly harmful amounts of B radiation, ground-level ozone concentrations are known to be sufficient to imperil human, animal, vegetation, and structural well-being Tropospheric ozone is shaped by invasions from the stratosphere, but it is additionally created in situ by photochemical responses (Di Renzo et al. 2004).

Ozone can be utilized in air or water for postharvest medicines of new products of the soil. In cool extra spaces it very well may be ceaselessly or irregularly added to the capacity of the air. The two strategies have, as of late, gotten significant role, particularly in light of the absence of build ups on the produce and new administrative issues (Tzortzakis 2016).

Ozone is a notable solid oxidizing specialist and is by and large more secure to use than numerous other options and can be cost-efficiently produced and controlled nearby. Financially, ozone should be produced nearby, at the mark of use. It exceptionally well may be made by submitting oxygen or ultraviolet (UV) radiation of 285 nm; in any case, various trade applications utilize the crown discharge strategy for age. At the point when a tall voltage trading current is connected over a discharge gap inside the container or oxygen, it brings around the cleavage of oxygen particles to shape ozone. Plum fruit has been assessed with ozone vaporous application in postharvest. Plums are, for the most part, absent at 0°C for multiple weeks, and as long as three months, at relative humidity of 95%. Plums were immunized with different inoculum piles of *Botrytis cinerea* (dim shape), moved to chilled capacity (13°C) and presented to low-level ozone advancement (0.1 ppm) for eight days. Ozone treatment made approximately no impact on sore headway in Dark Golden plums, and implied that there was a basic relationship ($P< 0.05$) between the inoculum center and damage advancement. Spore creation was decreased by around 20% in ozone-treated plums. Low ozone practicality may be credited to the low gas fixation, the brief treatment length and the high storing temperature used (Tzortzakis and Chrysargyris 2017).

10.4 IRRADIATION TREATMENT

Gamma light at present is utilized as a strategy for food conservation. Gamma illumination successfully defers the maturing, forestalls growing and decreases microbial tally prompts, thus expanding the time span of a plum's usability. Combinatory medicines have additionally generally been explored, as they frequently bring about synergistic impacts. Gamma light mixed with different medicines (e.g., heat, washing and altered environment stockpiling and edible covering measures) gives a viable outcome in broadening the time frame of realistic usability of the plum fruit (Salem et al. 2016).

Beneath both capacity circumstances, gamma-light treatment was basically effective in keeping up capacity quality while decreasing yeast and shape count. Up to 8–12 days of comprehensive stockpiling, no microbial burden was recognized in occasions lit at 1.0 kGy, 1.2 kGy, and 1.5 kGy. Measurements of 1.2 kGy and 1.5 kGy hindered microbial development for up to 28 days under refrigerated circumstances. Under encompassing amassing settings, a parcel scope of 1.2–1.5 kGy expanded the length of usage of the plum by 16 days. In differentiate to the 12.5% spoil in un-illuminated control tests, light in combination with refrigeration postponed the onset of decay by 35 days. After 35 days of chilling, plums with a portion extent of 1.2–1.5 kGy had an extra 8-day development amid additional encompassing storing. As a result, gamma illumination treatment of Santaroza plums will help in overcoming the confine prerequisites for sending out purposes and will empower the scattering of fruits to diverse business divisions, in this manner helping in incredible market returns to the cultivators all through seasons of overabundance (Hussain et al. 2013).

Plums (*Prunus domestica* L.) of the cultivars Avalon (prepared) and Victoria (prepared and partly prepared) were exposed to UV-B radiation and visible light. The impacts of the drugs on anthocyanins' imperative material qualities and substance were examined. The plums that were exposed to light and radiation misplaced more weight than the plums that were kept in the dark. In comparison to the outward component of the fruit tissue, the interior piece had more noteworthy dissolvable concentration and titratable fineness. The combination of visible light and UV-B treatment expanded the number of dissolvable materials within the outside component of the fruit tissue on the exposed side in Victoria plums. The plums' skin had the most noteworthy amount of redness after a comparable treatment. The anthocyanin substance of possibly ready plums was lower than that of prepared plums, but this contrast was misplaced, taking after clear light and UV-B treatment (Vangdal et al. 2009).

10.5 HIGH-PRESSURE PROCESSING

High-pressure processing (HPP) is one of the finest cooling technologies, with huge potential as a warm decontamination alternative. HPP has, of late, received expanded consideration from both food companies and examiners, who are contributing considerable time and money toward the advancement of this innovation (Knorr et al. 2011).

Pressure and temperature are both critical components in high-pressure food applications. The commitment of temperature all through treatment has not been considered in numerous studies. Temperature changes are critical since they influence food gelling, protein quality, fat migration, solidifying and other things. The nonappearance of data on thermo-physical qualities under strain is the essential issue with warm development in HPP estimations. High-pressure figures are generally utilized in soil-arrangement products to inactivate microorganisms and catalysts, as well as to expand the time allotment of practical utilization whilemaintaining higher organoleptic, unmistakable, and feeding highlights (Rastogi et al. 2007).

The microscopic organisms in Ruby Globe plum purées were adequately decreased by high-pressure medications. The combined treatments utilized (weight level and planning time) were incapable of lessening the waiting peel plum purée development. The pressure-treated plum purée appeared to have superior shading maintenance than the untreated plum purée. In plum purées, HPP at weights extending from 400 to 600 MPa altogether decreased anthocyanins and totally dispensed with cell fortifications, compared to untreated purées. High-pressure drugs, on the other hand, might keep up phenol substance. In common, HPP had no negative affect on the plum purée's healthfulness. In this respect, HPP seems to be a fruitful method for guaranteeing a plum purée of adequate dietary esteem (Gonzalez-Cebrino et al. 2013).

The purées treated by hydrostatic high pressure, on the other hand, contained bioactive combinations, such as carotenoids and polyphenols. HPP might get a Songold plum purée with unmistakable levels of bioactive combinations held all through cold storage (Gupta et al. 2010).

10.6 HIGH-PRESSURE THERMAL (HPT) TREATMENTS

High-pressure thermal (HPT) warm drugs are made up of a combination of high-pressure calculations (500–900 MPa) and temperatures (70–120°C) conveyed over a brief period of time. The quick temperature rises during pressure and rapid temperature drop after flattening may offer assistance to decrease the harshness of warm impacts in classic heated progressions (García-Parra 2014).

From a microbiological point of view, high-pressure warm medicines would be superfluous in exceptionally low-corrosive food sources like plum (pH=3.4). Regardless, this strategy of handling proved very compelling in inactivating polyphenoloxidase. This protein was safe to high pressures at low temperatures, which might restrain the duration in which handled plum purées may be utilized practically. Moreover, HPT treatment was more viable in protecting the beginning shade of plum purée as well as the bioactive blend fabric of handled purées, such as anthocyanin and the development of cancer resistance. The combination of pressing components at 600 MPa and a beginning temperature of 70°C brought about within the most noteworthy inactivation of polyphenoloxidase and the most noteworthy preservation of bioactive purée combinations (Raheem 2013).

10.7 EDIBLE COATINGS

Food preservation developments are presently confronting impressive challenges in amplifying the utility of momentary foodstuffs (e.g., meat, eggs, and an assortment of unrefined soil items) that help to fulfill everyday nutritional requirements (Diaz-Montes and Catro-Munoz 2021). Moreover, nourishment conservation has advanced past just preservation; present-day strategies are concentrated on the satisfaction of two extra targets: the reasonableness of preowned cycles and the age of environment friendly products with no impact on. Additionally, they are trying to find other health-promoting qualities. The utilizing of palatable films and coatings is overseen by one of these defensive standards (Eum et al. 2009).

Edible coatings are utilized to extend the quality of new harvests and amplify their utility by discouraging the restrain layer, which controls oxygen, carbon dioxide, and water loss encompassing the items. As a result, the impact of edible coatings on fruit maturation might show itself in a variety of ways, such as altered environment aggregation due to contrasts in gas arrangement in secured products (Kumar et al. 2018).

Plum is a short-lived fruit with a reasonable use period of three to four days. A couple of considerations have proposed that edible coatings may well be used to preserve the quality and extend the shelf life of fruits (Panahirad et al. 2020a). The 1.5% gelatin covering delivered the most excellent results for most anticipated limits. PEC-based consumable covers may well be utilized as a standard postharvest treatment on plum fruits to preserve their antioxidative capability. In most cases, the applied covering was compelling in completely assessed properties. Ascorbic destructive, phenolic compounds and anthocyanin, and flavonoid substance of fruits were all progressed by PEC-based coatings, coming about in a larger antioxidative boundary. The antioxidative qualities and utilization length of plum fruits are expanded by employing a PEC-based edible coating (Sohail et al. 2014).

When compared to other bundling materials, fruits stored in delicate board container showed high storage soundness, followed by wooden cartons. Agreeing to a number of research investigations about, CMC (carboxymethylcellulose) at 1% and Pec (gelatin) at 1.5% created the most noteworthy results. Besides, a combination of 0.5% Pec and 1.5% CMC may be a great choice for securing the plum's healthy advantage all through postharvest. As a result, utilizing CMC, Pec, or indeed their combinations as a covering technique for expanding and improving postharvest subjective qualities of plum fruits might be regarded a secure and comfortable choice (Panahirad et al. 2020a).

In plum cultivars, alginate edible-based medications may be utilized as ordinary postharvest drugs to delay the postharvest maturing cycle and keep up plum quality. The starting of the ethylene climacteric top in secured plums was restrained by alginate eatable-based treatment at 1%, which was particularly limited in those plums treated with alginate at 3%, the two solutions being successful in delaying weight and corrosiveness calamities, progressing, and shading changes (Valero et al. 2013).

10.7.1 Polysaccharide-Based Coatings

Plum has a shelf life of about three to four days due to its high moisture content and hence is designated as a highly perishable fruit. Several studies have shown that edible coatings might be used to retain the quality and extend the storability of perishable products. Edible coatings based on polysaccharides are usually impermeable to oxygen exchange due to their organized chemical structure, but they are still vulnerable to moisture loss. Commonly used edible coatings are oil-free, colorless and of low caloric value and act as rancidity and dehydration inhibitors, hence extending the shelf life of fruits, with the preservation of phytonutrients, i.e., antioxidants, flavonoids, anthocyanins and vitamin C, along with improvement in the enzymatic action of peroxidase while declining the enzymatic action of polygalacturonases and polyphenol oxidases. Several studies have depicted an extension in the storability of plums if coated with pectin and carboxymethylcellulose alone or in combination. A study investigated the application of pectin: carboxymethylcellulose at a ratio of 0.5:1.5 and found excellent results in the shelf-life extension of plums. Thus, investigators classified this combination as a safe edible coating for the improvement of qualitative and nutritive features of plum fruit (Kumar et al. 2018).

Another study revealed that during simulated export transportation conditions, edible coatings retained the quality of plum fruits. Plums coated with alginate and chitosan delayed peel shrinkage and slowed the maturation process and physicochemical variations in fruit during storage. Edible coating made up of gum arabic was promising, as it delayed shriveling and significantly reduced weight loss and provided better physicochemical properties and nutrient retention, allowing storage for up 20 days provided a transportation simulation of 5 weeks (Panahirad et al. 2020b).

10.7.2 Biopolymer-Based Coatings

Biopolymers, like proteins; lipids, specifically carbohydrates; and anti-microbial, anti-fungal or flavor-enhancing additives or preservatives, are used to develop edible coatings with the incorporation of bioactive components of fruit itself or may be extracted from other agricultural wastes to delay ripening and physiological changes in fruit and to minimize the loss of plum fruit weight and firmness (Fawole et al. 2020). Pectin is in high demand because of its structural variations and flexibility (Formiga et al. 2019). Edible coatings carrying biologically active ingredients signify a substitute preservation technique that could modify and maintain a modified atmosphere within the plum through regulating gaseous and volatile compounds exchange and restraining moisture ingression or loss. Starch-based edible coatings have an edge over other biopolymer-based edible coatings because these are available in high quantities at low costs, and the properties of the film do not contribute color or taste with a high barrier to oxygen. Starch-based edible coatings can be improvised to provide mechanical resistance to degradation and to enhance functional properties with incorporation of certain lipids, surfactants and plasticizers (Khan 2019).

10.7.3 Gum-Based Coating

Hydrocolloids (water-soluble gums) are utilized as packaging films, coating agents, thickeners, texture modifiers, emulsifiers and stabilizers. Potential uses of natural gums, such as alginate, chitosan, psyllium, xanthan, mesquite, gellan, basil seed, tragacanth, guar and arabic gums in edible coatings have been mentioned in several studies, signifying their usability in the shelf-life extension of fruits (Sapper and Chiralt 2018).

Edible coatings can be a great vehicle to protect agricultural produce from fungal, microbial and bacterial attacks and reducing synthetic fungicide applications, once have incorporated biologically active natural ingredients in edible coating formulations. In plums, mostly *Rhizopus stolonifer* attacks and causes postharvest loss. A double coating made of gum arabic in combination with natural antifungal constituents, such as oregano and rosemary essential oils, helps prevent its attack and controls *Rhizopus* soft rot in plums (*Prunus domestica*) during storage (Salehi 2020).

10.7.4 Starch-Based Coatings

Edible coating application on surfaces of fruits has recently been of interest in the fruit and vegetable preservation and storage enhancement industry as an innovative, cost-effective and energy-efficient technology. Starch-based edible coating application to whole fruits and vegetables or their extracts retains the products' nutritional quality as well as extends their shelf life due to the cohesive molecular waxy covering over the fruit surface, which monitors the exchange of gases and moisture loss, making it better for the improvement or retention of plum fruit's quality during storage (Andrade et al. 2017). The thin wax layer naturally present on plum skin is a complex of n-alkanes, ketones, n-alkyl esters, fatty acids, alcohols, aldehydes and other compounds, including penta-cyclic triterpenoids as well as flavonoids, basically is partly permeable for water vapors and other gases. The natural epi-cuticular wax layer on the plum itself contributes to extending the fruit's shelf life (Thakur et al. 2018).

10.7.5 Pectin-Based Edible Coatings

Pectin-based edible coatings are considerably effective in preserving ascorbic acid, anthocyanin and flavonoid contents, along with antioxidative capability in plum fruits. The activities of enzymes are affected by the coatings. The impacts of pectin-based edible coatings on antioxidative features of plum fruits can be hypothetically considered as a progressive method to augment the nutrient values of plums (Baisak et al. 2019).

10.7.6 Carboxymethylcellulose-Based Coating

Carboxymethylcellulose-based edible coatings are an acclaimed formulation in maintaining the firmness of plum flesh as well as for the retention of acidity and phytonutrients, like flavonoids, phenolic constituents and the antioxidant aptitude of plum fruits. The enzymatic activity of plums is significantly affected by the coating.

In general, the application of carboxymethylcellulose on plum fruit can be a potentially favorable method to improve the shelf life of the fruit (Panahirad et al. 2019).

10.8 PLUM POWDER

Consumer interest in a variety of food products from a single food source has led to diversity in processing technologies. Powdered food production is one such creation that meets and satisfies this consumer need, and the process that yields plum flesh into powder keeping quality of food powder and energy efficiency of process (Cozzolino et al. 2021a). Plum peel, flesh or waste is converted to powder and provided to industries, be it as food or nutraceuticals, as customers have inclined interest in purchasing foods with phenolics and antioxidant profiles. The food's color, taste and flavor can be improved with addition of plum powder (Kang et al. 2019).

Plants are a rich source of such natural components, especially fruits. Plums are fruits of special interest for growers and farmers due to their economic importance, as its production requires huge agricultural areas (Cozzolino et al. 2021b) and the trees give a greater volume of fruit. Powder production can be sped up by improvising processing conditions, i.e., application of lowering pressure in a drying facility, which speeds up the rate of moisture removal from fruit flesh or peel, in other words quickening the drying process. Vacuum drying is good example of such process modifications where high-temperature application is substituted with controlled temperature due to heating by conduction rather than convection (Seke et al. 2021). Reduced temperature in vacuum drying prevents heat-related loss to bioactive compounds in fruits. Freeze drying is considered the safest option for drying plums or any other fruit because it retains the quality as well as quantity of bioactive ingredients during the process of drying/dehydration (Michalska et al. 2017).

Plums are categorized as highly perishable fruits, as they contain 87% water content, and for better export earnings it is necessary to reduce transportation costs and risk of spoilage. Therefore, the best way to guarantee safety and storage life is to convert it to a powdered form; this also provides the opportunity to enjoy seasonal fruit year-round and makes it supple for industries that develop plum products (Sifat et al. 2021).

Conversion of plum fruit through freeze drying is still ranked the highest due to the retention of sensitive bioactive components along with nutritional value and the preservation of its natural color, flavor and taste because of low-temperature application (Michalska et al. 2017).

10.8.1 Lyophilization of Plum

Lyophilization calls for a technique that involves the direct conversion of frozen ice particles to the gaseous phase, i.e., vapors, without passing through the liquid phase, so basically it is about freezing, sublimation and secondary drying (Pansare and Patel 2019). Internal dissemination is abstained in lyophilization since sublimation starts at the top and grows to deeper levels gradually before frozen icy crystals are quickly converted into vapours/steam, skipping conversion to the water/liquid state (Antal et al. 2010).

A study represented the production of lyophilized plums where fresh plums, after removal of non-edible parts, were passed through a manual juicer to obtain plum pomace mixed with the exocarp of the plum/plum peel, which was further passed through phase of freezing at –20°C until it was dried. CHRIST Freeze Dryer Alpha 2–4 LD, which is also lyophilizer was utilized (48 hours) to conduct freeze drying. Then the lyophilized substance (plum pomace lyophilisate) was converted to a powder in a Knifetec 1095 Sample Mill (FOSS, Germany), keeping the moisture content up to 3.6%. Sieves set at 800, 500, 400 and 200 m were used to decide molecule size dispersion of the lyophilized plum pomace powder. From 200 to 400 m, the most, i.e., 92%, of the lyophilized plum pomace particle division was stored in plastic holders in a dark and dry storage area until future examination. This powder was further added to hydrocolloid emulsions to yield fruit spread, which provided a replacement of commercially available fruit-flavored spreads, with natural fruit spreads having the benefits of rich fiber content, bioactive compounds and an antioxidant profile (Bajić et al. 2020).

The exceptional impact of freeze drying is developing an inhibition of strain-dependent food-borne pathogenic attacks to dried powders/extract powders (Silvan et al. 2020).

10.9 EXTRACTION TECHNIQUES UTILIZED FOR BIOACTIVE PHYTONUTRIENT EXTRACTION FROM PLUM

To overcome the essential downsides of conventional warm procedures, a variety of novel choices have been created, comprising high pressing factor, ultrasound, microwave, supercritical fluid and electro-technologies (such as cold plasma, pulsed electric field and high voltage electric release) (Putnik et al. 2018; Barba et al. 2016).

10.9.1 Extraction by Ultrasounds

Ultrasound-assisted extraction (UAE) is an interesting interaction for getting high-value combinations and is especially advantageous for heat-sensitive compounds (Esclapez et al. 2011). The extraction of flavonoids and polyphenols from plums has been done utilizing UAE. Within the extraction medium, UAE produces advancement and weight cycles (DiNardo 2018). The Box–Behnken plan was utilized to effectively advance the antioxidant UAE strategy. The finest conditions were found to be a 21-minute extraction period, 20.5% (v/v) ethanol, a liquid-to-solid proportion of 27.7 cm^3 g–1 and an extraction temperature of 66°C. Ferulic acid was the foremost plentiful phenolic component within the ideal extract, while caffeic acid was the least plentiful. The presence of the detected phenolic chemicals likely contributed to the extract's antioxidant activity. Because of these properties, the ethanolic extricate of plum seed may be utilized within the culinary, makeup and pharmaceutical industries (Savic and Gajic 2021).

10.9.2 Microwave-Assisted Extraction

Microwave-assisted extraction (MAE) has been broadly utilized in the extraction of nutraceuticals from plant frameworks because of its warming instrument, moderate capital expense, more limited extraction times and less dissolvable qualities compared to other technologies (Orsat and Routray 2012). The utilization of a microwave to remove polyphenols from plants got a lot of attention (Haddadi-Guemghar et al. 2014). MAE found that a 60% concentration of ethanol and a solid-to-solvent proportion of 0.1 g/mL created the most elevated phenolic substance and antioxidant action in yellow European plums. The extraction of phenols from various genotypes of yellow European plums moreover appeared to be affected by time and temperature in this study. For the extraction of bioactive components from yellow European plums, MAE has demonstrated to be an attainable elective to conventional extraction strategies. Food and nutraceutical companies may utilize the extraction of these components from plums to fortify meals and incorporate bioactive substances into their items (DiNardo et al. 2019).

10.9.3 Extraction by Pulsed Electric Field

To extend the bioactive chemical substance of the natural product, direct-intensity pulse electric fields (PEF) were applied to plums. PEF innovation may be an unused non-thermal food-planning strategy that includes applying focused electric field beats to materials among two cathodes for a brief period of time. The objective of this strategy is to keep food secure by inactivating microscopic organisms while maintaining the fundamental qualities. The PEF extraction is carried out at 4°C for 24 hours at a recurrence of 0.1 Hz. As a dissolvable, water is used. The use of direct-quality PEFs on plums improved the levels of a few bioactive chemicals, like anthocyanins, and the antioxidant action of plum purées to some degree. The incorporation of amino acids during the fabricating of plum purée was a deciding factor for the ultimate quality of purées in terms of protecting the color and concentration of bioactive compounds (García-Parra et al. 2018). PEF-I increased the extraction of anthocyanins and phenols in plums, which was followed by a significant rise in the DPPH level (Medina-Meza and Barbosa-Canovas 2015).

10.9.4 Supercritical Fluid Extraction (SFE)

SFE is harmless to the ecosystem and is also economical and non-combustible. It has been utilized in the extraction of fundamental oils (Routray et al. 2013) and enjoys a few advantages over regular techniques; for example, it has high solvating power, which can be effectively controlled through a variety of temperature and pressure factors. Additionally, the diffusivity of supercritical liquid is superior to conventional solvents, it provides high extraction yield and it is environmentally friendly since no harmful solvents are used. Be that as it may, SFE is expensive and has numerous perplexing activity boundaries. It is additionally hard to execute because of the complex requirements to operate it (Singh and Orsat 2014). SFE is performed at 40°C for 5 hours at 300 bar pressure. The solvent is methanol (Vladic et al. 2020). SFE proved successful and practical, and it served as the foundation for high-throughput extraction, purification and industrialized production of volatile matter from dried plums.

10.10 CONCLUSION

Innovative processing technologies have improved quality retention of plum flesh, peel, waste and value-added products, especially powder and extracts. Innovation in extraction technologies from conventional extraction to supercritical fluid extraction has helped in achieving enhanced yield and structural stability of plum's bioactive phytonutrients. Edible coating improvisation from a simple dip method to highly classified biopolymers has improved storage life, especially in regard to physico-chemical properties. The conversion of plum into powder via vacuum drying, freeze drying and specifically lyophilization has improved plum powders' textural and physical characteristics, such as color, taste, flavor and flow ability, along with best retention of bioactive ingredients' structure and quality, keeping intact the fruit's nutritional and health benefits. Still, there are innovations being done to improve the quality of plum and its products during product development and the extraction of specialized components at laboratory and industrial levels to enjoy it off season and to earn a profit through exportation.

10.11 REFERENCES

Andrade, S.C., Baretto, T.A., Arcanjo, N.M., Madruga, M.S., Meireles, B., Cordeiro, Â.M., and Magnani, M. 2017. Control of Rhizopus Soft Rot and Quality Responses in Plums (Prunus domestica L.) Coated with Gum Arabic, Oregano and Rosemary Essential Oils. *Journal of Food Processing and Preservation* 41 (6): e13251.

Antal, T., Kerekes, B., and Sikolya, L. 2010. *Measurement of Quality Properties of Dried Plum Varieties*. SZTE Repository of Papers.

Bajić, A., Pezo, L.L., Stupar, A., Filipčev, B., Cvetković, B.R., Horecki, A.T., and Mastilović, J. 2020. Application of Lyophilized Plum Pomace as a Functional Ingredient in a Plum Spread: Optimizing Texture, Colour and Phenol Antioxidants by ANN Modelling. *LWT* 1 (130): 109588.

Barba, F.J., Zhu, Z., Koubaa, M., Sant'Ana, A.S., and Orlien, V. 2016. Green Alternative Methods for the Extraction of Antioxidant Bioactive Compounds from Winery Wastes and By-Products: A Review. *Trends in Food Science & Technology* 49: 96–109.

Basiak, E., Geyer, M., Debeaufort, F., Lenart, A., and Linke, M. 2019. Relevance of Interactions Between Starch-Based Coatings and Plum Fruit Surfaces: A Physical-Chemical Analysis. *International Journal of Molecular Sciences* 20 (9): 2220.

Cantín, C.M., Crisosto, C.H., and Day, K.R. 2008. Evaluation of the Effect of Different Modified Atmosphere Packaging Box Liners on the Quality and Shelf Life of 'Friar' Plums. *HortTechnology* 18 (2): 261–265.

Chakraborty, S., Rao, P.S., and Mishra, H.N. 2015. Kinetic Modeling of Polyphenoloxidase and Peroxidase Inactivation in Pineapple (Ananas comosus L.) Puree During High-pressure and Thermal Treatments. *Innovative Food Science & Emerging Technologies* 27: 57–68.

Chen, Z., and Zhu, C. 2011. Combined Effects of Aqueous Chlorine Dioxide and Ultrasonic Treatments on Postharvest Storage Quality of Plum Fruit (Prunus salicina L.). *Postharvest Biology and Technology* 61 (2): 117–123.

Cozzolino, D., Phan, A.D.T., Aker, S., Smyth, H.E., and Sultanbawa, Y. 2021a. Can Infrared Spectroscopy Detect Adulteration of Kakadu Plum (Terminalia ferdinandiana) Dry Powder with Synthetic Ascorbic Acid? *Food Analytical Methods* 14 (9): 1936–1942.

Cozzolino, D., Phan, A.D.T., Netzel, M., Smyth, H., and Sultanbawa, Y. 2021b. Assessing the Interaction Between Drying and Addition of Maltodextrin to Kakadu Plum Powder Samples by Two Dimensional and Near Infrared Spectroscopy. *Spectrochimica Acta Part A: Molecular and Biomolecular Spectroscopy* 247. https://doi.org/10.1016/j.saa.2020.119121

Díaz-Montes, E., and Castro-Muñoz, R. 2021. Edible Films and Coatings as Food-Quality Preservers: An Overview. *Foods* 10 (2): 249.

Díaz-Mula, H.M., Martínez-Romero, D., Castillo, S., Serrano, M., and Valero, D. 2011. Modified Atmosphere Packaging of Yellow and Purple Plum Cultivars. 1. Effect on Organoleptic Quality. *Postharvest Biology and Technology* 61 (2): 103–109.

DiNardo, A. 2018. *Investigation and Optimization of Methods for the Extraction of Antioxidants and Polyphenols from Yellow European Plums (Prunus Domestica L.)*. PhD diss.

DiNardo, A., Brar, H.S., Subramanian, J., and Singh, A. 2019. Optimization of Microwave-Assisted Extraction Parameters and Characterization of Phenolic Compounds in Yellow European Plums. *The Canadian Journal of Chemical Engineering* 97 (1): 256–267.

Di Renzo, C.G., Altieri, G., D'Erchia L., Lanza, G., and Strano, M.C. 2004. Effects of Gaseous Ozone Exposure on Cold Stored Orange Fruit. *V International Postharvest Symposium* 682: 1605–1610.

Elsayed, N., Hammad, K.S.M., and Abd El-Salam, E.A.E.S. 2020. Plum (*Prunus domestica* L.) Leaves Extract as a Natural Antioxidant: Extraction Process Optimization and Sunflower Oil Oxidative Stability Evaluation. *Journal of Food Processing and Preservation* 44 (10): e14813.

Enachi, E., Grigore-Gurgu, L., Aprodu, I., Stănciuc, N., Dalmadi, I., Bahrim, G., Râpeanu, G., and Croitoru, C. 2018. Extraction, Purification and Processing Stability of Peroxidase from Plums (*Prunus domestica*). *International Journal of Food Properties* 21 (1): 2744–2757.

Esclapez, M.D., García-Pérez, J.V., Mulet, A., and Cárcel, J.A. 2011. Ultrasound-Assisted Extraction of Natural Products. *Food Engineering Reviews* 3 (2): 108–120.

Eum, H.L., Hwang, D.K., Linke, M., Lee, S.K., and Zude, M. 2009. Influence of Edible Coating on Quality of Plum (Prunus salicina Lindl. cv.'Sapphire'). *European Food Research and Technology* 229 (3): 427–434.

Evrendilek, G.A., Baysal, T.A.N.E.R., Icier, F.İ.L.İ.Z., Yildiz, H., Demirdoven, A., and Bozkurt, H. 2012. Processing of Fruits and Fruit Juices by Novel Electrotechnologies. *Food Engineering Reviews* 4 (1): 68–87.

Fanning, K.J., Topp, B., Russell, D., Stanley, R., and Netzel, M. 2014. Japanese Plums (Prunus salicina L.) and Phytochemicals—Breeding, Horticultural Practice, Postharvest Storage, Processing and Bioactivity. *Journal of the Science of Food and Agriculture* 94 (11): 2137–2147.

Fawole, O.A., Riva, S.C., and Opara, U.L. 2020. Efficacy of Edible Coatings in Alleviating Shrivel and Maintaining Quality of Japanese Plum (Prunus salicina Lindl.) During Export and Shelf Life Conditions. *Agronomy* 10 (7): 1023.

Formiga, A.S., Junior, J.S.P., Pereira, E.M., Cordeiro, I.N.F., and Mattiuz, B.-H. 2019. Use of Edible Coatings Based on Hydroxypropyl Methylcellulose and Beeswax in the Conservation of Red Guava "Pedro Sato." *Food Chemistry* 290: 144–151. https://doi.org/10.1016/j.foodchem.2019.03.142

García-Parra, J., González-Cebrino, F., Cava, R., and Ramírez, R. 2014. Effect of a Different High Pressure Thermal Processing Compared to a Traditional Thermal Treatment on a Red Flesh and Peel Plum Purée. *Innovative Food Science & Emerging Technologies* 26: 26–33.

García-Parra, J., González-Cebrino, F., Delgado-Adámez, J., Cava, R Martín-Belloso, O., Élez-Martínez, P., and Ramírez, R. 2018. Effect of High-Hydrostatic Pressure and Moderate-Intensity Pulsed Electric Field on Plum. *Food Science and Technology International* 24 (2): 145–160.

González-Cebrino, F., Durán, R., Delgado-Adámez, J., Contador, R., and Ramírez, R. 2013. Changes After High-Pressure Processing on Physicochemical Parameters, Bioactive Compounds, and Polyphenol Oxidase Activity of Red Flesh and Peel Plum Purée. *Innovative Food Science & Emerging Technologies* 20: 34–41.

Gupta, R., Balasubramaniam, V.M., Schwartz, S.J., and Francis, D.M. 2010. Storage Stability of Lycopene in Tomato Juice Subjected to Combined Pressure– Heat Treatments. *Journal of Agricultural and Food Chemistry* 58 (14): 8305–8313.

Haddadi-Guemghar, H., Janel, N., Dairou, J., Remini, H., and Madani, K. 2014. Optimisation of Microwave-Assisted Extraction of Prune (Prunus domestica) Antioxidants by Response Surface Methodology. *International Journal of Food Science & Technology* 49 (10): 2158–2166.

Hussain, P.R., Dar, M.A., and Wani, A.M. 2013. Impact of Radiation Processing on Quality During Storage and Post-Refrigeration Decay of Plum (Prunus domestica L.) cv. Santaroza. *Radiation Physics and Chemistry* 85: 234–242.

Kang, K.Y., Hwang, Y.H., Lee, S.J., Jang, H.Y., Hong, S.G., Mun, S.K., and Yee, S.T. 2019. Verification of the Functional Antioxidant Activity and Antimelanogenic Properties of Extracts of Poria cocos Mycelium Fermented with Freeze-Dried Plum Powder. *International Journal of Biomaterials*. doi:10.1155/2019/9283207.

Karaca, H., Pérez-Gago, M.B., Taberner, V., and Palou, L. 2014. Evaluating Food Additives as Antifungal Agents Against Monilinia Fructicola in Vitro and in Hydroxypropyl Methylcellulose—Lipid Composite Edible Coatings for Plums. *International Journal of Food Microbiology* 179: 72–79.

Khan, A.S., Singh, Z., Abbasi, N.A., and Swinny, E.E. 2008. Pre-or Post-Harvest Applications of Putrescine and Low Temperature Storage Affect Fruit Ripening and Quality of 'Angelino' Plum. *Journal of the Science of Food and Agriculture* 88 (10): 1686–1695.

Khan, M. 2019. Optimization of Extraction Condition and Characterization of Low Methoxy Pectin from Wild Plum. *Journal of Packaging Technology and Research* 3 (3): 215–221.

Knorr, D., Froehling, A., Jaeger, H., Reineke, K., Schlueter, O., and Schoessler, K. 2011. Emerging Technologies in Food Processing. *Annual Review of Food Science and Technology* 2: 203–235.

Kumar, P., Sethi, S., Sharma, R.R., Srivastav, M., Singh, D., and Varghese, E. 2018. Edible Coatings Influence the Cold-Storage Life and Quality of 'Santa Rosa' Plum (Prunus salicina Lindell). *Journal of Food Science and Technology* 55 (6): 2344–2350.

Manganaris, G.A., Vicente, A.R., and Crisosto, C.H. 2008. Effect of Pre-Harvest and Post-Harvest Conditions and Treatments on Plum Fruit Quality. *CAB Reviews: Perspectives in Agriculture, Veterinary Science, Nutrition, and Natural Resources* 3 (9): 1–9.

Medina-Meza, I.G., and Barbosa-Cánovas, G.V. 2015. Assisted Extraction of Bioactive Compounds from Plum and Grape Peels by Ultrasonics and Pulsed Electric Fields. *Journal of Food Engineering* 1 (166): 268–275.

Michalska, A., Wojdyło, A., Łysiak, G.P., and Figiel, A. 2017. Chemical Composition and Antioxidant Properties of Powders Obtained from Different Plum Juice Formulations. *International Journal of Molecular Sciences* 18 (1): 176.

Nencetti, V., Peano, C., Palara, U., Pirazzini, P., Mezzetti, B., Capocasa, F., and Catalano, L. 2008. Plum Production in Italy: State of the Art and Perspectives. *IX International Symposium on Plum and Prune Genetics, Breeding and Pomology* 874: 25–34.

Orsat, V., and Routray, W. 2012. Microwave-Assisted Extraction of Flavonoids: A Review. *Food Bioprocess Technology* 5: 409–424.

Panahirad, S., Naghshiband-Hassani, R., Bergin, S., Katam, R., and Mahna, N. 2020a. Improvement of Postharvest Quality of Plum (Prunus domestica L.) Using Polysaccharide-Based Edible Coatings. *Plants* 9 (9): 1148.

Panahirad, S., Naghshiband-Hassani, R., Ghanbarzadeh, B., Zaare-Nahandi, F., and Mahna, N. 2019. Shelf Life Quality of Plum Fruits (Prunus domestica L.) Improves with Carboxy methylcellulose-Based Edible Coating. *HortScience* 54 (3): 505–510.

Panahirad, S., Naghshiband-Hassani, R., and Mahna, N. 2020b. Pectin-based Edible Coating Preserves Antioxidative Capacity of Plum Fruit During Shelf Life. *Food Science and Technology International* 26 (7): 583–592.

Pansare, S.K., and Patel, S.M. 2019. Lyophilization Process Design and Development: A Single-Step Drying Approach. *Journal of Pharmaceutical Sciences* 108 (4): 1423–1433.

Peano, C., Girgenti, V., Sottile, F., and Giuggioli, N.R. 2009. Improvement of Plum Storage with Modified Atmosphere Packaging. *X International Controlled and Modified Atmosphere Research Conference* 876: 183–188.

Putnik, P., Lorenzo, J.M., Barba, F.J., Roohinejad, S., Jambrak, A.R., Granato, D., Montesano, D., and Kovačević, D.B. 2018. Novel Food Processing and Extraction Technologies of High-Added Value Compounds from Plant Materials. *Foods* 7 (7): 106.

Raheem, D. 2013. Application of Plastics and Paper as Food Packaging Materials-An Overview. *Emirates Journal of Food and Agriculture* 177–188.

Rastogi, N.K., Raghavarao, K.S.M.S., Balasubramaniam, V.M., Niranjan, K., and Knorr, D. 2007. Opportunities and Challenges in High Pressure Processing of Foods. *Critical Reviews in Food Science and Nutrition* 47 (1): 69–112.

Rodrigues, S., and Fernandes, F.A.N., eds. 2012. *Advances in Fruit Processing Technologies*. CRC Press.

Routray, W., Orsat, V., Ramawat, K., and Mérillon, J.M. 2013. Preparative Extraction and Separation of Phenolic Compounds. In *Natural Products*, 1st ed., 2013–2045. Springer.

Salehi, F. 2020. Edible Coating of Fruits and Vegetables Using Natural Gums: A Review. *International Journal of Fruit Science* 20 (Sup 2): S570–S589.

Salem, A.E., Naweto, M., Mostafa, M., and Authority, A.E. 2016. Combined Effect of Gamma Irradiation and Chitosan Coating on Physical and Chemical Properties of Plum Fruits. *Journal of Nuclear Technology in Applied Science* 4: 91–102.

Sapper, M., and Chiralt, A. 2018. Starch-Based Coatings for Preservation of Fruits and Vegetables. *Coatings* 8 (5): 152.

Savic, I.M., and Gajic, I.M.S. 2021. Optimization Study on Extraction of Antioxidants from Plum Seeds (Prunus domestica L.). *Optimization and Engineering* 22 (1), 141–158.

Seke, F., Manhivi, V.E., Shoko, T., Slabbert, R.M., Sultanbawa, Y., and Sivakumar, D. 2021. Effect of Freeze Drying and Simulated Gastrointestinal Digestion on Phenolic Metabolites and Antioxidant Property of the Natal Plum (Carissa macrocarpa). *Foods* 10 (6): 1420.

Sifat, S.A., Trisha, A.T., Huda, N., Zaman, W., and Julmohammad, N. 2021. Response Surface Approach to Optimize the Conditions of Foam Mat Drying of Plum in Relation to the Physical-Chemical and Antioxidant Properties of Plum Powder. *International Journal of Food Science*. Doi:10.1155/2021/3681807.

Silvan, J.M., Michalska-Ciechanowska, A., and Martinez-Rodriguez, A.J. 2020. Modulation of Antibacterial, Antioxidant, and Anti-Inflammatory Properties by Drying of Prunus Domestica L. Plum Juice Extracts. *Microorganisms* 8 (1): 119.

Singh, A., and Orsat, V. 2014. Key Considerations in the Selection of Ingredients and Processing Technologies for Functional Foods and Nutraceutical Products. *Nutraceutical and Functional Food Processing Technology* 79–111. doi:10.1002/9781118504956.ch3.

Sharmin, M.R., Islam, M.N., and Alim, M.A. 2015. Shelf-Life Enhancement of Papaya with Aloe Vera Gel Coating at Ambient Temperature. *Journal of the Bangladesh Agricultural University* 13 (1): 131–136.

Sohail, M., Afridi, S.R., Khan, R.U., Ullah, F., and Mehreen, B. 2014. Combined Effect of Edible Coating and Packaging Materials on Post-Harvest Storage Life of Plum Fruits. *ARPN Journal of Agriculture and Biological Sciences* 9: 134–138.

Steffens, C.A., Vidal Talamini Do Amarante, C., De, E., Alves, O., Brackmann, A., Corrêa, T.R., and Pansera Espindola, B. 2014. Storage of 'Laetitia' Plums (Prunus salicina) under Controlled Atmosphere Conditions. *African Journal of Biotechnology* 13 (32): 3239–3243. https://doi.org/10.5897/AJB2014.13845

Sunmathi, M.C.D., Sudhakaran, N.K.S., and Arungandhi, K. 2019. Preservation Techniques of Plums—A Review. *International Journal for Innovative Research in Science & Technology* 6 (7): 9–12.

Terefe, N.S., Buckow, R., and Versteeg, C. 2014. Quality-Related Enzymes in Fruit and Vegetable Products: Effects of Novel Food Processing Technologies, Part 1: High-Pressure Processing. *Critical Reviews in Food Science and Nutrition* 54 (1): 24–63.

Thakur, R., Pristijono, P., Golding, J.B., Stathopoulos, C.E., Scarlett, C.J., Bowyer, M., and Vuong, Q.V. 2018. Development and Application of Rice Starch Based Edible Coating to Improve the Postharvest Storage Potential and Quality of Plum Fruit (Prunus salicina). *Scientia Horticulturae* 237: 59–66.

Tzortzakis, N. 2016. Ozone: A Powerful Tool for the Fresh Produce Preservation. In *Postharvest Management Approaches for Maintaining Quality of Fresh Produce*, 175–207. Springer.

Tzortzakis, N., and Chrysargyris, A. 2017. Postharvest Ozone Application for the Preservation of Fruits and Vegetables. *Food Reviews International* 33 (3): 270–315.

Valero, D., Díaz-Mula, H.M., Zapata, P.J., Guillén, F., Martínez-Romero, D., Castillo, S., and Serrano, M. 2013. Effects of Alginate Edible Coating on Preserving Fruit Quality in Four Plum Cultivars During Postharvest Storage. *Postharvest Biology and Technology* 77: 1–6.

Vangdal, E., Hagen, S., and Bengtsson, G. 2009. Effect of Postharvest UV-B Radiation and Visible Light on Health and Sensory Related Parameters in Plums (Prunus domestica L.). *VI International Postharvest Symposium* 877: 1325–1328.

Vladic, J., Gavaric, A., Jokic, S., Pavlovic, N., Moslavac, T., Popovic, L., Matias, A., Agostinho, A., Banozic, M., and Vidovic, S. 2020. Alternative to Conventional Edible Oil Sources: Cold Pressing and Supercritical CO2 Extraction of Plum (Prunus Domestica L.) Kernel Seed. *Acta Chimica Slovenica* 67 (3): 778–784.

11 Utilisation of Plum Peels and Seeds

Jessica Pandohee[1], *Jadala Shankaraswamy*[2], *Mohd Aaqib Sheikh*[3] *and Nisar A. Mir*[4]

[1] Centre for Crop and Disease Management, School of Molecular and Life Sciences, Curtin University, Bentley, WA 6102, Australia

[2] Department of Fruit Science, College of Horticulture, Sri Konda Laxman Telangana State Horticultural University, Wanaparthy, Telangana, 509382, India

[3] Department of Food Engineering and Technology, Sant Longowal Institute of Engineering & Technology, Longowal, Punjab, 148106, India

[4] Department of Biotechnology Engineering and Food Technology, University Institute of Engineering, Chandigarh University, Mohali, 140413, Punjab, India

CONTENTS

DOI: 10.1201/9781003205449-11

11.1 INTRODUCTION

The domestic plum (*Prunus domestica*) is a stone fruit belonging to the Rosaceae family. It is commonly consumed as a snack in its raw form or used as an ingredient in the manufacture of prunes (dried plums), jams and juices. During the production of plum products, its seed, peel and sometimes pulp form a large part of the waste products. Considering that over 11.7 million tons of plum are produced annually, a large amount of agri-food waste from plum is sent to landfills, where it causes pollution (Wojdyo et al. 2021).

Phytochemicals with recognized health advantages can be found in the seed and peel of plum fruit. As a result, the use of the seeds and peels may have additional uses in the food industry while simultaneously reducing pollution. Through the use of plum agri-waste, several UN Sustainable Goals can be achieved. For example, extracting bioactive compounds, such as phytochemical and phenolic compounds, from plum peels, seeds and skin can provide a bioactive ingredient for supplementation, therefore improving nutrition and promoting sustainable agriculture (Sustainable Goals 2 and 3).

Agri-food waste that does not go to landfills causes less pollution and ensures that our cities and settlements are sustainable and free of contamination (Sustainable Goal 11). Therefore, the use of agri-food waste from plums to make energy ensures that part of our energy requirement is being met through the constant availability of plum seeds and peels, therefore ensuring access to reliable and sustainable energy. This chapter presents the utilization of plum seeds and peels to provide value-added products and services.

11.2 PLUM-BASED PRODUCTS

11.2.1 Dried Plums

Dried fruits are considered an efficient source of energy. Besides their high caloric content, they are also rich in nutrients and minerals. Plum is an important source of polyphenols, including caffeic acid, chlorogenic acid, flavonol and their derivatives (Balasundram et al. 2006). Dried prunes are an abundant source of dietary fibre, particularly the polysaccharide pectin, which gives a feeling of satiety when consumed. Additionally, dried prunes give a mild laxative effect and help in lowering of blood cholesterol levels.

To prepare dried plum, plum fruits are harvested at the fully mature stage, when the total soluble solids (TSS) content in the fruit is a minimum of about 22%. Originally, dried plums were prepared by sun drying, but with the invention of new technologies, such as dehydrators, the drying process is now automated and more controlled. However, the cheapest method of preparation is sun drying, and it is still used by most plum-producing developing countries. A disadvantage of sun drying is that it may result in contamination of the fruit from the environment, which may lead to the production of unhygienic, poor tasting fruit. The moisture content during drying is lowered to below 18% to reduce water activity and prevent microbiological deterioration. The most common dehydrators are forced-air dehydrators, with drying times ranging from 24 to 36 hours. Temperature during the drying period is kept between 62°C and 74°C. After the drying period and moisture equilibration, the dried plums are packed and stored in a cool and dry place.

11.2.2 Plum Leather

Leather is a fruit-based confectionary product prepared from puréed pulp. It can be eaten as a snack or dessert. The texture of any fruit-based leather is soft and rubbery. To prepare fruit leather, the pulp of the fruit is puréed first, and then the ingredients are mixed to improve its sensory and other physicochemical properties. Fruit leather is made into sheets or rolls by drying puréed pulp of a fruit after mixing the necessary ingredients. The prepared fruit purée is either sun dried or oven dried (Madhav 2016).

11.2.3 Plum Jam and Jelly

Jams and jellies are homogenous processed products prepared from fresh, frozen or semi-processed fruits in addition to sugar, pectin and acid. Jams and jellies prepared from plums have not gained much commercial importance, but plums are usually mixed into other fruit jams that are deficient in phenolic components. Kim and Padilla-Zakour (2004) evaluated the influence of processing on the anthocyanin content, antioxidant capacity, and total phenolic content of several fresh fruits, concluding that while phenolic content is unaffected by processing, such as jam creation, antioxidant activity is reduced. Plums could be used with other fruits which are low in phenolic content for the preparation of jams and jellies. Viktorija et al. (2013) prepared jams from the Stanley plum, using various sweeteners like fructose, sorbitol and agave syrup instead of sucrose. The jams prepared were considered acceptable and microbiologically safe. The sensorial analysis of the jams showed that jam prepared from plum using fructose was best in terms of taste, smell, aroma and consistency.

11.2.4 Fresh-Cut Plums

For production of fresh-cut plums, the fruit is washed, trimmed, peeled and then cut into a usable form, or we can say that the physical appearance of the plum is altered without affecting its freshness. The product is then packaged properly to be supplied to the consumers. These products offer convenience, value and nutrients to the consumers without affecting the fresh state of the product. It has been shown that the quality of fresh-cut plum products remain good for about five days when refrigerated. The shelf life of these products depends on the variety as well as on the stage of ripeness. Cutting plums and applying hormones, including ethylene and methyl jasmonate, has been found to be a long-lasting strategy to improve the health-promoting antioxidant properties of these fresh foods (Cisneros-Zevallos and Heredia 2004). Fresh-cut products are 100% usable. Although the market for fresh-cut produce is developing, a very small percentage of the plum is marketed in this form because of volatile market conditions.

11.2.5 Plum Paste

Plum paste can be used in the preparation of many food formulations. Researchers have been pushed to manufacture a range of plum products, including plum paste, as a result of studies into the possible health advantages of polyphenols and dietary fibre. For the preparation of plum paste, fresh or frozen fruits are used. Frozen fruits should be thawed overnight at a temperature of 4°C. Plums taken are heated to a temperature of 95°C and then macerated in a steam-jacketed kettle for about 10 minutes. After maceration and cooling, the pits are filtered. Plum purée thus obtained is concentrated in a steam-jacketed kettle by continuous stirring at a temperature of 85–90°C to increase the TSS content between 25°C and 30°C. The packaged end product should be cooled to ambient room temperature prior to storage in dark conditions at 4°C (Wang et al. 1995).

11.2.6 Plum Juice

For the preparation of plum juice, the fruits are collected at a mature and firm stage of development. The fruits are stored at 4°C until further processing. Plum juice preparation by enzymatic hydrolysis involves the removal of pits from the pulp part of the fruit. The puréed pulp is then heated to about 90°C to inactivate the browning-causing enzymes, such as polyphenols oxidase. The fruit purée is then cooled to 50°C and subjected to enzymatic hydrolysis for a period of about two hours using food-grade enzymes, like multi-cellulase complex, pectinase, hemicellulase, arabinase, cellulose, pectin esterase, amongst others. When the enzymatic hydrolysis is over, the clarified fruit juice is obtained by the process of centrifugation at about 5000 rpm for five minutes. Apart from the products above discussed, plum can be formed into many other products, such as canned plum, plum powder, plum marmalade, amongst others. Different proportions of plum powder can be mixed with other ingredients for the preparation of products like chutney mixes and ready-to-serve drink mixes.

11.3 BY-PRODUCTS OF PLUM-PROCESSING INDUSTRIES

To improve the sustainability of the food chain, the focus of researchers has increased towards the valorization of by-products produced from the processing of agricultural commodities. Among the agricultural produce, fruits and vegetables play a very important role in the human diet. Dietary habits are constantly evolving, and nowadays there is an increased demand for fruits and vegetables in order to consume a diverse range of phytochemicals (Vilariño et al. 2017).

The active agents present in by-products like organic acids, polyphenols, flavonoids, vitamins and minerals, amongst others, help in the management of various degenerative diseases in humans. Unfortunately, the by-products obtained from horticultural products like fruits and vegetables and their bioactive content have been ignored until now, when we have seen an increase in processing to recover highly valuable biomolecules from what was previously waste materials. During the processing of plum into foodstuffs such as alcoholic beverages, juices, jams, jellies, nectars and dried fruit, amongst others, to provide an alternative to fresh consumption, a large amount of residue in the form of peels, pomace and stones, which comprises about 10–25% of the raw material, are generated (González-García et al. 2014).

A large amount of by-products are produced during the processing of plums on a large scale that in turn will lead to a loss of nutrients and energy and cause a number of environmental problems, like pollution of water and soil (Damiani et al. 2012). Although the food-processing industries have developed significantly during the past few decades, the losses and wastes during processing are hard for researchers to ignore. The problem related to the generation of by-products has attracted scientific, political, economic as well as commercial interest and thus has forced the European Union towards becoming a zero-waste economy by the year 2025. Moreover, waste produced in the plum industry can be used as a source of energy and thus has the potential to be used as a substitute for the substrates used as conventional sources of energy.

The utilization of industrial waste also helps to reduce costs. As the wastes produced are additional challenges, more sustainable methods should be developed for their appropriate re-use or disposal (Shalini and Gupta 2010). Plum production is increasing throughout the world, and as a result the amount of by-product from the fruit is also increasing. These by-products, such as peel and pomace, are rich in dietary fibre and are usually either discarded or fed to animals. These by-products are a valuable source of health-promoting compounds, thus the focus of researchers is to extract the benefits from them. To convert the by-products into value-added products, it is necessary to have a complete knowledge of the components responsible for their quality, such as pigment, pectin content and carbohydrates.

11.3.1 Pomace

Pomace is a plum by-product that is normally discarded, transported to landfills or fed to animals. Increased interest in consuming cloudy plum juice has resulted in pomace from plum to be re-used and re-added to the juice prior to packaging. This form of plum product is recognized as being high in dietary fibre, polyphenols, pigments, and other beneficial health-promoting ingredients. This by-product has proved to be an efficient source of pectin that acts as a thickening, gelling and stabilizing agent in many food products (Gil et al. 2002). Plum pomace is a low-cost source of dietary fibre (64.5%), with the amount of dietary fibre varying depending on the method of juice extraction and the kind of fruit.

The growing interest in the utilization of plum pomace is because of its suitable functionality owing to the presence of a balanced percentage of soluble and insoluble dietary fibre with proven antioxidant properties. Soluble dietary fibre provides a range of health benefits to humans, including lowering the risk of cardiovascular disease and cholesterol levels. Dietary fibre that is insoluble reduces the risk of colorectal cancer. Rather than being viewed as a waste, the pomace collected should be viewed as a raw material for the production of a variety of healthier supplements, including dietary fibres, bicolor and bioactive chemicals with demonstrated antioxidant potential (Milala et al. 2013).

The consumption of dietary fibre and an individual's weight increase have an inverse connection Dwivedi et al. (2014). Plum pomace is a prospective source of pigments, particularly anthocyanins, with high economic potential, according to reports, and might be used to produce bicolour for the food industry. Dietary fibre, according to Elleuch et al. (2011), improves the physical, structural, textural and sensory aspects of a variety of food preparations. Sahni and Shere (2018) reported that the incorporation of pomace in bakery products improves the sensorial attributes and storage quality due to its associated antioxidants, polyphenols and dietary fibres. Plum pomace, being a cheap source of dietary fibre, could be used for designing new functional foods and could allow for effective waste management for food industries.

11.3.2 Plum Kernels

The amount of stone remaining after plum processing is quite large. Plum pits are by-products of the plum canning industry. The plum stones left after processing are thrown out as waste. The use of plum pits could be extremely beneficial to both the food-processing industry and the environment (Kostic et al. 2016). Plum kernels are found inside the pits of the fruit. *Prunus domestica* is not a traditional staple or commonly used ingredient in routine cuisine, and when dry and mature it can be regarded as a non-traditional source of nutrients. Plum kernels are high in lipids, proteins, vitamins, minerals and fibre, amongst other nutrients. As a result, they represent a low-cost source of chemicals that could be used to make a variety of value-added products (Górna et al. 2015).

Plum kernels contain a large amount of dietary proteins, oils, fibres, minerals and phenolic and cyanogenic compounds (Dulf et al. 2016). Of those, the cyanogenic glycoside (amygdalin) is very distinctive and determines the healthy function of the plum kernel to some extent (Zhang et al. 2019). The characteristics of plum seed oil are similar to that of apricot kernel oil. In some cases, fruit stones have been employed in the manufacturing of cosmetics. The use of plum kernels as a source of protein, polyphenols and oil with a high content of unsaturated fatty acids could be advantageous.

11.3.2.1 Chemical Composition of Plum Kernels

Within the plum species, varietal diversity produces in variances in compositional features. Plum kernels have a higher concentration of oils, dietary proteins, fibres, vitamins, minerals and bioactive substances, making them a good source of quick energy (Górnas et al. 2017).

11.3.2.2 Oils

Regardless of species and variety, plum kernels contain a considerable amount of oils. The oil content ranges from 36.5% to 50%, which could be extracted and used for consumption. Górnas et al. (2017) found that plum kernels produce oils rich in unsaturated fatty acids, phytosterols and lipophilic antioxidants, as well as imparting distinct naturally nutty tastes. Oleic acid, linoleic acid, palmitic acid, stearic acid, palmitoleic acid and eicosanoic acid are the most important fatty acids (Kamel and Kakuda 1992; Kostic et al. 2016; Górnas et al. 2017). In addition to a high quantity of unsaturated fatty acids, the larger amounts of oleic and linoleic acids found in plum kernel oil indicate that it is stable and resistant to rancidity. The presence of these acids indicate that plum kernel oil may also be ideal for food oils, cosmetic preparation, moisturizing creams and biodiesel manufacturing. Plum kernels have been shown to have high levels of essential fatty acids (linoleic acid), which are necessary for human metabolism but cannot be produced by the body. It is commonly known that meals high in unsaturated fatty acids aid in the management of a number of physiological and biological activities, as well as the prevention of degenerative diseases, such as cardiovascular disease, atherosclerosis and inflammatory disorders (Abd Aal et al. 1987; Zhao et al. 2007). Because of the large levels of unsaturated fatty acids, exceptional stability and antioxidant components, plum kernel oil is nutritionally appealing and regarded as a significant source of manufacturing high-value goods for food and nutraceutical supplements.

11.3.2.3 Proteins

Plant proteins are now recognized as significant sources of bioactive chemicals and are gaining popularity as a result of their digestibility and nutritional value (Garcia et al. 2013). Demand for relatively cheap sources of protein is increasing globally, and plum kernels contain a substantial amount of dietary proteins that are not usually used and characterized. Plum kernel meal has a high protein content, ranging from 28 g/kg to 41.3 g/kg (Gonzalez-Garcia et al. 2014; Xue et al. 2018). The extraction of protein from plum kernels using high-intensity focused ultrasounds yielded 40% (dry basis) in less than an hour. The plum kernels presented a relatively full array of essential amino acids, which indicates that the kernels could play a vital role in health maintenance and as a supplement designed to maintain muscle mass. Plum kernel proteins might also be an attractive source of bioactive peptides that have a positive influence on body function (Gonzalez-Garcia et al. 2014).

11.3.2.4 Amygdalin

Many plants are able to synthesize cyanogenic glycosides, which is related to bitterness. Cyanogenic glycosides are water-soluble, heat-stable secondary metabolites. Cyanogenic glycosides are not toxic when intact, but they become toxic when enzymes (endogenous or exogenous, such as β-glycosidase and α-hydroxynitrilel yases) come in contact with them, producing hydrogen cyanide, which causes tissue damage after bruising or chewing. The predominant cyanogenic glycoside in Rosaceae species is amygdalin ($C_{20}H_{27}NO_{11}$), and plum kernels contain a relatively high amount (0.1–17.5%) (Gonzalez-Garcia et al. 2014). The plant toxicant amygdalin (D-mandelonitrile-β-D- gentiobioside), or vitamin B17, also known as nitrilosides, is generated from phenylalanine and consists of two glucose molecules, one hydrogen cyanide molecule and one benzaldehyde molecule (Zhang et al. 2019).

Amygdalin is classified as a cyanogenic glycoside because it contains a cyanide group between a glycoside and a benzene ring that can be released after hydrolysis, and each molecule includes a nitrile group, which releases hydrogen cyanide (Cho et al. 2006). Depending on the variety of plum and environment conditions, plum seeds contain different levels of amygdalin. For example, green, black, purple, yellow and red plum kernels contain 17.5, 10, 2.16, 1.54 and 0.44 mg/g of amygdalin respectively (Gonzalez-Garcia et al. 2014). Consumption of cyanogenic seeds by humans can be harmful since they can be hydrolyzed to form hydrogen cyanide, resulting in sub-acute cyanide poisoning with symptoms such as anxiety, headaches, dizziness and confusion (Shi et al. 2019).

The anti-fibrosis, anti-inflammation analgesia, auxiliary anti-cancer, immune-regulation, anti-atherosclerosis and potential to treat neurodegenerative disease advantages of amygdalin have been thoroughly documented throughout the years (He et al. 2020). However, while amygdalin is innocuous in and of itself, its metabolic product, hydrocyanic acid, when degraded in the presence of an enzyme (-glucosidases or -hydroxynitrile lyases), has the potential to be poisonous. To be more specific, a low dose is beneficial to one's health, whereas a high dose is toxic and can cause symptoms similar to cyanide poisoning, such as headache, nausea, vomiting and dizziness (Shi et al. 2019).

To reduce cyanide toxicity, processing methods such as roasting, soaking, peeling, crushing, fermenting, grating, grinding, boiling and drying have been employed for years to reduce cyanide concentration by removing water-soluble glycosides before ingestion (Gonzalez-Garcia et al. 2014). To minimize the potential for toxicity to tolerable levels, traditional procedures, such as soaking, autoclaving, fermenting and boiling, have been used (Bolarinwa et al. 2016). When kernels are macerated or crushed, hazardous chemicals are degraded by enzymes, resulting in the generation of hydrogen cyanide. During grinding and consecutive soaking, -glycosidase activity causes acceptable amygdalin breakdown (Tuncel et al. 1990). In general, finer particles result in faster glycoside breakdown. After 30 minutes of soaking, finely powdered seeds contain no glycosides. Soaking reduced amygdalin but had no effect on limiting amino acids (El-Adawy and Kadousy 1995).

For cyanogenic glycoside reduction, heat treatments such as autoclaving, extrusion and microwave roasting are commonly utilized (Wu et al. 2008). Microwave heating has recently emerged as the most versatile option for eliminating anti-nutritional elements. After soaking for one hour, microwave heating at 500 W for nine minutes is enough to completely degrade the hazardous chemical (Ahmed et al. 2015). In several common beans, microwave cooking efficiently reduces anti-nutrients (tannins, phytic acid and trypsin inhibitor activity) while also improving protein quality (Shimelis and Rakshit 2005). A reduction in hydrogen cyanide levels under permissible limits (230 mg/kg of linseed) was achieved using a microwave power of 400 W for 4 minutes and 50 seconds (Ivanov et al. 2012). Beans have been heated in a variety of ways, including microwaving, to boost their nutritional content (Hernande-Infante et al. 1998).

11.4 TRANSFORMING FOOD WASTE INTO VALUABLE RESOURCES

There has been increasing concern about environmental degradation as a result of the enormous amounts of industrial post-production waste coming from plum fruits in recent years. As a result, the eventual disposal cost of wastes coming from agriculture, food manufacturing and even households has risen. Plum fruits that do not match the fresh market's specifications are transferred to enterprises that produce organic waste, which accounts for around 60% of all processed fruit. Given the environmental risks associated with the accumulation of high-organic-content fruit waste solid residue after processing, it is interesting to investigate plum trash as a viable bioresource for producing energy. Effective and sustainable use of biowaste as essential bioresources for slow pyrolysis and gasification of trash to produce fuel and biochar generation using sustainable approaches to achieve a near-zero-waste scenario. The fundamental issue is the wide range of waste generated by different types of production in different regions, as well as different substrate specifications (Mishra et al. 2011). However, there are a number of strategies that can help minimize not only the quantity of waste from plum fruit but also the entire expense of running a business (Lipiński et al. 2018).

11.5 ENERGY RECOVERY FROM FOOD WASTE

11.5.1 Post-Processing Waste of Plum for Bioenergy Resource

Fruit processing represents a substantial amount of solid waste, which is either thrown in landfills or rivers, causing environmental problems, or recycled to a lesser extent through livestock as food supplies. The plum fruit's involvement in the production and processing of the fruit differs by country. The majority of stone fruits are processed in processing plants, implying that a significant amount of waste in the form of stones is used as a source material for energy production.

The majority of the machinery used in fruit-processing units are powered by technical steam, which means the average annual quantity of coal burned is significant, and the biggest load is in aseptic packing, as well as powering offices and social areas and lighting the entire production plant. In this regard, efficient exploitation of waste from plum processing to meet required energy demand in processing plants can help reduce external energy demand.

Many countries have taken steps to boost the use of cleaner, more reliable energy sources to lessen the rising level of energy demand and the problems that come with it (Sirin and Ege 2012). The use of plum fruit wastes for heat and energy production reduces the quantity of waste that ends up in landfills, slows the depletion of fossil resources, and thereby mitigates a variety of climate change effects.

11.5.2 Contribution of Plum Towards Bioenergy Resources

Plum stone (Kernel+shell) is a waste and by-product of the plum-processing industry, accounting for 2.8% of the total weight of the fruit (Kamel and Kakuda 1992).

Bioactive peptides (González-Garca et al. 2014), active carbons (Nowicki et al. 2010b) and carbonaceous adsorbents (Kostic' et al. 2016) are all found in the solid outer shell (Nowicki et al. 2010a, 2010b). Unconventional oils have gotten a lot of press in recent years due to their useful properties in a range of industries (Górnas and Rudzinska 2016; Górnas et al. 2013).

11.5.3 Characteristics of Biomass Derived from the Plum

Several features of biomass should be known before deciding on the conversion technique. When selecting plum biomass, the calorific value, ash content or volatile matter content and chemical component composition are all crucial factors to consider (Ramosa et al. 2008). Plum stones contain 51.3% carbon, with a heat of combustion of 21.12 $MJ.kg^{-1}$ and have a low ash content (0.9%). As a result, detailed research on the optimization of pellet manufacturing and their subsequent usage in thermal/thermochemical technologies are required to manage waste in the form of plum stones for heating reasons (Cagnon et al. 2009).

11.5.4 Factors Affecting Bioenergy Production from Plum Biomass

11.5.4.1 Levels of Carbon, Hydrogen, Oxygen, Nitrogen and Sulphur

Solid fuels are mostly made up of carbon, nitrogen and oxygen. During combustion, carbon and oxygen interact in an exothermic reaction to produce carbon dioxide and water, contributing to the higher heating value of the fuel and the combustion process (Obernberger and Thek 2004). Carbon is one of the most significant ingredients in the combustion process, and biomass with a high carbon content has a higher heating value (Obernberger and Thek 2004). Prune biomass has 48–52% carbon, while plum fruit stones contain 51–55% carbon and fruit pits contain 52.11–55.29% carbon. Because hydrogen, along with carbon, is essential for defining the energy properties of solid biofuels, a deficit could be a concern (Obernberger and Thek 2004). Trimmed biomass has 6.07–6.74% hydrogen, while unpruned biomass contains 6.21–6.94% hydrogen (stone).

Sulphur is the least abundant gas in biomass, but it is the most important ingredient in terms of environmental impact, along with nitrogen. Research was conducted to determine nitrogen and sulphur levels in trimmed wood (0.74–0.91% and 0.16–0.18% respectively) and plum stone (0.36–1.60% and 0.03–0.04% respectively). Nitrogen and sulphur maximum permitted values for trimmed biomass are 0.1–0.5% and 0.01–0.05% respectively and for fruit stone are 0.2–0.3% and 0.05–0.5% respectively, based on the criteria employed in this research. Because nitrogen, along with sulphur, influences the emissions of hazardous gases (NOx and SO2) during biomass combustion (Sáez Angulo and Martnez Garca 2001; Garcia et al. 2012), these gases should have the lowest feasible concentration. During burning, sulphur in biomass is oxidised to sulphur dioxide (and in small amounts to sulphur [VI] SO3 oxide), generating alkali and sulphates (Obernberger et al. 2006). The emissions of oxides of sulphur from wood biomass burning are usually minimal. Low emission levels were also obtained for plum stone granulate combustion with the addition of rye bran (Zaj da et al. 2017).

11.5.4.2 Total Ash Content

Total ash content is one of the most studied aspects of biomass, and it is created during biomass combustion from inorganic, organic and aqueous components that are both natural and technogenic (Vassilev et al. 2010). Furthermore, ash is an undesirable component of biomass because of its catalytic effect on thermal breakdown; additionally, a higher ash concentration leads to greater carbon and gas concentrations. Because biomass ash has a low melting temperature, when it melts during the heating process, it generates "slag." Slag builds up in furnaces and boilers, obstructing energy transfer and lowering complete combustion (Hodgson et al. 2010). In pruned plum biomass, 6.6% of the ash content is present. Furthermore, the ash level of plum stone is 1.8%. The varying amount of mineral nutrients in soil explains the variability in ash content in plum biomass. Furthermore, the amount of ash in plum fruit biomass is affected by the climate conditions in places where the biomass is harvested. The ash content of biomass can sometimes be linked to the typical temperatures of the sites where the biomass was collected. This, however, should not be regarded as a rule of thumb.

11.5.4.3 Plum Biomass Treatment Strategies for Energy Production

Thermo-chemical operations, such as combustion, gasification and pyrolysis, are recommended for biomass with an ash concentration of less than 1%. (Kenney et al. 2013). Pyrolysis is a key technology for making renewable biofuels and chemicals from biomass. It produces solid, liquid and gaseous fuels by heating solid biomass in the oxygen-free environment. Pyrolytic runaway is a new and nearly uncontrollable conversion regime in this biomass conversion. This is associated with rapid heating of the entire sample, up to temperatures nearly 300K higher than a bench-scale pyrolyser's external heating temperature, and the nearly instantaneous emission of a whole volatile matter content. A packed-bed bench-scale device is used to conduct pyrolysis research. The particles in the packed-bed pyrolysis samples were created to maintain the characteristic thickness of the residue, ranging from 0.5 mm to 4 mm with widths of 2–6 mm.

To preserve the material microstructure, pulverisation to prepare pellets of equivalent density was avoided in this method. As a result, the material determines the sample mass. Different conditions that favour the development of secondary pyrolysis increase the size of the reaction exothermicity: a) microstructures with a low porosity, b) extraordinary particle thicknesses that preserve the original material structure, c) considerable alkali chemical concentration and d) molecules of organic water vapour with high intra-particle pressures. Following conversion, there is likely to be a high level of structural stability. Despite their comparable physical attributes and microstructure, particle size and chemical characteristics are important determinants for the worldwide exothermicity of fruit nut shell and stone pyrolysis from a quantitative standpoint (Di Blasi et al. 2017).

Inorganic impurities present in gasification-generating gas includes sulphur compounds, nitrogen compounds, alkali metals (mainly potassium as well as sodium) and hydrogen chloride. Each pollutant has its own set of issues that tend to range from surface corrosion fouling to catalyst deactivation that is both rapid and permanent, depending on the downstream uses. Pollution levels in the production gas based on biomass-bound inorganic contaminants vary greatly and are inversely related to the inorganic content of the starting solid feedstocks. As a result, the syngas cleaning process is critical. Depending on downstream technology and/or emission limits, different levels of cleaning are required. Pollutants originating from these species have been linked to several health problems. Nitrogen, sulphur and chlorine, in particular, are the most important non-metal biomass inorganics that are monitored during gasification.

In future applications, nitrogen from biomass is primarily transformed to ammonia in producer gas, with miniscule amounts of hydrogen cyanide. The nitrogen-based impurities can even be oxidised to produce nitrogen oxides; both are controlled pollutants. The majority of sulphur in biomass is converted to hydrogen sulphide, carbonyl sulphide and other sulphur-containing compounds. Organically linked sulphur is released during devolatilisation as the organic fuel matrix disintegrates. Sulphur is liberated in this process via the production of sulfhydryl radicals which come from the thermal breakdown of the sulphur-bound organic compounds. The extremely reactive hydrogen, carbon and oxygen may be removed from the char by SH radicals, yielding hydrogen sulphide and sulphur dioxide.

Currently, direct combustion of biomass produces 95–97% of the world's bioenergy. As a result, if the world's burned biomass is estimated to be seven billion tonnes, around 480 million tonnes (480 MT) of biomass ash might be generated yearly. Challenges with these inorganics during combustion include agglomeration alkali deposits, slagging, fouling and corrosion (Gudka et al. 2016).

11.5.5 Advantages of Plum Biomass Utilisation in Energy Production

1. The availability of plum kernels, shells and stones contributes greatly to national energy supply and is the best way to promote low-cost energy and the self-sufficiency of the processing industry.
2. Social and economic benefits from the establishment of new jobs in rural products.
3. Important environmental benefits through use of biomass resources is an alternate for minimising environmental issues, such as carbon dioxide emissions into the atmosphere generated by the use of fossil fuels, and these bioenergies possess very little sulphur, therefore they do not really emit sulphur dioxide.

11.6 CONCLUSION

Plum is a fruit rich in phytochemicals and fibres; therefore, it has several positive impacts on human health. While the enormous amount of by-products and waste from the processing of plum fruits used to be a source of unwanted pollution, nowadays new applications and uses have been proposed. The re-use of pomace in fruit juice is a successful way to reduce food waste going to landfills. Moreover, the agriwaste from plum has great potential to be used as raw material to produce energy. The supply of electricity from plum biomass is becoming increasingly important due to its capacity to produce energy in a cost-effective manner. As a result, the management of plant origin post-production waste should be increasingly associated with the use of plum waste and its utilisation as raw material for energy.

11.7 REFERENCES

Abd Aal, M.H.E., Gomaa, E.G., and Karara, H.A. 1987. Bitter Almond, Plum and Mango Kernels as Sources of Lipids. *Lipid/Fett* 89 (8): 304–306.

Ahmed, A., Rahman, A.E., Abdalla, E., Hadary, E., and Aleem, M.I.A.E. 2015. Detexofication and Nutritional Evaluation of Peach and Apricot Meal Proteins. *Journal of Biological Chemistry and Environmental Science* 10: 597–622.

Balasundram, N., Kalyana, S., and Samir, S. 2006. Phenolic Compounds in Plants and Agri-Industrial By-Products: Antioxidant Activity, Occurrence, and Potential Uses. *Food Chemistry* 99 (1): 191–203.

Blasi Di, C., Branca, C., and Galgano, A. 2017. On the Experimental Evidence of Exothermicity in Wood and Biomass Pyrolysis. *Energy Technology* 5: 19–29.

Bolarinwa, I.F., Sulaiman, A.O., Sogo, J.O., Feyisayo, T.A., and Ifasegun, A.O. 2016. Effect of Processing on Amygdalin and Cyanide Contents of Some Nigerian Foods. *Journal of Chemical and Pharmaceutical Research* 8 (2): 106–113.

Cagnon, B., Py, X., Guillot, A., Stoeckli, F., Chambat, G. 2009. Contributions of Hemicellulose, Cellulose and Lignin to the Mass and the Porous Properties of Chars and Steam Activated Carbons from Various Lignocellulosic Precursors. *Bioresour Technology* 100: 292–298.

Cho, A.-Y, Yi, K.S., Rhim, J.-H., Kim, K.-I., Park, J.-Y., Keum, E.-H., Chung, J., and Oh, S. 2006. Detection of Abnormally High Amygdalin Content in Food by an Enzyme Immunoassay. *Molecules & Cells (Springer Science & Business Media BV)* 21 (2).

Cisneros-Zevallos, L., and Heredia, J.B. 2004. *Antioxidant Capacity of Fresh-Cut Produce May Increase After Applying Ethylene and Methyl Jasmonate.* Proceedings of the 2004 Institute Food Technologists Annual Meeting.

Damiani, C., da Silva, F.A., Cândido Rodovalho, E., Becker, F.S., Asquieri, E.R., Oliveira, R.A., and Lages, M.E. 2012. Aproveitamento De Resíduos Vegetais Para Produção De Farofa Temperada Utilization of Waste Vegetable for the Production of Seasoned Cassava Fl our. *Alimentos e Nutrição Araraquara* 22 (4): 657–662.

Dulf, F.V., Vodnar, D.C., and Socaciu, C. 2016. Effects of Solid-state Fermentation with Two Filamentous Fungi on the Total Phenolic Contents, Flavonoids, Antioxidant Activities and Lipid Fractions of Plum Fruit (Prunus domestica L.) By-products. *Food Chemistry* 209: 27–36.

Dwivedi, S.K., Joshi, V.K., and Mishra, V. 2014. Extraction of Anthocyanins from Plum Pomace Using XAD-16 and Determination of Their Thermal Stability. *Journal of Scientific & Industrial Research* 73: 57–61.

El-Adawy, T.A., and El-Kadousy, S.A. 1995. Changes in Chemical Composition, Nutritional Quality, Physico-chemical and Functional Properties of Peach Kernel Meal During Detoxification. *Food Chemistry* 52 (2): 143–148.

Elleuch, M., Bedigian, D., Roiseux, O., Besbes, S., Blecker, C., and Attia, H. 2011. Dietary Fibre and Fibre-Rich by-Products of Food Processing: Characterisation, Technological Functionality and Commercial Applications: A Review. *Food Chemistry* 124 (2): 411–421.

García, M.C., Puchalska, P., Esteve, C., and Marina, M.L. 2013. Vegetable Foods: A Cheap Source of Proteins and Peptides with Antihypertensive, Antioxidant, and Other Less Occurrence Bioactivities. *Talanta* 106: 328–349.

García, R., Pizarro, C., Lavín, A.G., and Bueno, J.L. 2012. Characterization of Spanish Biomass Wastes for Energy Use. *Bioresource Technology* 103: 249–258.

Gil, M.I., Tomás-Barberán, F.A., Hess-Pierce, B., and Kader, A.A. 2002. Antioxidant Capacities, Phenolic Compounds, Carotenoids, and Vitamin C Contents of Nectarine, Peach, and Plum Cultivars from California. *Journal of Agricultural and Food Chemistry* 50 (17): 4976–4982.

González-García, E, Marina, M.L., and Concepción García, M. 2014. Plum (Prunus Domestica L.) by-Product as a New and Cheap Source of Bioactive Peptides: Extraction Method and Peptides Characterization. *Journal of Functional Foods* 11: 428–437.

Górnaś, P., Mišina, I., Grāvīte, I., Lācis, G., Radenkovs, V., Olšteine, A., Segliņa, D., Kaufmane, E., and Rubauskis, E. 2015. Composition of Tocochromanols in the Kernels Recovered from Plum Pits: The Impact of the Varieties and Species on the Potential Utility Value for Industrial Application. *European Food Research and Technology* 241 (4): 513–520.

Górnaś, P., Rudzínska, M., Raczyk, M., Misina, I., Soliven, A., and Seglin, A.D. 2016c. Composition of Bioactive Compounds in Kernel Oils Recovered from Sour Cherry (Prunus cerasus L.) By-Products: Impact of the Cultivar on Potential Applications. *Industrial Crops and Products* 82: 44–50.

Górnaś, P., Rudzińska, P., and Soliven, A. 2017. Industrial By-Products of Plum Prunus Domestica L. and Prunus cerasifera Ehrh. as Potential Biodiesel Feedstock: Impact of Variety. *Industrial Crops and Products* 100: 77–84.

Górnaś, P., Siger, A., and Seglin, A.D. 2013. Physicochemical Characteristics of the Cold-Pressed Japanese Quince Seed Oil: New Promising Unconventional Bio-Oil from by-Products for the Pharmaceutical and Cosmetic Industry. *Ind Crops Products* 48: 178–182.

Gudka, B., Jones, J.M., Lea-Langton, A.R., Williams, A., and Saddawi, A. 2016. A review of the mitigation of deposition and emission problems during biomass combustion through washing pre-treatment. *Journal of the Energy Institute* 89(2): 159–171.

He, X.-Y, Wu, L.-J., Wang, W.-X., Xie, P.-J., Chen, Y.-H., and Wang, F. 2020. Amygdalin-A Pharmacological and Toxicological Review. *Journal of Ethnopharmacology* 254: 112717.

Hernandez-Infante, M., Sousa, V., Montalvo, I., and Tena, E. 1998. Impact of Microwave Heating on Hemagglutinins, Trypsin Inhibitors and Protein Quality of Selected Legume Seeds. *Plant Foods for Human Nutrition* 52 (3): 199–208.

Hodgson, E.M., Fahmi, R., Yates, N., Barraclough, T., Shield, I., Allison, G., Bridgwater, A.V., and Donnison, I.S. 2010. Miscanthus as a Feedstock for Fastpyrolysis: Does Agronomic Treatment Affect Quality? *Bioresource Technology* 101: 6185–6191.

Ivanov, D., Bojana, K., Tea, B., Radmilo, C., Đuro, V., Jovanka, L., and Slavica, S. 2012. Effect of Microwave Heating on Content of Cyanogenic Glycosides in Linseed. *Ratarstvo i Povrtarstvo* 49 (1): 63–68.

Kamel, B.S., and Kakuda, Y. 1992. Characterization of the Seed Oil and Meal from Apricot, Cherry, Nectarine, Peach and Plum. *Journal of the American Oil Chemists' Society* 69 (5): 492–494.

Kenney, K.L., Smith, W.A., Gresham, G.L. et al. 2013. Westover. *Biofuels* 4: 111–127.

Kim, D.-O., and Padilla-Zakour, O.I. 2004. Jam Processing Effect on Phenolics and Antioxidant Capacity in Anthocyanin-rich Fruits: Cherry, Plum, and Raspberry. *Journal of Food Science* 69 (9): S395–S400.

Kostić, M.D., Veličković, A.V., Joković, N.M., Stamenković, O.S., and Veljković, V.B. 2016. Optimization and Kinetic Modeling of Esterification of the Oil Obtained from Waste Plum Stones as a Pretreatment Step in Biodiesel Production. *Waste Management* 48: 619–629.

Lipiński, A.J., Lipiński, S., and Kowalkowski, P. 2018. Utilization of Post-Production Waste from Fruit Processing for Energetic Purposes: Analysis of Polish Potential and Case Study. *Journal of Material Cycles and Waste Management* 20: 1878–1883.

Madhav, K. 2016. Studies on Development of Tomato Leather Prepared for Geriatric Nutrition. *Journal of Nutrition & Food Sciences* 6: 446.

Milala, J., Kosmala, M., Sójka, M., Kołodziejczyk, K., Zbrzeźniak, M., and Markowski, J. 2013. Plum Pomaces as a Potential Source of Dietary Fibre: Composition and Antioxidant Properties. *Journal of Food Science and Technology* 50 (5): 1012–1017.

Mishra, N., El-Aal Bakr, A.A., Niranjan, K., and Tucker, G. 2011. Environmental Aspects of Food Processing. In *Food Processing Handbook*, edited by J.G. Brennan and A.S. Grandison, vol. 2, 2nd ed., 571–592. Wiley. ISBN: 978-3-527-32468-2.

Nowicki, P., Skrzypczak, M., and Pietrzak, R. 2010a. Effect of Activation Method on the Physicochemical Properties and NO2 Removal Abilities of Sorbents Obtained from Plum Stones (Prunus domestica). *Chemical Engineering Journal* 162: 723–729.

Nowicki, P., Wachowska, H., and Pietrzak, R. 2010b. Active Carbons Prepared by Chemical Activation of Plum Stones and Their Application in Removal of NO2. *Journal of Hazardous Materials* 181 (1–3): 1088–1094.

Obernberger, I., Brunner, T., and Barnthaler, G. 2006. Chemical Properties of Solid Biofuels-Significance and impast. *Biomass Bioenergy* 30: 973–982.

Obernberger, I., and Thek, G. 2004. Physical Characterisation and Chemical Composition of Densified Biomass Fuels with Regard to Their Combustion Behaviour. *Biomass Bioenergy* 27: 653–669.

Ramosa, M.A., de Medeirosa, P.M., de Almeidaa, A.L.S., Felicianob, A.L.P., and de Albuquerque, U.P. 2008. Can Wood Quality Justify Local Preferences for Firewood in an Area of Caatinga (Dryland) Vegetation? *Biomass and Bioenergy* 32: 503–509.

Sáez Angulo, F., and Martínez García, J.M. 2001. Emisiones en la Combustión de Biomasa y el Medio Ambiente. *Energia* 161: 75–83.

Sahni, P., and Shere, D.M. 2018. Utilization of Fruit and Vegetable Pomace as Functional Ingredient in Bakery Products: A Review. *Asian Journal of Dairy & Food Research* 37 (3): 202–211.

Shalini, R., and Gupta, D.K. 2010. Utilization of Pomace from Apple Processing Industries: A Review. *Journal of Food Science and Technology* 47 (4): 365–371.

Shi, J., Chen, Q., Xu, M., Xia, Q., Zheng, T., Teng, J, Li, M., and Fan, L. 2019. Recent Updates and Future Perspectives about Amygdalin as a Potential Anticancer Agent: A Review. *Cancer Medicine* 8 (6): 3004–3011.

Shimelis, E.A., and Rakshit, S. 2005. Effect of Microwave Heating on Solubility and Digestibility of Proteins and Reduction of Antinutrients of Selected Common Bean (Phaseolus Vulgaris L.) Varieties Grown in Ethiopia. *Italian Journal of Food Science* 17 (4): 407–418.

Sirin, S.M., and Ege, A. 2012. Overcoming Problems in Turkey's Renewable Energy Policy: How can EU Contribute. *Renew Sust Energ Rev* 16: 4917–4926.

Tunçel, G., Nout, M.J.R., Brimer, L., and Göktan, D. 1990. Toxicological, Nutritional and Microbiological Evaluation of Tempe Fermentation with Rhizopus Oligosporus of Bitter and Sweet Apricot Seeds. *International Journal of Food Microbiology* 11 (3–4): 337–344.

Vassilev, S.V., Baxter, D., Andersen, L.K., and Vassileva, C.G. 2010. An Overview of the Chemical Composition of Biomass. *Fuel* 89: 913–933.

Viktorija, S., Karakashova, L., Babanovska-Milenkovska, F., Delchev, N., Nakov, G., and Necinova, L. 2013. The Quality Characteristics of Plum Jams Made with Different Sweeteners. *Technology* 18–19.

Vilariño, M.V., Franco, C., and Quarrington, C. 2017. Food Loss and Waste Reduction as an Integral Part of a Circular Economy. *Frontiers in Environmental Science* 5: 21.

Wang, W.M, Siddiq, M., Sinha, N.K., and Cash, J.N. 1995 Effect of Processing Conditions on the Physicochemical and Sensory Characteristics of Stanley Plum Paste. *Journal of food Processing and Preservation* 19 (1): 65–81.

Wojdyło, A., Nowicka, P., Tkacz, K., and Turkiewicz, I.P. 2021. Fruit Tree Leaves as Unconventional and Valuable Source of Chlorophyll and Carotenoid Compounds Determined by Liquid Chromatography-Photodiode-Quadrupole/Time of Flight-Electrospray Ionization-Mass Spectrometry (LC-PDA-qTof-ESI-MS). *Food Chemistry* 349: 129156.

Wu, M., Li, D., Wang, L.-J., Zhou, Y.-G., Brooks, M.S.-L., Chen, X.D., and Mao, Z.-H. 2008. Extrusion Detoxification Technique on Flaxseed by Uniform Design Optimization. *Separation and Purification Technology* 61 (1): 51–59.

Xue, F., Zhu, C., Liu, F., Wang, S., Liu, H., and Li, C. 2018. Effects of High-Intensity Ultrasound Treatment on Functional Properties of Plum (Pruni domesticae semen) Seed Protein Isolate. *Journal of the Science of Food and Agriculture* 98 (15): 5690–5699.

Zajac, G., Szyszlak-Bargłowicz, J., Słowik, T., Wasilewski, J., and Kuranc, A. 2017. Emission Characteristics of Biomass Combustion in a Domestic Heating Boiler Fed with Wood and Virginia Mallow Pellets. *Fresenius Environ Bull* 26: 4663–4670.

Zhang, N., Zhang, Q.A., Yao, J.L., and Zhang, X.Y. 2019. Changes of Amygdalin and Volatile Components of Apricot Kernels During the Ultrasonically-Accelerated Debitterizing. *Ultrasonics Sonochemistry* 58: 104614.

Zhao, X., Wang, H., You, J., and Suo, Y. 2007. Determination of Free Fatty Acids in Bryophyte Plants and Soil by HPLC with Fluorescence Detection and Identification by Online MS. *Chromatographia* 66 (3): 197–206.

12 Plum and Its Products
Properties and Health Benefits

Xian Lin[1,2], *Baojun Xu*[2] *and Jessica Pandohee*[3]

[1] Sericultural & Agri-Food Research Institute, Guangdong Academy of Agricultural Sciences/Key Laboratory of Functional Foods, Ministry of Agriculture and Rural Affairs/Guangdong Key Laboratory of Agricultural Products Processing, Guangdong, 510610, China

[2] Food Science and Technology Program, BNU-HKBU United International College, Zhuhai, Guangdong 519087, China

[3] Centre for Crop and Disease Management, Curtin University, Western Australia 6151, Australia.

CONTENTS

DOI: 10.1201/9781003205449-12

12.1 INTRODUCTION

To date, more than 40 types of plums have been reported. Despite the botanical diversity of plums, *Prunus domestica* (originating from Europe) and *Prunus salicina* and its hybrids (from Japan) are the main species that are commercially grown and consumed around the world (Wallace 2017). Other species of plum in the Middle East and in some regions of Europe are considered as wild fruits. Prune is one of the most popular plum products, and it usually refers to dried plums made from processing varieties of *Prunus domestica.* L. Besides prune, other forms of plum products include canned plums, plum sauce, plum juice, dried plum purée and plum powder, amongst others. As natural products, prunes are known to comprise a plethora of biochemical compounds beneficial for human health. The two plum species *Prunus domestica* and *Prunus salicina* are well-studied in food science and nutrition. This chapter provides an update on the health benefits of plum and its products.

12.2 PROPERTIES

12.2.1 Major Nutritional Ingredients

12.2.1.1 Carbohydrates

Carbohydrate composition of plum and its products as reported in the SR-Legacy (USDA 2019) are shown in Table 12.1. As seen in Table 12.1, plum and its products have a large proportion of carbohydrates, and dried plums (uncooked) displayed the highest carbohydrate content, at 63.9 g/100g FW. Amongst the carbohydrate components, plum and its products are rich in sugars and dietary fibre. Plum sugars and its products are mainly composed of glucose, fructose and sucrose. Especially in dried plums (uncooked), the glucose and fructose content reached 25.5 and 12.4 g/100g FW respectively.

Moreover, Dikeman et al. (2004) found that plum and its products also contained sorbitol and inositol. The sorbitol content levels were in high and reached 2.7, 1.1 and 9.1 g/100g FW in fresh prune plum, prune juice and dried plums respectively. As previous studies have evidenced, sorbitol may have beneficial health effects, including laxative, anti-cancer, anti-diabetic and anti-bacterial properties (Stacewicz-Sapuntzakis 2013). Table 12.1 also presents the oligosaccharide contents, which were only in the range of 0.1–1.4 g/100g FW.

TABLE 12.1
Carbohydrate Composition of Plum and Its Products

Component (g/100g FW)	Fresh plums	Dried plums (uncooked)	Dried plums (stewed)	Canned plums (water pack)	Plum sauce
Carbohydrate, by difference	11.4	63.9	28.1	11	42.8
Fibre dietary	1.4	7.1	3.1	0.9	0.7
Sugars	9.92	38.1	25	10.1	
Sucrose	1.57	0.15			
Glucose	5.07	25.5			
Fructose	3.07	12.4			
Lactose	0	0			
Maltose	0.08	0.06			
Galactose	0.14	0			
Starch	0	5.11			

12.2.1.2 Organic Acids

Organic acids provide plums with their characteristic tart taste. Quinic acid and malic acid have been detected as the primary organic acid in plums, with contents of 668 mg/100 mL and 104 mg/100 mL in prune juice respectively (van Gorsel et al. 1992). In prune-making plums, the quinic acid and malic acid contents were 1.1 g/100g and 0.29 g/100g respectively. The quinic acid and malic acid contents in dried plums were 4.3 g/100g and 1.1 g /100g respectively. Moreover, citric and fumaric acids were also detected in fresh plums (García-Mariño et al. 2008). Quinic acid, as a bioactive component, was reported to have potential beneficial cardiovascular, anti-diabetic and bone-health effects. Additionally, moderate amounts of oxalic acid were also reported in dried plums (Stacewicz-Sapuntzakis 2013).

12.2.1.3 Vitamins

Vitamin composition of plum and its products reported in the SR-Legacy (USDA 2019) are shown in Table 12.2. As Table 12.2 shows, the vitamin C content in fresh plum was 9.5 mg/100g FW, which was significantly higher than its products. This was attributed to the heat-sensitive antioxidant property of vitamin C. Factors such as high-temperature processing and long duration of storage could lead to the degradation of vitamin C. For vitamin B compounds, the richest component in fresh plums was choline. Riboflavin, niacin and vitamin B6 were well preserved in plum products, especially in dried plums. However, no folic acid or vitamin B12 were detected in plum and its products. Betaine was only detected in dried plums (uncooked), which was probably added during processing.

Fresh plums had a vitamin A content of 17 μg/100g FW based on retinol activity equivalents, but no retinol was detected, suggesting the existence of other vitamin

TABLE 12.2
Vitamin Composition of Plum and Its Products

Component (mg/100g FW)	Fresh plums	Dried plums (uncooked)	Dried plums (stewed)	Canned plums (water pack)	Plum sauce
Vitamin C, total ascorbic acid	9.5	0.6	2.9	2.7	0.5
Thiamine	0.028	0.051	0.024	0.021	0.018
Riboflavin	0.026	0.186	0.1	0.041	0.084
Niacin	0.417	1.88	0.723	0.37	1.01
Pantothenic acid	0.135	0.422	0.107	0.072	0.059
Vitamin B6	0.029	0.205	0.218	0.027	0.078
Folate, total	5 000	4 000	0	3 000	6 000
folic acid	0	0	0	0	0
Choline, total	1.9	10.1	4.4	1.3	
betaine		0.4			
Vitamin B12	0	0	0	0	0
Vitamin A, RAE	17000	39000	17000	46000	2000
Retinol	0	0	0	0	0
β-Carotene	190000	394000	173000	502000	26000
α-Carotene	0	57000	25000	0	0
β-Cryptoxanthin	35000	93000	41000	93000	0
Lycopene	0	0	0	0	
Lutein + zeaxanthin	73000	148000	65000	49000	
Vitamin E (α-tocopherol)	0.26	0.43	0.19	0.18	
β-Tocopherol	0	0			
γ-Tocopherol	0.08	0.02			
δ-Tocopherol	0	0			
α-Tocotrienol	0.04	0.01			
β-Tocotrienol	0	0			
γ-Tocotrienol	0.01	0			
δ-Tocotrienol	0	0			
Vitamin D (D2 + D3)	0	0	0	0	
Vitamin K (phylloquinone)	6400	59500	26100	4300	
Vitamin K (Dihydrophylloquinone)	0	0			

A compounds. Some carotenoids are provitamin A, such as α-carotene, β-carotene, γ-carotene and β-cryptoxanthin. The carotenoids detected in plum included β-carotene, lutein, β-cryptoxanthin and zeaxanthin, which were also well maintained in the products. α-Carotene was only detected in dried plums and canned plums. As Table 12.2 displays, α-tocopherol was the richest vitamin component in plum and its products, with contents ranging from 0.18 mg/100g FW to 0.26 mg/100g FW. Dried plums exhibited high levels of vitamin K (phylloquinone), which was 59.5 μg/100g FW and 2.6.1 μg/100g FW in dried plums (uncooked) and dried plums (stewed) respectively.

12.2.1.4 Amino Acids

Table 12.3 lists the amino acid composition of fresh plum and its products as reported in the SR-Legacy (USDA 2019). As shown, fresh plums, dried plums and canned plums contained total amino acid contents of 0.604 g/100g, 1.668 g/100g and 0.258 g/100g respectively. Aspartic acid accounted for the largest proportion in fresh plum and its products, which was around half of the total amino acid content. Van Gorsel et al. (1992) have also analyzed the amino acids of fresh prune and various prune juices. Results showed that processed prune juices contained high levels of citrulline, α-aminobutyric acid, *O*-phosphoethanolamine, and taurine. However, frozen juice from fresh prunes had a very little amount of these amino acids. There were higher levels of threonine, proline, serine, isoleucine, valine and γ-aminobutyric acid in frozen juice from fresh prunes than in processed prune juices. Frozen juice from fresh prunes had much higher cysteine and glutamine contents than those

TABLE 12.3
Amino Acid Composition of Plum and Its Products

Component (g/100g FW)	Fresh plums	Dried plums (uncooked)	Canned plums (water pack)
Tryptophan	0.009	0.025	–
Threonine	0.01	0.049	0.008
Isoleucine	0.014	0.041	0.008
Leucine	0.015	0.066	0.01
Lysine	0.016	0.05	0.008
Methionine	0.008	0.016	0.003
Cystine	0.002	0.011	0.002
Phenylalanine	0.014	0.052	0.008
Tyrosine	0.008	0.021	0.003
Valine	0.016	0.056	0.009
Arginine	0.009	0.037	0.006
Histidine	0.009	0.027	0.006
Alanine	0.028	0.066	0.014
Aspartic acid	0.352	0.801	0.122
Glutamic acid	0.035	0.114	0.018
Glycine	0.009	0.047	0.006
Proline	0.027	0.13	0.017
Serine	0.023	0.059	0.01
Total amino acid content	0.604	1.668	0.258

TABLE 12.4
Mineral Composition of Plum and Its Products

Component (mg/100g FW)	Fresh plums	Dried plums (uncooked)	Dried plums (stewed)	Canned plums (water pack)	Plum sauce
Ca	6	43	19	7	12
Fe	0.17	0.93	0.41	0.16	1.43
Mg	7	41	18	5	12
P	16	69	30	13	22
K	157	732	321	126	259
Na	0	2	1	1	538
Zn	0.1	0.44	0.19	0.08	0.19
Cu	0.057	0.281	0.123	0.039	0.078
Mn	0.052	0.299	0.131	0.033	0.114
Se	0	300	100	0	400

of prune juices processed from dried prunes. It is noted that all kinds of prune juices presented higher GABA content than stone fruit juices (van Gorsel et al. 1992).

12.2.1.5 Minerals

Table 12.4 lists the mineral composition in fresh plum and its products as reported in the SR-Legacy (USDA 2019). As is shown, plum and its products are good sources of copper, potassium and manganese. The mineral composition of plum and its products varied in large scales from various studies, as the results depend on many factors, including cultivar, growing environment and agricultural practices.

12.2.2 Phenolic Compounds

The key bioactive compounds group in plum has been shown to be phenolic compounds. Sahamishirazi et al. (2017) assessed the total phenolic and individual anthocyanin contents of 178 cultivars of plum and found the total phenolic contents ranged between 38.45 mg/100g FW and 841.50 mg/100g FW (based on gallic acid equivalents). The great variety in the total phenolic contents are attributed to several factors, such as cultivar, cultivating and environmental conditions, fruit ripening and post-harvest practices.

Forty individual phenolic compounds have been identified (Yu et al. 2021). Amongst them, chlorogenic acid is the major phenolic acid detected in the dried plums, including 5-O-caffeoylquinic acid, 3-O-caffeoylquinic acid and 4-O-caffeoylquinic acid (Stacewicz-Sapuntzakis 2013). Another important class of phenolic compounds in plums are anthocyanins, including cyaniding-3-glucoside and cyanidin-3-rutinoside (Sahamishirazi et al. 2017). For proanthocyanidins, studies have shown high amounts of proanthocyanidins, which are polymeric forms of flavan-3-ols. The contents of proanthocyanidins were found to be 0.79 mg/g DW and 0.59 mg/g DW in the pulp and skin of fresh plums respectively. However, the actual value would be higher, as some of the proanthocyanidins may be challenging to be extracted

(Stacewicz-Sapuntzakis 2013). Yu et al. (2021) investigated the free and bound phenolic fractions in plums. Results showed the free phenolic fraction mainly contained epicatechin, neochlorogenic acid and procyanidin B2, and the bound phenolic fraction were mostly composed of catechin and epicatechin.

Processing could significantly affect the phenolic compounds of plums. Evidence showed that long drying (60–72 hours) of plums at 60°C led to the degradation of phenolic compounds, but shorter drying (38–44 hours) at higher temperature of 85°C preserved the phenolic compounds. Osmotic dehydration of plums in a sucrose solution could also lead to the loss of phenolic compounds (Stacewicz-Sapuntzakis 2013).

12.2.3 Volatile Compounds

The aroma of plum and its products is largely dependent on their volatile compounds and contributes to the acceptance of the consumers. Studies showed that the aromatic volatile compounds of plums were complex, including various classes of compounds, such as esters, alcohols, aldehydes, terpenes, alkanes, acids, lactones, ketones and phenols. Among them, esters, aldehydes and alcohols were reported to be the main contributors to plum aroma (González-Cebrino et al. 2016).

Aroma of plum products could be significantly different with that of the fresh plum, as processing conditions have great impacts on the volatile compounds. González-Cebrino et al. (2016) detected the volatile compounds of plum purée and identified 40 volatile compounds, with major compounds identified as aldehyde, hexan-1-ol and (Z)- hex-3-en-1-ol. The researcher also found that a high-pressure processing treatment had mild effects on the aroma of plum purée, only leading to 1.8% of total aroma changes.

12.3 HEALTH BENEFITS

Due to its unique nutritional and bioactive profile, plum and its products are endowed with a series of beneficial health effects (Figure 12.1).

12.3.1 Antioxidant Capacity

Antioxidant capacity is closely related toa series of diseases, including cardiovascular diseases, immune dysregulation, neurodegenerative diseases, cancer and even anxiety and depression (Bouayed et al. 2007). Plum utilization has beneficial health properties, largely due to their antioxidants. The high level of phenolic content is the main contributor to the great antioxidant capacity of plums. Studies have demonstrated that the antioxidant capacity of plum was positively correlated with its phenolic content (Yu et al. 2021). The antioxidant property of plums is often attributed to lipid-like molecules, such as vitamin C, carotenoids and α-tocopherol.

The antioxidant capacity of plum and its products have been assessed by various chemical assays, using different radicals or oxidants. The antioxidant capacity of phenolics in plums were reported to be 121.74 mg TE/g DW, 95.63 mg TE/g DW, 116.27 mg TE/g DW and 443.98 mg TE/g DW, as determined by the DPPH, FRAP, ABTS and ORAC assays, respectively (Yu et al. 2021). The range of total ORAC of dried plum and prune juice was 2127–8578 μmol TE/100g and 2036–2127 μmol

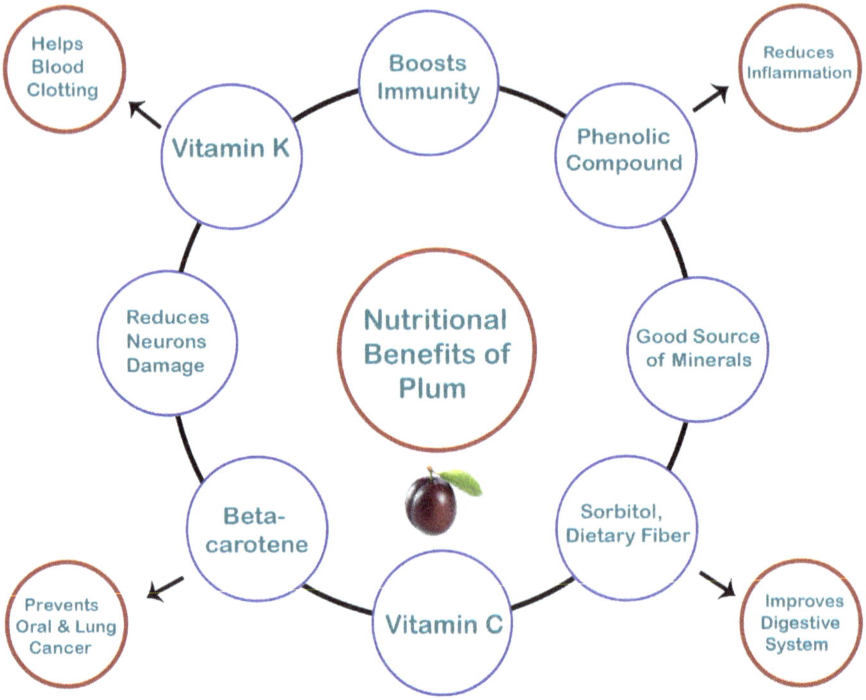

FIGURE 12.1 Nutritional benefits of plum fruit.

TE/100g respectively. The antioxidant capacity of dried plums were 1482, 2300 and 6054 μmol TE/100g, as analyzed by ABTS, TRAP and FRAP assays, respectively (Stacewicz-Sapuntzakis 2013). Interestingly, studies showed that thermal drying preserved or even increased the antioxidant capacity of plums, while osmotic dehydration led to different degrees of reduction (Stacewicz-Sapuntzakis 2013). However, the results obtained from the chemical assays were hard to compare, and the biological relevance needed further in vivo investigation.

In cell studies, Kumar et al. (2009) found that the production of malondialdehyde was reduced by 32% when the dried plum phenolics extract was at a concentration of 1000 μg/mL, indicating the potent antioxidant capacity of dried plum phenolics extract, after macrophage cells were stimulated with a mixture of $FeSO_4$ and H_2O_2 to induce lipid peroxidation, and malondialdehyde was measured as an indicator of lipid peroxidation. In human trials, Ko et al. (2005) reported that consumption of 150 mL of plum juice for 30 minutes displayed significant antioxidant activities in human plasma by inhibiting reactive oxygen species generation. Another study also confirmed the antioxidant activity of plums. After consumption of plums (195 g, twice a day) for five days, urinary 6-sulfatoxymelatonin and the total urinary antioxidant capacity levels significantly increased (Igwe and Charlton 2016). However, Prior et al. (2007) found consumption of died plums didn't affect the plasma antioxidant activity. It is noted that many of the health benefits of plum and its products mentioned in the following parts were also functioned based on their potent antioxidant activity.

12.3.2 Modulation of Cardiovascular Health

Negishi et al. (2007) showed that five weeks of prune supplementation suppressed blood pressure in stroke-prone spontaneously hypertensive rats. The anti-hypertension effects of prune extract were probably attributed to its antioxidant property. In clinical studies, Tinker et al. (1991) found that plasma low-density lipoprotein cholesterol of adult men with mild hypercholesterolemia was effectively reduced by the consumption of dried plums, as compared to the groups consuming grape juice. The lithocholic acid, as one of the bile acids in stool, was effectively decreased as well. Ahmed et al. (2010) studied the impacts of daily consumption of prunes on pre-hypertensive patients. Patients consumed a small dosage of prunes (11.5 g), a high dosage of prunes (23 g) or water (control group) for eight weeks. Results showed that patients who consumed a small dosage of prunes or a glass of water daily displayed a significant reduction in systolic and diastolic blood pressure in the morning before diets, but those who consumed a high dosage of prunes only experienced a decrease in systolic blood pressure. Moreover, the consumption of prunes significantly prevented the increase of serum high-density lipoprotein compared to the control group.

The anti-atherosclerosis effect of dried plum powder was evaluated in apolipoprotein E-deficient mice models, which developed atherosclerotic lesions when fed with cholesterol (Gallaher and Gallaher 2009). Results showed that the consumption of dried plums could prevent the development of atherosclerosis. After consuming diets with 4.75% supplementation of dried plum for five months, the atherosclerotic lesion area was significantly decreased in the entire arterial tree and the aortic arch. A marker of inflammation in mice, serum amyloid P-component was also significantly reduced by comparison with that of the control group. But groups treated with 9.5% supplementation of dried plum did not exhibit obvious anti-atherosclerosis effects.

The functional capacity was highly related to the bioactive phytochemicals in the materials; thus, some cardiovascular health benefits of plum and its products could be deduced according to the major bioactive phytochemicals. Chlorogenic acid is the major bioactive component in plums. Rodriguez de Sotillo and Hadley (2002) showed that after three weeks of treatment with intravenous injection of chlorogenic acid (5 mg/Kg body weight/day), the weight gain of insulin-resistant and obese Zucker rats was significantly prevented, and the fasting plasma triacylglycerol and cholesterol concentrations were decreased.

Plum and its products also contain high amounts of pectin, which may endow the plums with anti-atherosclerosis effects. Wu et al. (2003) conducted an atherosclerosis clinical study with 500 participants to investigate the relationship between dietary fibre intake and cardiovascular disease risk. A significant relationship between the evolution of atherosclerosis and the consumption of pectin was observed. In addition, it was found that the total fibre intake was negatively related with the plasma triglyceride levels as well as the ratio of total cholesterol to HDL-cholesterol.

12.3.3 Modulation of Immune System

Rendina et al. (2012) investigated the effects of dried plum on the immune response related to ovarian hormone deficiency. Ovariectomized female mice were fed with supplementations of 0%, 5%, 15% or 25% of dried plum in their diets. After four weeks of treatment, supplementation with 15% or 25% dried plum significantly inhibited splenocytes to produce TNF-α ex vivo, responding to concanavalin A stimulation. The results indicated that dried plum had positive effects on suppressing splenocyte activation in the situation of ovarian hormone deficiency.

Hooshmand et al. (2015) analyzed the effects of dried plum polyphenols on the production of nitric oxide and cyclooxygenase-2 mouse macrophage cell lines, which were stimulated with bacterial liposaccharide. Results showed dried plum polyphenols with a concentration of 1000 μg/mL significantly decreased the NO production by 43% after 12 hours of treatment. Moreover, expression of COX-2 induced by LPS was also significantly decreased, indicating that the dried plum polyphenols possessed immunomodulatory properties.

Karasawa et al. (2012) used the mite-sensitized BALB/c mice model to investigate the effects of prune on allergic responses. Prunes were extracted by hot water and consumed through oral ingestion. Results showed that, by supplementation of prune extract, sneezing events as well as immunoglobulin E levels were effectively reduced. The results indicated that the prune extract may alleviate type I allergic symptoms in mice by adjusting the type 1 helper T cell/type 2 helper T-cell balance and inhibiting mast cell degranulation.

In an animal study, chickens were raised with supplementation of 1% freeze-dried plum powder in their diet and infected with a coccidiosis parasite to evaluate the immunomodulatory property of dried plum. Results showed that chickens with the supplementation of dried plum grew better than those fed with a standard diet and fewer oocysts were observed in the stool (Lee et al. 2008). Supplementation of dried plum also led to an increase in spleen lymphocyte proliferation and the expression of mRNA for interferon-γ and interleukin-15 in intestinal lymphocytes. However, an in vitro study reported that freeze-dried plum powder (10–500 μg/mL) did not promote the proliferation of mouse splenocytes (Lin and Tang 2007).

Mishra et al. (2012) reported the antiulcer effect of plums. In the study, the antiulcerogenic activity of ethanolic fruit extract of plums was evaluated using pyloric ligation model. Wistar albino rats were fed with 200 mg/kg of plum extract for seven days and a 70.58% inhibition of ulcers was observed, indicating the excellent antiulcer effect of plums (Mishra et al. 2012).

12.3.4 Neurologic and Psychiatric Effects

The cognitive decline that accompanies aging may be the result of long-term accumulation of oxidative stress on neurologic processes. Some studies have demonstrated the beneficial health effects of plum on cognition improvement related to aging (Igwe and Charlton 2016). The effects on cognition are mainly endowed by the antioxidant capacity of plums. Kuo et al. (2015) studied the effects of plums rich in polyphenols on cognitive function using mice fed with high-cholesterol diets. Mice were fed with supplementations of plum powder for five months. Results showed that mice that were fed ahigh-cholesterol diet showed a significant increase in mRNA expression of Cyp46, BACE1, Ab and 24-hydroxycholesterol in the brain cortex and hippocampus, indicating damage in cognitive function. However, the index exhibited no significant change when mice were fed with a supplementation of 5% plum powder in the diet, suggesting that plum could ameliorate the symptoms of neurodegenerative conditions caused by a high-cholesterol diet.

In another study, effects of supplementation of 2% dried plum powder and 100% plum juice on age-related deficits were compared using aged rats. The results showed that plum juice has exerted positive effects in cognition, but the dried plum powder has not. The differences could be attributed to much lower levels of antioxidants in dried plum powder, which was caused by processing (Shukitt-hale et al. 2009).

12.3.5 Bone-Protective Effects

The role of plums in bone health has gained great attention. Researchers have evaluated the bone-protective effects of plums in cell, animal and clinical studies. In their cell study, Bu et al. (2008) investigated the osteoclast differentiation and the osteoblast function of dried plum using RAW 264.7 murine macrophage cells and MC3T3-E1 cells respectively. Results demonstrated the polyphenols of dried plum effectively inhibited osteoclastogenesis, thus decreasing osteoclast activity by down-regulating NFATc1 and inflammatory mediators. Polyphenols of dried plum could also increase mineralized nodule formation and enhance the osteoblast activity (Bu et al. 2008).

Evidence from animal studies showed that dried plum can prevent bone loss and alter the situation of pre-existing bone loss. Post-menopausal in particular women are concerned about accelerated bone loss due to decreased calcium absorption. Ovariectomized female rat models were used for the study of hormone deficiency–induced bone loss. Results indicated that a diet supplemented with 25% dried plum powder after ovariectomy inhibited the decrease of bone mineral density (BMD) of the lumbar vertebrae and femur (Arjmandi et al. 2001). Another study was carried out pre-existing bone loss. After 40 days of bone loss, ovariectomized rats were treated with a diet supplementation of dried plum powder for 60 days (Deyhim et al. 2005). Results showed that, when rats were treated with a 5% dried plum powder supplement, the BMD of their femurs and tibiae were restored. BMD of the femur/tibia and lumbar vertebrae were restored when the supplementation of dried plum powder was 5% and 25%, respectively. Moreover, supplementation of dried plum powder in the diet also improved the bone structural and biomechanical properties (Rendina et al. 2012). In another similar study, results showed that the supplementation of 15% or 25% dried plum had positive effects on the restoration of body and femoral BMD, as well as improvement of trabecular bone volume and cortical thickness (Stacewicz-Sapuntzakis 2013).

Aging is another factor that increases chances of bone loss, thus osteoporosis becomes a concern for old people, though men experience it at a slower rate than women. In a study, castrated rats were used as models for male osteoporosis and the therapeutic effects of supplementation of dried plum powder was investigated. When supplemented with 15% and 25% dried plum powder, the loss of BMD was prevented after 90 days of treatment. When supplemented with 5% dried plum powder in their diets, effective results were observed in the BMD of the lumbar vertebrae and femur. In another study, castrated rats were used as models and treated with a supplementation of 25% dried plum powder after 90 days of bone loss. Results indicated that dried plum powder supplementation effectively reversed BMD loss in the whole skeleton with positive changes in bone structure. Age-related mouse models have also been used to study the effects of dried plum on age-related bone loss. When fed with a diet containing 25% dried plum for six months, the bone volume of adult and elderly mice increased by 50% and 40% respectively. But the control groups fed with normal diets suffered over 20% of bone loss. Therefore, it can be concluded that plum and its products are good sources for the prevention and restoration of bone loss related to aging (Stacewicz-Sapuntzakis 2013). To reveal the mechanism of bone-sparing effects of plums, researchers found that the consumption of dried plum significantly decreased bone resorption. Moreover, the consumption of dried plum may increase the insulin-like growth factor in both male and female rats, thus stimulating bone formation (Stacewicz-Sapuntzakis 2013).

Clinical studies are mainly focused on post-menopausal women. By intervention of 100g/day dried plum for three months, serum levels of bone-specific alkaline phosphatase activity (BSAP) and insulin-like growth factor-I (IGF-I) were significantly increased when compared to the group that consumed 75 g/day of dried apple. The increase of BSAP and IGF-I indicated the greater rates of bone formation (Arjmandi et al. 2002). The effects of 100g/day of dried plum with a duration of one year were also investigated. Results indicated a significant increase in BMD at the spine and ulna as compared to the control group. Moreover, by consumption of dried plum, serum sclerostin and osteoprotegerin levels were decreased and increased, respectively. These indicated dried plum exerted positive effects to the bone formation partly by inhibition of sclerostin and promotion of osteoprotegerin (Hooshmand 2013). In another study, results showed that 50 g/day and 100g/day of dried plum consumption for six months prevented the loss of total body BMD but had no effect on the BMD of the hip, spine orulna. Hence, the consumption of dried plums is a good choice for post-menopausal women when it comes to bone health (Stacewicz-Sapuntzakis 2013).

The beneficial effects of dried plum on bone health may be attributed to some bioactive compounds. The phenolics present in plums may be a bioactive compound, as a diet rich in phenolics may facilitate bone formation and inhibit bone resorption (Wallace 2017). Moreover, the high level of vitamin K in plum and its products may also contribute to bone health by enhancing calcium balance. Vitamin K is involved in the regulation of the growth of hydroxyapatite crystals in bone as a cofactor in γ-carboxylation of osteocalcin, thus promoting bone mineralization accordingly. Additionally, boron, magnesium and potassium may also facilitate to maintain bone health by reducing urinary calcium excretion and increasing oestradiol levels in post-menopausal women or maintaining BMD (Stacewicz-Sapuntzakis 2013).

12.3.6 Modulation of Metabolic Syndrome

12.3.6.1 Diabetes

Glycemic index (GI) is a measure of the glycemic potential per gram of carbohydrate. Low-GI foods induce satiety better than high-GI foods and do not cause a significant increase or decrease in blood insulin after a meal. Therefore, it is recommended to use low-GI foods to control diabetes and body weight. Dried plums have been classified as a low-GI food, with aglycemic index of 40 ± 6 (Atkinson et al. 2008). The high level of sorbitol in prunes may be part of the reason for their low GI. Sorbitol has a low GI of 9, and the glycemic load of a diet consisting of glucose and sorbitol is less than predicted based on the sum of the individual loads. This may be due to reduced glucose absorption in the presence of sorbitol (Livesey 2003).

Another component of dried plums—chlorogenic acid—has been evidenced in in vivo, animal and clinical studies (Stacewicz-Sapuntzakis 2013). In a clinical study, Rodriguez de Sotillo and Hadley (2002) investigate the impacts of chlorogenic acid on fasting plasma glucose using obese, hyperlipidemic and insulin-resistant (fa/fa) Zucker rats. Rats were treated with intravenous infusion of chlorogenic acid (5 mg/Kg body weight/day) for three weeks. Results showed that the intravenous infusion of chlorogenic acid effectively reduced the post-prandial peak response to glucose, suggesting chlorogenic acid could improve glucose tolerance. A trial was carried out on healthy volunteers who consumed glucose with caffeinated or decaffeinated coffee, both of which contained 354 mg of chlorogenic acid. The results of plasma glucose, insulin and gastrointestinal hormone profiles suggested that chlorogenic acids had an antagonistic effect on glucose transport, but caffeine may damage glucose tolerance (Johnston et al. 2003). Similar results were also observed by Thom (2007), who conducted a clinical study on 12 healthy volunteers. Results showed that instant coffee rich in chlorogenic acid significantly diminished the absorption of glucose compared to the control.

12.3.6.2 Obesity

The consumption of dried fruit is related to less abdominal obesity and a reduced risk of being overweight in adults (Keast and Jones 2009). Results of animal studies have demonstrated the anti-obesity effects of plum. Rodriguez de Sotillo and Hadley (2002) used hyperlipidemic, insulin-resistant and obese Zucker rats as models to investigate the impacts of chlorogenic acid on body weight. The results indicated that the intravenous infusion of chlorogenic acid (5 mg/Kg body weight/day) for three weeks effectively prevented further increased body weight. Noratto et al. (2015) fed obese and lean Zucker rats with plum and peach juice and found that only plum juice significantly suppressed weight gain. This effect was probably due to the phenolics, as the phenolic content in plums was three times that of peaches. The clinical studies suggested that consuming dried plums as snack could significantly exert satiety effects, thus leading to smaller energy intake. An important factor for the obesity effects of plum or its products is attributed to its satiety effects (Stacewicz-Sapuntzakis 2013).

12.3.7 Gastrointestinal Health Effects

12.3.7.1 Laxative Effects

Dried plums are generally recognized as a laxative food. Studies have evidenced that the consumption of plum or its products helps relieve constipation. In a clinical trial, 54 volunteers with mild gastrointestinal (GIT) symptoms were subjected to two weeks of prune juice consumption (125 mL twice a day). The results showed that the consumption of prune juice decreased the amount of difficulty in defecation but increased flatulence (Piirainen et al. 2007). Similarly, patients with chronic constipation symptoms who consumed 237 mL of plum juice everyday for two weeks or 50 g of dried plums twice a day with meals for three weeks both found improved constipation symptoms (Cheskin et al. 2012; Attaluri et al. 2011). The present evidence indicated that plums had mild laxative effects.

The mild laxative effects of plum could be attributed to the synergistic effect provided by dietary fibre, sorbitol and polyphenols (Attaluri et al. 2011). Studies showed that the consumption of sorbitol had a stool-softening effect in small doses but could quickly change the fluid balance in the colon due to its osmotic effects in high doses (McRorie et al. 2000). Sorbitol could also affect gut motility because of its prebiotic effect. By fermentation of colonic bacteria, sorbitol could be transformed into a short-chain fatty acid, especially butyric acid, which could have positive effects on the colon environment, thus alleviating constipation (Livesey 2003). Phenolics in plums, such as chlorogenic acid, caffeic acid and ferulic acid, were also shown to contribute to increased gastrointestinal motility (Igwe and Charlton 2016; Badary et al. 2006). Moreover, a study reported that fresh plums of unknown varieties were rich in the neurotransmitter serotonin and could promote intestinal fluid secretion and gut motility (Stacewicz-Sapuntzakis 2013).

12.3.7.2 Satiety Effect

A short-term clinical study has been conducted on normal-weight individuals to investigate the satiety effect of dried plums. Dried plums were served as pre-meal snacks and the energy and satiety were recorded. As compared to a bread product, a pre-meal snack of dried plums led to decreased energy intake at later meals, and increased satiety was maintained between the snack and meal (Stacewicz-Sapuntzakis 2013).

12.3.8 Antimicrobial Properties

The antimicrobial activities of fresh and dried plums on food-borne pathogens have been analyzed (Stacewicz-Sapuntzakis 2013). The crude ethanol extract of dried plum, along with ethyl acetate and chloroform fraction, showed varying degrees of antimicrobial activity against five Gram-positive bacteria, including *Staphylococcus aureus*, *Bacillus cereus*, *Streptococcus intermedius*, *Bacillus pumilus*, and four Gram-negative bacteria, including *Escherichia coli*, *Salmonella typhi*, *Proteus mirabilis Shigella flexneri* and *Klebsiela pneumoniae* (Yaqeen et al. 2013). Cevallos-Casals et al. (2006) selected a plum genotype (*Prunus salicina* Erhr. and hybrids) with rich phenolic compounds and found that they showed strong antibacterial activity against *Salmonella Enteritidis* and *Escherichia coli*. The strong antimicrobial activity of plum may be attributed to the antimicrobial properties of phenolics, including anthocyanins and chlorogenic acid. Phenolic compounds may exhibit their antimicrobial activity by the reactive portion of free hydroxyl groups (Cevallos-Casals et al. 2006). Phenolic compounds could also kill *Helicobacter pylori* by interacting with nitrites in the stomach and producing nitric oxide. In addition, a class of phenolic polymers (proanthocyanidin) was capable of inhibiting the adhesion of *Helicobacter pylori* to the stomach walls (Stacewicz-Sapuntzakis 2013).

12.3.9 Anti-Tumour Effects

The anti-tumor effects of plum and its products are highly associated with plums'antioxidant capacity. Previous studies indicated that a lower risk of cancer is related to a diet that is rich in antioxidants, as antioxidants play a significant role in the prevention of reactive oxygen species-mediated carcinogenesis.

Considerable cell culture studies have been carried out to evaluate the anti-tumor effects of plum and its products. Fujii et al. (2006) found that 250 μg/mL of concentrated prune juice was reported to be capable of reducing the viability of Caco-2 (a human colon cancer cell line) and KATO III (the stomach carcinoma cells) and exerted increased apoptosis in Caco-2 cells but had no impact on normal colon fibroblasts. Olsson et al. (2004) found that fresh plum extracts with a concentration in the range of 0.05–0.5% could suppress the proliferation of HT29 (human colon cancer cells) and MCF-7 (estrogen-dependent breast cancer cell line). Noratto et al. (2009) investigated the antiproliferative effects of plum extract and various phenolic fractions on MCF-710A (estrogen-dependent breast cancer cells) and MDA-MB-435 (estrogen-independent breast cancer cells). Results showed that plum extract exhibited antiproliferative activities on both cell lines. The dose range effective for cell growth suppression was in the range of 40–70 and 700–1000 mg of chlorogenic acid equivalent/L for MDA-MB-435 and MCF-7 respectively, suggesting better inhibitory effects against estrogen-independent breast cancer cells. Among various phenolic fractions of plum extract, proanthocyanidin fractions and flavonol, including quercetin glucoside and rutinoside, displayed stronger antiproliferative activities. The phenolic acid fraction also suppressed the proliferation of normal breast epithelial cells, though with a milder effect than other fractions. In particular, chlorogenic and neo-chlorogenic acids showed a relative higher degree of antiproliferation on MDA-MB-435 cells but low toxicity against the normal MCF-10A cells. Moreover, plum was reported to exert anti-tumor effects by reducing the carcinogenesis caused by heterocyclic amines, which were products during cooking meat and are considered to be carcinogenic in animals tests. Edenharder et al. (2002) found that the application of fresh plum homogenate reduced DNA damage induced by one of the heterocyclic amines in Chinese hamster lung fibroblast cultures at similar concentrations of red wine or red grape homogenate.

For animal studies, Yang and Gallaher (2005) used rat models injected with a carcinogen compound (azoxymethane) to investigate the effects of dried plum supplementation on colon cancer risk factors. Rats were fed with 4.75% and 9.5% of dried plum powder supplementation in diets for ten days. Results showed that supplementation with dried plums did not decrease the number of aberrant crypt foci but positively changed other colon cancer risk factors. With the supplementation of dried plums, the fecal bile acids, deoxycholic acid and hyodeoxycholic acid concentrations were significantly decreased. Moreover, cecal 7α-dehydroxylase activity, nitroreductase activities and supernatant oxygen radical absorbance capacity were reduced, increased and increased respectively.

Furthermore, inferring from active ingredients, dried plums may have protective effects against colon cancer, as they are rich in fibre, sorbitol and phenolic compounds. Fibre and sorbitol could shorten the gut transit time and reduce fecal bile acid concentration, while phenolic compounds could prevent DNA damage in colonic epithelium. Dietary fibre could also bind cancer-promoting secondary bile acids in the intestinal tract, thus reducing plasma cholesterol and facilitating the removal of secondary bile acids (Stacewicz-Sapuntzakis 2013).

12.4 CONCLUSION

Plums have been a go-to fruit to help with cases of decreased bowel movements to severe constipation due to its high fibre content. However, in recent years, plum and its products have been shown to modulate more than just the gastrointestinal system in the human body; other important benefits include the modulation of cardiovascular, immune and metabolic systems. These benefits are attributed to bioactive compounds present in plum and its products. While this chapter details the major nutritional, phytochemical and volatile organic constituents in plums and their role in human health, much work remains to be done in understanding the mode of action of the phytochemical in plum in improving our health.

12.5 REFERENCES

Ahmed, T., Sadia, H., Batool, S., Janjua, A., and Shuja, F. 2010. Use of Prunes as a Control of Hypertension. *Journal of Ayub Medical College, Abbottabad: JAMC* 22 (1): 28–31.

Ahmed, T., Sadia, H., Khalid, A., Batool, S., and Janjua, A. 2010. Prunes and Liver Function: A Clinical Trial. *Pakistan Journal of Pharmaceutical Sciences* 23 (4): 463–466.

Arjmandi, B.H., Lucas, E.A., Juma, S., Soliman, A., Stoecker, B.J., and Khalil, D.A. 2001. Dried plums prevent ovariectomy-induced bone loss in rats. *Journal of the American Nutraceutical Association* 4: 50–56.

Arjmandi, B.H., Khalil, D.A., Lucas, E.A., Georgis, A., Stoecker, B.J., Hardin, C., Payton, M.E., and Wild, R.A. 2002. Dried Plums Improve Indices of Bone Formation in Postmenopausal Women. *Journal of Women's Health and Gender-Based Medicine* 11, 61–68.

Atkinson, F.S., Foster-Powell, K., and Brand-Miller, J.C. 2008. International Tables of Glycemic Index and Glycemic Load Values: 2008. *Diabetes Care* 31 (12): 2281–2283.

Attaluri, A., Donahoe, R., Valestin, J., Brown, K., and Rao, S.S.C. 2011. Randomised Clinical Trial: Dried Plums (prunes) vs. Psyllium for Constipation. *Alimentary Pharmacology and Therapeutics* 33 (7): 822–828.

Badary, O.A., Awad, A.S., Sherief, M.A., and Hamada, F.M.A. 2006. In Vitro and in Vivo Effects of Ferulic Acid on Gastrointestinal Motility: Inhibiton of Cisplatin-Induced Delay in Gastric Emptying in Rats. *World Journal of Gastroenterology* 12 (33): 5363–5367.

Bouayed, J., Rammal, H., Dicko, A., Younos, C., and Soulimani, R. 2007. Chlorogenic acid, a polyphenol from Prunus domestica (Mirabelle), with coupled anxiolytic and antioxidant effects. *Journal of the Neurological Sciences* 262 (1–2): 77–84.

Bu, S.Y., Lerner, M., Stoecker, B.J., Boldrin, E., Brackett, D.J., Lucas, E.A., and Smith, B.J. 2008. Dried plum polyphenols inhibit osteoclastogenesis by downregulating NFATc1 and inflammatory mediators. *Calcified Tissue International* 82 (6): 475–488.

Cevallos-Casals, B.A., Byrne, D., Okie, W.R., and Cisneros-Zevallos, L. 2006. Selecting new peach and plum genotypes rich in phenolic compounds and enhanced functional properties. *Food Chemistry* 96 (2): 273–280.

Cheskin, L., Mitola, A., Ridoré, M., Kolge, S., Hwang, K., and Clark, B. 2012. A Naturalistic, Controlled, Crossover Trial of Plum Juice versus Psyllium versus Control for Improving Bowel Function. *The Internet Journal of Nutrition and Wellness* 7 (2): 1–10.

Deyhim, F., Stoecker, B.J., Brusewitz, G.H., Devareddy, L., and Arjmandi, B.H. 2005. Dried Plum Reverses Bone Loss in an Osteopenic Rat Model of Osteoporosis. *Menopause* 12 (6): 755–762.

Dikeman, C.L., Bauer, L.L., and Fahey, G.C. 2004. Carbohydrate Composition of Selected Plum/Prune Preparations. *Journal of Agricultural and Food Chemistry* 52 (4): 853–859.

Edenharder, R., Sager, J.W., Glatt, H., Muckel, E., and Platt, K.L. 2002. Protection by Beverages, Fruits, Vegetables, Herbs, and Flavonoids Against Genotoxicity of 2-Acetylaminofluorene and 2-Amino-1-Methyl-6-Phenylimidazo[4,5-b]pyridine (PhIP) in Metabolically Competent V79 cells. *Mutation Research—Genetic Toxicology and Environmental Mutagenesis* 521 (1–2): 57–72.

Fujii, T., Ikami, T., Xu, J.W., and Ikeda, K. 2006. Prune extract (Prunus domestica L.) suppresses the proliferation and induces the apoptosis of human colon carcinoma caco-2. *Journal of Nutritional Science and Vitaminology* 52 (5): 389–391.

Gallaher, C.M., and Gallaher, D.D. 2009. Dried plums (prunes) reduce atherosclerosis lesion area in apolipoprotein E-deficient mice. *British Journal of Nutrition* 101 (2): 233–239.

García-Mariño, N., De La Torre, F., and Matilla, A.J. 2008. Organic acids and soluble sugars in edible and nonedible parts of Damson Plum (Prunus domestica L. subsp. insititia cv. Syriaca) fruits during development and ripening. *Food Science and Technology International* 14 (2): 187–193.

González-Cebrino, F., García-Parra, J., and Ramírez, R. 2016. Aroma profile of a red plum purée processed by high hydrostatic pressure and analysed by SPME-GC/MS. *Innovative Food Science and Emerging Technologies* 33: 108–114.

Hooshmand, S., Brisco, J., Elam, M., Chai, S.C., and Arjmandi, B.H. 2013. Bone Protective Effects of Dried Plum is Through Increasing Osteoprotegerin and Suppressing Sclerostin Levels. *The FASEB Journal* 27 (S1): 371.2–371.2.

Hooshmand, S., Kumar, A., Zhang, J.Y., Johnson, S.A., Chai, S.C., and Arjmandi, B.H. 2015. Evidence for anti-inflammatory and antioxidative properties of dried plum polyphenols in macrophage RAW 264.7 cells. *Food and Function* 6 (5): 1719–1725.

Igwe, E.O., and Charlton, K.E. (2016). Review. *A Systematic Review on the Health Effects of Plums (Prunus domestica and Prunus salicina). Phytotherapy Research* 30 (5): 701–731, January.

Johnston, K.L., Clifford, M.N., and Morgan, L.M. 2003. Coffee acutely modifies gastrointestinal hormone secretion and glucose tolerance in humans: Glycemic effects of chlorogenic acid and caffeine. *American Journal of Clinical Nutrition* 78 (4): 728–733.

Karasawa, K., Miyashita, R., and Otani, H. 2012. Anti-allergic Properties of a Fruit Extract of Prune (prunus domestica L.) in Mite-sensitized BALB/c Mice. *Food Science and Technology Research* 18 (5): 755–760.

Keast, D.R., and Jones, J.M. 2009. Dried Fruit Consumption Associated with Reduced Overweight or Obesity in Adults: NHANES, 1999-2004. *The FASEB Journal* 23.

Ko, S.H., Choi, S.W., Ye, S.K., Cho, B.L., Kim, H.S., and Chung, M.H. 2005. Comparison of the Antioxidant Activities of Nine Different Fruits in Human Plasma. *Journal of Medicinal Food* 8 (1): 41–46.

Kumar, A., Hooshmand, S., and Arjmandi, B.H. 2009. Dried Plum Polyphenols Decreased Markers of Inflammation and Lipid Peroxidation in RAW264.7 Macrophage Cells. *The FASEB Journal* 27 (S1).

Kuo, P.H., Lin, C.I., Chen, Y.H., Chiu, W.C., and Lin, S.H. 2015. A High-Cholesterol Diet Enriched with Polyphenols from Oriental Plums (*Prunus salicina*) Improves Cognitive Function and Lowers Brain Cholesterol Levels and Neurodegenerative-Related Protein Expression in Mice. *British Journal of Nutrition* 113 (10): 1550–1557.

Lee, S.H., Lillehoj, H.S., Lillehoj, E.P., Cho, S.M., Park, D.W., Hong, Y.H., Park, H.J. et al. 2008. Immunomodulatory properties of dietary plum on coccidiosis. *Comparative Immunology, Microbiology and Infectious Diseases* 31 (5): 389–402.

Lin, J.Y., and Tang, C.Y. 2007. Determination of total phenolic and flavonoid contents in selected fruits and vegetables, as well as their stimulatory effects on mouse splenocyte proliferation. *Food Chemistry* 101 (1): 140–147.

Livesey, G. 2003. Health potential of polyols as sugar replacers, with emphasis on low glycaemic properties. *Nutrition Research Reviews* 16 (2): 163–191.

McRorie, J., Zorich, N., Riccardi, K., Bishop, L., Filloon, T., Wason, S., and Giannella, R. 2000. Effects of olestra and sorbitol consumption on objective measures of diarrhea: Impact of stool viscosity on common gastrointestinal symptoms. *Regulatory Toxicology and Pharmacology* 31 (1): 59–67.

Mishra, N., Gill, N.S., Mishra, A., Mishra, S., Shukla, A., and Upadhayay, A. 2012. Evaluation of antioxidant and antiulcer potentials of prunus domestica fruit methanolic and extract on wistar albino rats. *Journal of Pharmacology and Toxicology* 7: 305–311.

Negishi, H., Onobayashi, Y., Xu, J.W., Njelekela, M.A., Kobayakawa, A., Yasui, N., Yamamoto, J., Ikami, T., Ikeda, K., and Yamori, Y. 2007. Effects of Prune Extract on Blood Pressure Elevation in Stroke-Prone Spontaneously Hypertensive Rats. *Clinical and Experimental Pharmacology and Physiology* 34: 47–48.

Noratto, G., Martino, H.S.D., Simbo, S., Byrne, D., and Mertens-Talcott, S.U. 2015. Consumption of polyphenol-rich peach and plum juice prevents risk factors for obesity-related metabolic disorders and cardiovascular disease in Zucker rats. *Journal of Nutritional Biochemistry* 26 (6): 633–641.

Noratto, G., Porter, W., Byrne, D., and Cisneros-Zevallos, L. 2009. Identifying peach and plum polyphenols with chemopreventive potential against estrogen-independent breast cancer cells. *Journal of Agricultural and Food Chemistry* 57 (12): 5219–5226.

Olsson, M.E., Gustavsson, K.E., Andersson, S., Nilsson, Å., and Duan, R.D. 2004. Inhibition of cancer cell proliferation in vitro by fruit and berry extracts and correlations with antioxidant levels. *Journal of Agricultural and Food Chemistry* 52 (24): 7264–7271.

Piirainen, L., Peuhkuri, K., Bäckström, K., Korpela, R., and Salminen, S. 2007. Prune juice has a mild laxative effect in adults with certain gastrointestinal symptoms. *Nutrition Research* 27 (8): 511–513.

Prior, R.L., Gu, L., Wu, X., Jacob, R.A., Sotoudeh, G., Kader, A.A., and Cook, R.A. 2007. Plasma antioxidant capacity changes following a meal as a measure of the ability of a food to alter in vivo antioxidant status. *Journal of the American College of Nutrition* 26 (2): 170–181.

Rendina, E., Lim, Y.F., Marlow, D., Wang, Y., Clarke, S.L., Kuvibidila, S., Smith, B.J. et al. 2012. Dietary supplementation with dried plum prevents ovariectomy-induced bone loss while modulating the immune response in C57BL / 6J mice ☆. *The Journal of Nutritional Biochemistry* 23 (1): 60–68.

Rodriguez de Sotillo, D.V., and Hadley, M. 2002. Chlorogenic acid modifies plasma and liver concentrations of: Cholesterol, triacylglycerol, and minerals in (fa/fa) Zucker rats. *Journal of Nutritional Biochemistry* 13 (12): 717–726.

Sahamishirazi, S., Moehring, J., Claupein, W., and Graeff-Hoenninger, S. 2017. Quality assessment of 178 cultivars of plum regarding phenolic, anthocyanin and sugar content. *Food Chemistry* 214: 694–701.

Shukitt-hale, B., Ph, D., Kalt, W., Ph, D., Carey, A.N., A, M., Ph, D. et al. 2009. Plum juice, but not dried plum powder, is effective in mitigating cognitive deficits in aged rats. *NUT* 25 (5): 567–573.

Stacewicz-Sapuntzakis, M. 2013. Dried Plums and Their Products: Composition and Health Effects-An Updated Review. *Critical Reviews in Food Science and Nutrition* 53 (12): 1277–1302.

Thom, E. 2007. The effect of chlorogenic acid enriched coffee on glucose absorption in healthy volunteers and its effect on body mass when used long-term in overweight and obese people. *Journal of International Medical Research* 35 (6): 900–908.

Tinker, L.F., Schneeman, B.O., Davis, P.A., Gallaher, D.D., and Waggoner, C.R. 1991. Consumption of prunes as a source of dietary fiber in men with mild hypercholesterolemia. *American Journal of Clinical Nutrition* 53 (5): 1259–1265.
van Gorsel, H., Chingying, L., Kader, A.A., Kerbel, E.L., and Smits, M. 1992. Compositional Characterization of Prune Juice. *Journal of Agricultural and Food Chemistry* 40 (5): 784–789.
Wallace, T.C. 2017. Dried plums, prunes and bone health: A comprehensive review. *Nutrients* 9 (4): 1–21.
Wu, H., Dwyer, K.M., Fan, Z., Shircore, A., Fan, J., and Dwyer, J.H. 2003. Dietary fiber and progression of atherosclerosis: The Los Angeles Atherosclerosis Study. *American Journal of Clinical Nutrition* 78 (6): 1085–1091.
Yang, Y., and Gallaher, D.D. 2005. Effect of dried plums on colon cancer risk factors in rats. *Nutrition and Cancer* 53 (1): 117–125.
Yaqeen, Z., Naqvi, N.U.H., Sohail, T., Zakir-Ur-Rehman, Fatima, N., Imran, H., and Atiq-Ur-Rehman. 2013. Screening of solvent dependent antibacterial activity of Prunus domestica. *Pakistan Journal of Pharmaceutical Sciences* 26 (2): 409–414.
Yu, J., Li, W., You, B., Yang, S., Xian, W., Deng, Y., Yang, R. et al. 2021. Phenolic Profiles, Bioaccessibility and Antioxidant Activity of Plum (*Prunus salicina* Lindl). *Food Research International* 143, March.
U.S. Department of Agriculture, Agricultural Research Service. FoodData Central. 2019. www.fdc.nal.usda.gov.

13 Effects of Pre- and Postharvest Processing Technologies on Bioactive Compounds of Plum

Sabeera Muzzaffar and Munazah Sidiq
Department of Food Science and Technology,
University of Kashmir, Srinagar, Jammu and
Kashmir, India

CONTENTS

DOI: 10.1201/9781003205449-13

13.1 INTRODUCTION

Plum, a stone fruit, is a member of the Rosaceae family, which is a dominant family of fruits that includes apples, blackberries, cherries, nectarines, peaches, pears, raspberries and strawberries. Approximately 19–40 different plum species exist; however, only the two species of genus *Prunus*, i.e., European plums (*Prunus domestica* L.), which is a hexaploid, and Japanese plum (*Prunus salicina* L.), a diploid, are recognized for their commercial significance. It is presumed that the European plum has its origin somewhere near the Caspian Sea. It was discovered about 2,000 years ago and is believed to have been introduced by the Pilgrims into the United States during the 17th century. China is believed to be the origin of the Japanese plum; however, the name was derived from Japan, where it was developed and planted (Ezinne et al. 2016). The berries of this family are recognized by diverse physical and dietary features. The polyphenolic compounds are present in sufficient quantities in these fruits, and such features have led to an increased focus on their functional properties (Paliyath et al. 2008).

Different plum cultivars exhibit major differences in their quality parameters viz., total soluble solids (TSS), total phenols and anthocyanins. Such differences are observed in fresh as well as processed products. Quality parameters of the fruit, like sugars, organic acids, color, etc., change during ripening (Usenik et al. 2008). Consumer choice for fresh and processed plum are affected by the stage of maturity, the variety, horticultural practices, place of origin, season of growth, storage conditions and also procedures followed during processing (Donovan et al. 1998; Usenik et al. 2008).

The systematic classification of plums is:

Kingdom: *Plantae*
Division: *Magnoliophyta*
Class: *Magnoliopsida*
Order: *Rosales*
Family: *Rosaceae*
Subfamily: *Maloideae* or *Spiraeoideae*
Genus: *Prunus*
Subgenus: *Prunus*

13.2 PLUM PHENOLICS AND ANTIOXIDANTS

Plum fruit is rich in different phytochemicals, which are primarily phenolic acids and derivatives of flavonols (Tomás-Barberán et al. 2001; Nakatani et al. 2000; Auger et al. 2004). Such components, besides imparting organoleptic properties like perception, hue and essence in fruits, vegetables and beverages, have strong antioxidant capacities as well (Cao et al. 1997; Kim et al. 2003b; Vinson et al. 2001). The radical scavenging activity of plum fruits is particularly higher than that of oranges, apples and strawberries (Kayano et al. 2002). Although apples are a more popular commercial product, plums show 4.4 times higher radical scavenging activity (Wang et al. 1996). Redox imbalance often induces the formation of free radicals associated with chronic diseases. The reactive free radicals, including superoxide anion (O^{2-}), hydroxyl radical (OH^-), and peroxy radical (ROO^-), cause damage to biological systems in terms of the disruption of lipid peroxidation, protein denaturation, cell membranes, membrane fluidity, oxidative DNA damage and adjustment of the functioning of platelets (Fridovich 1978; Kinsella et al. 1993). Such free radicals are scavenged by the dietary oxidants, thereby providing an efficient defense mechanism (Elmegeed et al. 2005; Gutteridge 1993; Harris et al. 1992).

Phenolic components in plants perform various diverse biological functions, which includes their contribution in growth, development and protection against stress caused due to various biotic and abiotic condition. Besides influencing pigmentation, they also exhibit antimicrobial and antifungal functions. Further, phenolics protect against insects, UV radiation and free radicals, which may be produced in the process of photosynthesis. They also have a role in the chelation of toxic heavy metals (Parry et al. 2005). Although nearly every organism has developed antioxidant resistance and repair systems to defend against the production of free radicals, they do not provide complete protection (Nwanna and Oboh 2007). The most convenient and realistic manner to counter retrogressive disease is to enhance internal antioxidant capacity by consuming more fruits and vegetables (Oboh et al. 2008). These types of plant-based foods are rich in naturally occurring antioxidants that manipulate free radicals. Bouayed et al. (2007), Shukitt-Hale et al. (2009), Zaidi et al. (2012) and Nile and Park (2014) reported that the phenolics present in plum provides anti-inflammatory, cytoprotective, antioxidant, anti-carcinogenic, antimutagenic, antimicrobial and neuroprotective characteristics.

13.2.1 Anthocyanins

Anthocyanins comprise an anthocyanidin (aglycone form) bonded to one or more sugars at different hydroxylation sites (Usenik et al. 2009). There are six common anthocyanidins present in different fruits, including pelargonidin, peonidin, delphinidin, cyanidin, petunidin and malvidin (Kotepong et al. 2011). The sugars generally associated with anthocyanidins include glucose, galactose, rhamnose and arabinose. Further modifications in these structures are brought by the addition of other compounds, such as methyl groups, acetic acid, propionic acid, caffeic acid and malonic acid, bonded to sugar moieties. The most abundant anthocyanin in edible plants is aglycone, constituting about 90 per cent of anthoc yanin-containing fruit (Prior and Cao 2000). Several factors, like temperature, pH, light, oxygen and the presence of other compounds, influence anthocyanins (Rein 2005) and thus will affect the color displayed. For instance, a change in pH will alter the structure of anthocyanin and lead to a change in color (Stintzing et al. 2002). Co-pigmentation between anthocyanins and other compounds, like colorless flavonols or phenolic acids, or association with metals may affect the intensity of the color displayed (Boulton 2001). Plums possess high levels of anthocyanins (Proteggente et al. 2002). The chief anthocyanins in plums are cyanidin-3-glucoside and cyanidin-3-rutinoside (Wu and Prior 2005), but further cyanidin derivatives, including cyanidin 3-(6″-acetoyl) glucoside, cyanidin 3-(6″-malonyl) glucoside and cyanidin-3-galactoside and peonidin derivatives, have also been quantified. The anthocyanin content of dark-fleshed and dark-peeled varieties makes them substantial sources of dietary anthocyanins. Tomás-Barberán et al. (2001) revealed that the anthocyanin was concentrated predominantly in the peel of the fruit for a variety of plum samples. The phenolic composition will differ between cultivars. A literature review suggests that fruits high in cyanidin, peonidin or pelargonidin glycosides shows more anti-inflam matory effects than fruits containing mostly delphinidin, malvidin and petunidin glycosides, like blueberries. One of the research investigations suggests that the total per-day average intake of flavonoids in consumer fruits, beverages, vegetables and berries contributes 67, 25, 5 and 3 per cent anthocyanins, respectively (Prior and Cao 2000). Further, anthocyanin-rich foods and derived products are believed to protect against hypertension, inflammation, cardiovascular disease and oesophageal cancer (Cassidy et al. 2000; Chen et al. 2007). As anthocyanins have generated importance among researchers in the last decade, several breeding programs were conducted to develop anthocyanin-rich hybrids. In one such program funded by the Government of Australia, the Queen Garnet plum, a hybrid of the Japanese plum was established. This plum has remarkably enhanced anthocyanin levels, reaching up to 277 mg per 100g of fruit (Fanning et al. 2013), which is supposed to be higher than the total anthocyanin content of regular plums (5–173 mg per 100g) across harvest years (Miletić et al. 2012). Studies conducted using this hybrid plum have showed antithrombotic activity in humans (Santhakumar et al. 2015a), and its useful impact on metabolic syndrome in rat models, in vivo and in vitro bioactivity has also been demonstrated (Bhaswant et al. 2015).

13.2.2 Flavonols

Among the flavonols, quercetin is most often found in plant material; however, kaempferol and myricetin also exist in this group (Hertog et al. 1992). Flavonols are mainly found in the peel or leaves of plants, as their synthesis is promoted by exposure to light (Awad et al. 2000). In plums, quercetin-3-O-rutinoside was frequently reported as the primary flavonol (Kim et al. 2003a). Several quercetin glycosides may also be present in plums, including quercetin pentosyl-hexoside, quercetin-3-glu coside, quercetin-3-rutinoside, quercetin pentosyl-pentoside, quercetin-3-xyloside, quercetin-3-arabionside, quercetin acetyl-hexoside and quercetin-3-rhamnoside (Tomás-Barberán et al. 2001; Chun et al. 2003). Quercetins were found to be the most available polyphenols in over two-thirds of the plum selections tested, while rutin was the most abundant glycoside; however, it was not detected in all selections (Mubarak et al. 2012).

Plum fruits contain various important secondary metabolites, including flavonoids and phenolic acids (Tomás-Barberán et al. 2001), which shows a strong antioxidant capacity (Cao et al. 1997). Phenolic phytochemicals have an ability to reduce oxidative stress due to their capability to act as foragers of electrophiles and free radicals. Oxidative stress occurs when a well-maintained balance between pro-oxidants and antioxidants is disturbed, preferably favoring the pro-oxidants. This concept has helped explain the differences in the development of various diseases, aging and senescence of humans and animals. Free radicals, antioxidants and co-factors are the three main areas that contribute to aging (Rahman et al. 2006). Fruits and vegetables acts as a basic food source by providing important nutrients for supporting life; moreover, they also possess a diversity of phytochemicals, such as phenolics and flavonoids (Tomás-Barberán et al. 2001). Free radicals are responsible for inducing oxidative stress, leading to membrane lipid oxidation, oxidative damage to protein and DNA, enzyme inactivation and gene mutation that may be a cause of carcinogenesis.

Various reactive oxygen species (ROS) that include hydroxyl radical, peroxyl radical and superoxide radical anion are physiologically generated while performing metabolic activity in living organisms. Vast intensities of reactive oxygen species beyond the normal defense capacity of antioxidant systems may result in oxidative damages, which are associated with many age-related degenerative diseases, like cancers and Alzheimer's disease. Several reports have proposed that the phytochemical content contributing to the antioxidant activity of fruits and vegetables are associated with reduced risks of cancers and cardiovascular disease. Phenolic compounds (flavonoids), nitrogenous compounds (chlorophyll derivatives), tocopherols, carotenoids and ascorbic acids are the major phytochemicals present in plants. These antioxidants exert their action either by scavenging free radical species or impeding the fabrication of reactive species, thereby inhibiting destruction to lipids, proteins and nucleic acids, which eventually leads to cell damage and death (Ames et al. 1993). Moreover, the antioxidants present in fruits and vegetables have also been found to show anticarcinogenic and antimutagenic activity. Plum phenolics, such as anthocyanins have an ability to scavenge superoxide radical anions, an initiator of an ROS generation system (Chun et al. 2003). Thus, oxidative damage could be decreased by ingesting anthocyanin-rich fruits and their products, proving their higher antioxidant capacity than apples, one of the most widespread fruits in the market (Kim et al. 2003b).

13.3 INFLUENCE OF PREHARVEST FACTORS ON PLUM FRUIT QUALITY AND BIOACTIVE COMPOUNDS

Variation in ripening and storage periods strongly influence fruit quality. Non-uniform fruit quality compromises consumer satisfaction (Murray et al. 2005). The ripening process of plum takes place from the top of the tree to the bottom, and as a result the lowest fruit matures 10–14 days later than the uppermost fruits that are better exposed to light. Compared to other fruits, like peaches and nectarines, the first harvest of plum fruit is usually the largest, and for color-filled varieties the produce is usually limited to a certain portion of tree, which is segregated by exposure to light; for example, for the first harvest the top third of the tree is harvested and for the second harvest the middle third is picked (Manganaris et al. 2008).

"Bioactive compounds" may be defined as extra-nutritional compounds that are usually present in smaller amounts in food. They are widely studied owing to their nutraceutical potential (Kris Etherton et al. 2002). Plums are repositories of such composites, such as phenolics and antioxidants. Among the phenolics present in plum, neo-chlorogenic acid, *p*-coumaroyl, quinic acid, cholorogenic acid and rutin predominate (Singh and Khan 2010). These compounds have an important role in defense mechanisms against chronic diseases including cardiovascular diseases and cancers.

13.3.1 Environmental Factors

Commonly, fruits and vegetables are grown in an open field and hence are affected by natural environmental fluctuations, including cultivar/genotype, light, temperature, irrigation, fertilizers and so on.

13.3.1.1 Cultivar/Genotype

The fruit cultivar or rootstock plays a major role in determining postharvest fruit quality. The Soluble solid content (SSC) and Titrable acidity (TA) concentration shows significant variation among various cultivars. Studies demonstrated that the Black Amber plum exhibits a higher respiration and ethylene production rate than that of the Amber Jewel variety, suggesting that the Black Amber cultivar might have a shorter shelf life than the Amber Jewel (Khan 2016). Similarly, for bioactive compounds, it was researched that the red plum has a higher concentration of antioxidant and flavonoid compounds, but the sugar plum cultivar exhibits higher antioxidant activity (FRAP and ABTS). In addition, the resveratrol concentration of red plum was higher than other genotypes (Murathan et al. 2020).

13.3.1.2 Light

Light is an important factor for photosynthesis, hence it has a profound effect on the growth and development of plants, but its entire spectrum is not useful for plants.

Usually only the visible part of the light spectrum is utilized by living organisms. Moreover, light also plays a role in controlling flowering time and morphogenesis. For morphology and the development of plants, two major photoreceptors—phyto chromes (absorbs red/far-red light) and cryptochromes (absorbs blue/ultraviolet A [UV-A] light)—are involved (Quail et al. 1995).

The quality of the light disturbs the photo-oxidative properties of plants by modifying their antioxidant defense systems, causing an increase in the activity of antioxidant enzymes. In vegetables like tomatoes, kale, peas and Chinese cabbage the usage of single spectral or combined red light (625–630 mm) and blue lights (465–470 mm) results in more antioxidant activity than the usage of a white light source (Johkan et al. 2010; Lee et al. 2016). Furthermore, treatment of green (510 nm), yellow (595 nm) or even mixed red-white LEDs also enhances the potential of scavenging free radicals and the accumulation of anthocyanin (Dong et al. 2014). This increase in antioxidant activity might be due to the generation of free radicals, glucosinolates, ß-carotene, reactive oxygen species, scavenging enzymes, like superoxide dismutase, and also vitamin C (Wu et al. 2007; Muneer et al. 2014). Therefore, it would be of interest to discover the health benefits of ingesting LED-treated crops (Hasan et al. 2017).

Apart from single-spectral red light, the merging of blue and red LEDs also enhances the buildup of prime metabolites, anthocyanin, total polyphenols and flavonoids. Further, red LEDs have a distinct effect on anthocyanin accumulation as opposed to blue LEDs, perhaps because of the enhanced expression of the MdMYB10 and MdUFGT genes responsible for anthocyanin biosynthesis (Lekkham et al. 2016). Environmental light augmented with red, blue and green LEDs also results in the enhancement of organic acids, phenolic compounds, α-tocopherol, vitamin C and nitrate in different crops (Bantis et al. 2016; Lin et al. 2013). The role of LEDs in the generation of all these secondary metabolites lies within the activity of the PAL enzyme, which is involved in the phenyl propanoid pathway (Heo et al. 2012). In ginseng plants, the major secondary metabolites produced by the isoprenoid pathway are ginsenosides, which ehave high medicinal values. It has been reported that under the influence of blue LEDs (450 nm and 470 nm) there was an increase in concentration (from 2 to 74 per cent) of total ginsenosides than that under dark conditions. Therefore, it is likely possible that LEDs can be used as elicitors, activates the appearance of key enzymes (like squalene synthase) in the isoprenoid pathway, or may also generate ROS, which on the other hand may enhance the activity of defense-related genes, and in this way increases the synthesis of ginsenosides (Park et al. 2012). Formerly, it has been assumed that the increased accumulation of prime metabolites in crops might be because of an inhibition of photosynthetic products caused by LEDs. Additionally, light also affected signal transduction pathways, which also includes enzymes, metabolites and secondary messengers. The data strongly proposes that light can also be used for the generation of medicinally important secondary metabolites in plants. For the enhancement of the nutritional value of crops under controlled cultivation conditions, blue LEDs or a combination of red/blue LEDs might be the best choice (Kozai et al. 2016). The effect of sunlight has been shown to be linked with enhanced activity of bioactive compounds, like that of ascorbic acid (Lee and Kader 2000) and anthocyanins. As per the results of Murray et al. (2005), poor color development takes place in plums that are kept under shade. Further exposure of plums toward UV radiation resulted in the accumulation of anthocyanins (Arakawa 1993) and enhanced the activity of enzymes involved in phenyl propanoid metabolism, including PAL, chalcone synthase and dihydro flavonol reductase (Tomás-Barberán and Espin 2001). Moreover, differences in temperature during the day and at night also affect anthocyanin accumulation in plums (Tsuji et al. 1983), and the regulated deficiency of water results in decreased levels of vitamin C and carotenoids in fruit peels, whereas anthocyanins and procyanidin increase (Buendía et al. 2008).

13.3.1.3 Temperature

Temperature has a profound effect on the bioactive compounds of plum. The intermediate temperature changes the phytochemical composition, which is correlated to the antioxidant activity and also to the taste of the fruit. It has been suggested that the storage of Friar plums at transitional temperatures of 5°C and 15°C causes greater and faster accumulation of anthocyanins, including cyanidine-3-o-glucoside, and less accumulation of cyanidin-3-o-rutinoside, but at 0°C and 25°C no anthocyanin accumulation takes place. In the case of plums stored at 5°C, the antioxidant activity is greatly due to anthocyanins. The combination of anthocyanin with phenolics like syringic, protocatechuic, trans-p-coumaric and caffeic acids at 15°C contributes to total antioxidant activity. Storing plums at 15°C delays the breakdown of sucrose to its monomer-glucose and fructose, whereas storage at 5°C maintained accumulated fructose. The altered composition at transitional temperatures changes the composition of organic acids, which in turn generates varied tastes and flavors with reasonable SSC and TA. Hence, storing plums at intermediate temperatures can be an effective method for the postharvest changes of naturally occurring bioactive compounds in Friar plums. This method would provide another choice of fruits in the market for consumers who are concerned with promoting good health (Wang et al. 2018).

13.3.1.4 Irrigation

Of the total global water consumption, 80–85 per cent of water accounts for agricultural use, regulating physiological processes including growth, hormone signaling, exocytosis, nutrient collection and so on. However, in dry or semi-arid regions, water deficit is a major environmental factor affecting the quality of fruit, in particular secondary metabolites. At present, for various fruit tree species, moderate water stress conditions are measured not only for the purpose of saving water or controlling vegetative growth but also for enhancing the value of the fruit (Behboudian et al. 2011). It has been researched that stress caused by a dearth of water induces a series of enzymes involved in the phenyl propanoid pathway, such as flavonone-3-hydroxylase (F3H), C4H, PAL, 4CL, and CHS (Deluc et al. 2009).

In the case of stone fruits, the appropriate period for regulated deficit irrigation (RDI) is during the second phase of growth (Naor 2006b). Deficit irrigation (DI) indicates a water stress period. During this period, fruit growth is minimal and hence is not affected by water deficits. For early cultivars, this growth phase is short and the postharvest period is long; therefore, it may be a suitable time for irrigation. However, precautions should be taken for the possible negative postharvest effects of water deficiency on the development of flower buds (Johnson and Handley 2000). Water stress during postharvest can also result in fruit disorders (Naor et al. 2005). The interaction between water and plants has garnered specific attention in present years. Generally, it is known that pressure generated by water enhances the quality of plum fruit (Fereres et al. 1990). In the case of stone fruits, growth follows a typical double-sigmoid pattern, with exponential growth taking place in the first stage (cell division), followed by a short period of slow growth because of pit hardening and the embryo development phase and finally rapid growth prior to harvest. The length of each individual stage differs by location and cultivar (DeJong and Goudriaan 1989). The reaction of stone fruit trees to DI is based on the growth stage at which it was used (Handley and Johnson 2000). For plums, the application of DI after pit hardening results in smaller fruit and reduced yield, while the composition, color and quality of the fruit is improved (Maatallah et al. 2015; Naor 2006). The process also results in an increase in anthocyanin accumulation in the case of peaches (Girona et al. 2003). The studies have revealed that DI also improves TSS of peach fruit and reduces acidity. During the second stage of growth, the sensitivity of fruit to water stress is minimum, owing to the slow growth rate during that period (Lawrence and Melgar 2020).

13.3.1.5 Fertilizers

The term "fertilizer" refers to a material that provides the necessary chemical elements required for the proper growth and development of a plant. It is an important tool used by farmers to increase the yield and quality of a crop. The main fertilizers used are chemical or mineral fertilizers, plant residues and manures. Today, the use of fertilizers is necessary to interchange the elements removed from the soil in the form of food and agricultural products. The labels on fertilizers consist of three bold numbers. The initial number indicates the nitrogen (N) content, the second refers to the amount of phosphate (P_2O_5) and the third number indicates the amount of potash (K_2O). For instance, a bag of 10–10–10 fertilizer represents that the bag has 10 percent nitrogen, 10 per cent phosphorous and 10 per cent potassium. The amount of fertilizer added to a plant influences the plant's quality. The principal nutrients plants need are nitrogen and potassium (Marschner 1995). Although nitrogen deficiency has been suggested to cause a decrease in the size of fruits and also result in poor flavor, its excessive use also causes some adverse effects on fruit. As reported in the case of stone fruits, the excessive use of nitrogen as a fertilizer can cause a reduction in flesh firmness and sweetness (Rettke et al. 2006), the fading of its red color (Crisosto et al. 1997) and also increased exposure to postharvest diseases (Daane et al. 1995). Potassium deficiency in stone fruits has a negative impact on productivity, quality and storage (Chatzitheodorou et al. 2004; Ruiz 2006).

The usage of chemical fertilizers enhances the yield and quality of a crop, but on the other hand has a negative influence on the environment, so to save our environment from the adverse effects of chemical fertilizers, a shift toward environmentally friendly fertilizers (EFFs) was taken. EFFs provide an appropriate method for enhancing nutrient efficiency, reducing leaching and volatilization loss of enrichers and most importantly decreasing environmental hazards. These EFFs cause a decrease in environmental pollution by reducing or regulating the release of nutrients into the soil and are therefore are also called "enhanced efficiency fertilizers" (EEFs) (Chalk et al. 2015; Timilsena et al. 2015).

Organic fertilizers are a type of EFF consisting of plant derivatives ranging from new or dehydrated plants to animal manure and also waste from agriculture (Wohlfarth and Schroeder 1979; Das and Jana 2003; Kumar et al. 2004). The nutritional composition of biological fertilizers differs by source, and biodegradable materials provide a better nutrient source. The nitrogen and phosphorous concentration is less in comparison to that of chemical fertilizers. In addition, moisture content is another parameter responsible for diluting the nitrogen or phosphorus content of organic fertilizers. Besides these two elements, the concentration of carbon is either equal to or greater than that of nitrogen and phosphorous (Schroeder 1978; Qin et al. 1995; Barkoh et al. 2005). Plums grown by organic fertilizer show more antioxidant and total phenolic activity than plums grown by conventional methods. Overall, the content of bioactive compounds in organic plums was higher and had less adverse effects on the environment (Cuevas et al. 2015).

13.3.2 Chemical Treatments

13.3.2.1 Calcium Spray

The bivalent ions of cationic calcium are required for basic processes associated with changes involved during ripening, such as changes in cell wall structure, membrane reliability, functionality and the activity of specific enzymes. Its deficiency results in physiological disorders, including cracking, bitter pit, etc. (White and Broadley 2003). To overcome all these disorders, calcium treatment is typically given to plants, which delays maturation and senescence and also confirms positive effects on the value and storage of the product. Therefore, diverse methods have been followed for the postharvest calcium treatment of fruit. The application of calcium can also be done before harvest to provide an extra supply of minerals. As calcium absorption from the soil to the shoot portion of plant is restricted, the calcium is sprayed directly onto the plant, allowing for an effective increase in the calcium concentration of the respective fruit (Ferguson and Boyd 2002).

Lopez et al. (2003) found that a preharvest calcium spray in combination with titanium shows an increase in fruit size, reduces weight loss, maintains firmness and also improves fruit surface color during storage. Further management of orchard soil also laid its impact on accumulation of secondary metabolites. A study by Lombardi-Boccia et al. (2004) revealed that fruit emerging in orchards where the soil had been enveloped with a natural meadow assembled more ß-carotene, tocopherol and total phenolic compounds compared to fruits that were grown in soil covered with *Trifolium*. Moreover, the application of synthetic auxins during pit hardening resulted in cell enlargement and increased fruit size (Stern et al. 2007).

13.3.2.2 Methyl Jasmonate

Methyl jasmonate (MJ) is a natural growth regulator phytohormone that is distributed among plants and determined by plant species, variety and developmental stage. The exposure of plants to stress increases the synthesis of phenols (Cantos et al. 2003). MJ—a methyl ester derivative of jasmonic acid—plays an important role in plant maintenance, including growth, fruit bearing, defense and stress-like injury, shock, drought, insect attack and so on (Ahmad et al. 2019). MJ treatment in fruits like pomegranates (Sayyari et al. 2011), loquats (Cao et al. 2009), raspberries (Wang and Zheng 2005) and grapes (Ruiz-Garcia et al. 2012) during preharvesting were reported to cause an increase in total phenolics. Similarly, a study by Ozturk et al. (2015) reported that MJ treatment results in a significant increase in water-soluble phenolics in plums. Further, a study by Fang et al. (2002) suggested cholorogenic acid as the most predominant polyphenol in plums. All these studies suggest that preharvest treatment of MJ results in increasing the antioxidant activity of plums by increasing its phenolic content and also provides better antimutagenic and anti-cancerous activity. Several studies have reported that MJ encouraged secondary metabolites in various fruits, including apples, grapevines, strawberries, plums and peaches. Moreover, its application also resulted in the accumulation of anthocyanin in apples, strawberries and beta-carotene in the case of tomatoes. A study by Khan and Singh (2007) reported that MJ treatments clearly influenced the bioactive compounds of plums during the process of ripening.

13.3.2.3 Aminoethoxyvinylglycine

In addition to MJ, another preharvest treatment of aminoethoxyvinylglycine (AVG) has been found to decrease the phenolics in Black Amber (Ozturk et al. 2012) and Fortune plums (Karaman et al. 2013). Further, it has also been stated that preharvest AVG usage decreases the antioxidant activity in Braeburn apples (Ozturk et al. 2013). As mentioned earlier, plums are an excellent source of antioxidants compared to other Mediterranean fruits and have various health benefits (Sahamishirazi et al. 2017). AVG behaves as a competitive inhibitor for converting S-adenosylmethionine (SAM) to 1-Aminocyclopropane carboxylic acid (ACC), a precursor for ethylene synthesis. Because of its ability to block the synthesis of ethylene, pre- and postharvest application of AVG has been used to interrupt the ripening process and increase the storage capacity of climacteric fruits. Generally, it has been suggested that preharvest treatment of AVG alters the ripening process significantly by decreasing the production of ethylene.

13.3.2.4 Salicylic and Oxalic Acid

Salicylic acid is a plant-growth hormone and generator of a wide variety of metabolic and physiological responses in plants to affect the growth and development of the plant. The acid acts as an accepted and safe phenolic compound with great ability to monitor postharvest losses of the respective crop. The preharvest treatment of salicylic acid results in enhanced resistance to pathogens, controls postharvest deterioration, maintains the quality of the fruit, as reported in peaches (Wang et al. 2006), plums (Davarynejad et al. 2015) and sweet cherries (Xu and Tian 2008). It is also noteworthy that the preharvest treatment of salicylic acid results in increased levels of total phenolics and carotenoids in the golden Japanese plum cultivar. Furthermore, the same level of increase was observed by treating plums with putrescine and moringa leaf extract (Fatma et al. 2020). Similarly, preharvest treatments of plum trees (Black Splendor and Royal Rosa) with oxalic acid results in the production of fruit with decreased postharvest ethylene production, amplified content of bioactive compounds, like that of carotenoids, anthocyanins, phenolics and also an increased activity of antioxidant enzymes, such as superoxide dismutase (SOD), peroxidase (POD), catalase (CAT) and ascorbate peroxidase (APX). These enzymes, in combination with antioxidant compounds, are utilized in the scavenging of free radicals usually produced during the process of ripening and senescence. The treatment may also lead to an increase in the capability of plum tissues to scavenge the reactive oxygen species produced during fruit ripening and hence delays the postharvest ripening and senescence process. Apart from this, quality parameters and the concentration of bioactive compounds with neutraceutical potential were enhanced during long-term storage (Martínez- et al. 2018).

13.4 INFLUENCE OF POSTHARVEST TREATMENTS ON PLUM FRUIT QUALITY AND BIOACTIVE COMPOUNDS

Harvest time is commonly determined by changes in skin color, but for fruits with darker skin colors, harvest time is determined by fruit firmness. Regardless of early harvest and allowing the fruit to ripen off the tree, plums are picked at a more advanced ripening stage (Crisosto et al. 2004). They are usually handpicked and then transported to a packing house. For the removal of field heat, a delay in cooling should be avoided to prevent fruit from softening (Crisosto 2000).

13.4.1 Cold Storage

Temperature and humidity control play an important role in extending the shelf life of fresh produce (Kader 2003). Since plums are sensitive to cold temperatures, cold storage is limited to only a few weeks so as to prevent chilling injury (Iglesias-Fernandez et al. 2007). Typically, the market life of plums is dependent on the postharvest handling of the commodity and generally varies from one to six weeks (Crisosto et al. 1999). For plum storage, temperatures of 0–71.1°C and relative humidity of 90–95% are recommended (Crisosto 2000).

Cold storage is one of the most common postharvest tools for maintaining the organoleptic quality of a product. There have been very few reports on the outcome of cold storage on bioactive compounds. During harvest and cold storage, the evolution of bioactive compounds, i.e., phenolics, carotenoids and anthocyanins of colored varieties (red, purple and yellow), of plum was assessed. At the time of harvesting, both peel and pulp showed a difference in bioactive compounds and radical scavenging activity, the concentration being higher in the peel than in the flesh. Furthermore, for the period of cold storage, total phenolics, anthocyanins and carotenoids showed a general increase, which ultimately influenced antioxidant potential. The hydrophilic antioxidant (H-AOC) was interrelated to phenolics and anthocyanins while lipophilic was related to carotenoids; therefore, the carotenoids and phenolics could be the main hydrophilic and lipophyllic compounds contributing to antioxidant activity. So, overall cold storage does not have any negative impact on plum bioactives.

13.4.2 Controlled and Modified Atmospheres

Besides temperature, atmospheric storage together with controlled atmosphere (CA) and modified atmosphere (MA) storage have also proven to be operative in prolonging the life span and maintenance of fruit quality. In CA, the concentration of CO_2 is increased and that of O_2 is decreased in order to stop ethylene production and hence delay the ongoing ripening of the fruit, reduce biological syndromes and hinder postharvest decay (Ali et al. 2016). The application of storage has not been fruitful for all fruits and vegetables, but for plums the use of atmospheric storage has been beneficial, as it helps in maintaining fruit firmness during storage and shipment (1–2 per cent O_2+3–5 per cent CO_2) (Crisosto 2000). For Japanese plums, it was reported that CA consisting of 2.5–10 per cent CO_2 and 1–5 per cent O_2 at 0°C shows satisfactory effects. Moreover, it was also reported that for 80 days of CA storage there was a reduction in brown rot, internal breakdown and loss of weight and firmness (Menniti et al. 2006). In the same way, Japanese Black Amber plums were treated with 0.6 μL L^{-1} 1-MCP and kept in an atmosphere containing 2.5 per cent O_2 and 3 per cent CO_2 for 56 days and exhibited a delay in ripening, a slow down of lipid peroxidation and a reduction in incidents of chilling injury (Singh and Singh 2013). Further, the benefits of atmospheric storage are variety-dependent; for example, in Friar plums it was observed that the quality parameters, including acidity, pH and firmness, were not affected by modified atmosphere packaging (MAP) (Cantin et al. 2008).

In MAP, atmosphere is modified within the package of the product, usually in polymeric films having permeability to H_2O, O_2 and CO_2. The response of commodity usually depends on storage temperature, interchange of prestored chemicals and maturity stage (Saltveit 2003). It has been found that use of MAP lengthens the storage life of fruits and also preserves the quality attributes of the commodity, as was reported in the case of European plums. Besides, MAP also reduces weight loss in fruits during storage (Khan and Singh 2008). It was also observed that combining MAP with 1-MCP in blackberries and autumn giant plums results in a reduction in ethylene production, fruit softening, flesh browning, loss in weight and an increase in acidity, soluble solid content and firmness of the fruit (Erkan and Eski 2012).

Similarly, storage of Laetitia plums in low-density polyethylene MAP bags with CO_2 absorber increased storage life for a period of 60 days at 0.5°C (Stanger et al. 2017).

13.4.3 Chemical Treatments

The postharvest treatment of AVG has the ability to interrupt the fruit ripening, as was observed in Friar plum fruits. Usages of 300 mg L^{-1} AVG and 200 mg L^{-1} AVG to plum fruit can be considered a good postharvest tool, as its usage results in good maintenance and also delays the loss of flesh firmness, total anthocyanin and total phenolic content. AVG extensively inhibits the respiration rate and reddening of flesh in plums. Additionally, during cold storage, the use of AVG reduces the prevalence of chilling injury and maintains significantly higher antioxidant activity (Bal 2019).

Anthocyanins play an important role in the formation of the red color in plums, their accumulation in skin and flesh may increase the bioactive compounds and hence consumer acceptability (Wang et al. 2020). The synthesis of anthocyanins via the phenyl propanoid pathway is well known in the case of plants. For anthocyanin synthesis, five regulatory enzymes, phenylalanine ammonia lyase (PAL), chalcone isomeeraase (CHI), dihydro flavonol 4-reductase, anthocyanin synthase and flavo noid-*O*-glycosyltransferase, are involved. The PAL enzyme is directly linked to the synthesis of anthocyanin. Previous studies have reported that anthocyanin accumulation in grapes leads to an increase in PAL activity (Yang et al. 2013).

Besides AVG, benzothiadiazole (BTH) is also used as a postharvest tool. BTH-treated plum also exhibit an increase in PAL activity and anthocyanin concentration during storage, as reported in the study of Huan et al. (2020). Moreover, coming to total phenolics, it was found that postharvest treatment of plums with salicylic acid led to an increase in total phenolics, and this increase was maintained during cold storage. The increase in total phenolics might be attributed to an increase in the activity of PAL, an enzyme responsible for the biosynthetic phenolic pathway (Martínez-Esplá et al. 2017). Furthermore, a study by Davarynejad et al. (2015) found that using 4 mmol/L of putrescine and 4 mmol/L salicylic acid on Santa Rosa plums results in a significant decrease in total phenolic content while storing at 4°C.

Various researchers have assessed the outcome of an ethylene inhibitor 1-MCP on several plum varieties (Martinez-Romero et al. 2003). Applying the 1-MCP treatment to plums results in a delay in fruit softening and also reduces the activity of some cell wall–degrading enzymes, like that of polygalacturonase, galactosidase and glucosidase (Dong et al. 2001). These effects of 1-MCP are concentration dependent and are usually effective at concentrations near 0.5 ml/l (Manganaris et al. 2008). A study by Menniti shows that similar results were obtained for both low temperatures and room temperature at a concentration of 0.5 ml/l 1-MCP (Menniti et al. 2004). Further, it has also been reported that exposure of fruit to 1-MCP for 12 hours delays its softening. Moreover, the use of MCP has also proven to be effective in reducing chilling injury disorders in nonsuppressed climacteric plums (Candan et al. 2007).

MJ treatment to plum after harvest significantly affects the antioxidant level during ripening. A study by Khan and Singh (2007) suggests that treatment of 10^{-5}M MJ shows maximum levels of antioxidants compared to untreated plum varieties, which then declines, while treatment of 10^{-3} M MJ depicts a continuous increase in antioxidant level throughout the ripening period (Khan and Singh 2007). A decrease in antioxidant level might be due to the initiation of the senescence process (Srilaong and Tatsumi 2003). Similar results were also obtained in the case of Angelino fruit for antioxidant level where antioxidant level was increased by treating with MJ. Furthermore, MJ has been found to increase the antioxidant level in apples (Rudell et al. 2002).

13.4.4 Edible Coating

Plum is a stone fruit that is rich in phenolic compounds, as mentioned earlier. It has been reported that during the ripening and maturation process, the total phenolic content and antioxidant activity of stone fruits increases both in the peel and flesh, with concentrations being four to five times higher in peels than in flesh (Díaz-Mula et al. 2012; Kumar et al. 2017; Martínez-Romero et al. 2017). However, as senescence takes place, cell structure breaks down and ultimately results in a decrease of phenolic compounds (Gol et al. 2013; Thakur et al. 2018). The postharvest treatment of edible coatings has been stated to reduce the loss of phenolic compounds in plums and thereby increases antioxidant activity throughout postharvest storage (Kumar et al. 2017; Thakur et al. 2018) because the edible coatings are able to reduce the breakdown of cell structure and delay senescence in fruits. Moreover, the coatings also reduce the respiration, decreasing the oxygen availability within fruit for metabolic activities and hence decreasing the activity of phenol oxidase and peroxidase (Maftoonazad et al. 2008). The same increase in antioxidant level was found in cherries (Díaz-Mula et al. 2012) and apricots (Ghasemnezhad et al. 2010). Moreover, phenolic compounds, ascorbic acid and carotenoids also show increased antioxidant activity (Ahmed et al. 2009). Ascorbic acid has the ability to forage free radicals and also stimulate the generation of α-tocopherol (Ghasemnezhad et al. 2010). A study by Kumar et al. (2017) stated that ascorbic acid content was maintained in plums coated with chitosan stored for 35 days at 1 ± 1°C, 90 ± 5 per cent RH (relative humidity) in comparison to that of the control fruit throughout storage. Further, it has also been reported that edible coatings reduce the synthesis of carotenoids in plums by decreasing respiration rate and ethylene production, which subsequently decreases the enzyme activity for carotenoid synthesis (Marty et al. 2005). A study by Valero et al. (2013) reported an increase in total carotenoids in peels of four plum cultivars, among which two were purple skinned (Black Amber and Larry Ann) and two were yellow skinned (Golden Globe and Songold), and when these cultivars were coated with alginate and stored at 2°C, 90 per cent RH for 35 days, followed by 20°C, 65 per cent RH for 3 days, there was a delay in the increase of carotenoids.

Another edible coating postharvest treatment of CMC and pectin (Pec) shows an increase in vitamin C content, total phenolics, anthocyanins and flavonoids. The enzyme activity of peroxidase (POD) was increased and that of Polyphenol oxidase (PPO) and Polygalacturonase (PG) was decreased, as reported in a study of Panahirad et al. (2020) on the improvement of the postharvest quality of plum using polysaccharide-based edible coatings. The best results were obtained at a concentration of 1 per cent CMC and 1.5 per cent Pec. Moreover, a combination of 0.5 per cent Pec and 1.5 per cent CMC is good preparation for preserving the nutritive value of plum during postharvest. Therefore, the use of CMC and Pec or their combination could be regarded as a satisfactory and nontoxic coating method for extending and refining the postharvest quality features of plum fruit.

13.5 EFFECT OF PROCESSING

13.5.1 Thermal Processing

During heat treatment, the degradation of polyphenols takes place, resulting in significant changes in spectra position and fluorescent intensities. A dynamic study explored this influence of high temperatures on polyphenolic content (Turturica et al. 2015). The degradation of total polyphenolic content, total anthocyanin, monomeric compounds and flavonoid content follows a first-order kinetic model. The degradation rate constants showed that high temperature has a hastening effect on the degradation of the aforementioned compounds. This temperature dependence breakdown of anthocyanin, followed by total polyphenols and flavonoids is revealed by activation energy values. The values of activation energy reveal that these compounds are sensitive to temperature, and hence even small changes in temperature results in functionality loss.

13.5.1.1 Blanching

Blanching is one of the thermal processes that enhances the quality of food by disabling the harmful enzymes present inside the food. The timing for blanching is critical and therefore should be selected properly as per the size and type of food. Under-blanching fuels enzymatic activity and enhances the degradation rate, whereas over-blanching results in loss of texture, color and flavor. During the blanching process, bioactive compounds (phytochemicals) undergo a change as well. The extent of the changes depends on the process of blanching, phytochemical stability, enzymatic activity and the position of phytochemicals in the product. Among the various methods of blanching, water and steam blanching are most frequently employed because of their low operating costs. During water blanching, phytochemical loss into the water is a risk, as heat from the boiled water results in breaking down the cellular structure, which ultimately causes an increase in the release of phytochemicals into the water (Rungapamestry et al. 2007). On the other hand, steam blanching retains the bioactive compounds of food (Volden et al. 2009; Goodrich et al. 1989). A study by Volden et al. (2009) conveyed that 30–52 per cent glucosinolate damage takes place in cauliflower during water blanching compared to 18–22 per cent in steam blanching. Further, it was also observed that blanching before drum drying enhances antioxidant activity (Nayak et al. 2011). The stability of different bioactive compounds during blanching is a crucial aspect that controls the degradation rate. Although polyphenols are heat sensitive, some reports still have specified its degradation during blanching, while some authors have reported enhanced retention. This increase has been attributed to increased extractability, which also leads to enhanced bioavailability (Tibäck et al. 2009; Colle et al. 2010; Svelander et al. 2010). The stability of phytochemicals varies within the same food. For example, in the case of tomatoes, the two different phytochemicals—lycopene and beta-carotene—have different stability to heat. Lycopene is stable and beta-carotene is delicate to heat (Nguyen et al. 2001); hence, a greater loss of beta-carotene is observed during blanching (Svelander et al. 2010).

Another critical point for blanching is the timing of the treatment, as this can increase the loss of phytochemicals. In addition to time, enzymatic activity is also considered responsible for producing changes in bioactive compounds. The chief aim of blanching is to disable enzymes liable to cause changes in texture, color and flavor prior to various processes like drying, freezing and so on. Inactivation of those enzymes that facilitate the oxidation of polyphenolic compounds are proposed to increase their holding during blanching and other thermic processing (Brambilla et al. 2011). For example, if an innate PPO enzyme is not inactivated before food processing, it can cause polyphenolic oxidation during heat treatment and storage (Kader et al. 1997) and hence add to a greater loss of polyphenols (Rossi et al. 2003).

13.5.1.2 Drying

According to a study by Miletić et al. (2019), the process of drying has a profound impact on the chemical composition of plums. According to this study, drying at a temperature of 70°C or 90°C results in a slight decrease in total sugar content of three varieties of plum. The process also shows a decrease in saccharose in terms of glucose and fructose. In the case of anthocyanin, pigments and polyphenols, the convection drying process depicts a decrease of about 82 per cent for anthocyanin and 41 per cent for total polyphenols. For pigments, the loss ranges from 80 per cent for cyanidin-3-rutoside to 100 per cent for peonidin-3-rutinoside and peonidin-3-glucoside. But for anthocyanin, the loss proved less sensitive than that of pigments. For phenolic acids, the loss was 34 per cent. For chlorogenic acid, maximum loss of about 69 per cent was observed in drying. In terms of flavonoids, flavan-3-ols, a loss of 51 per cent was observed, and for flavanols a loss of 61 per cent was observed. The conventional drying process resulted in the complete disappearance of quercetin 3-o-galactoside.

For anthocyanin, dehydrating by vacuum or freeze drying proved to be more favorable, especially for cyanidin-3-rutinoside, and for cyanidin-3-glucoside smaller losses were observed in this process of drying. Moreover, it was also suggested that no significant changes were observed for individual polyphenolic content upon vacuum drying and lyophilization. The loss of polyphenolic compounds might be because of the oxidative reactions taking place between enzyme polyphenol oxidase and peroxidase. The enzyme polyphenol oxidase shows its effect at an optimum temperature of 40°C.

For antioxidant potential, a decrease was observed in the antioxidant capacity on lyophilization and microwave-assisted drying, whereas conventional drying shows an increase in antioxidant activity. Similarly another study by Piga et al. (2003) also suggests an upturn in scavenging free radicals while dehydrating at a temperature range of 60–80°C. At a temperature of 60°C, an increase of 65 per cent was observed, and at 80°C the antioxidant capacity increased from 90 per cent to 250 per cent compared to fresh fruit (Piga et al. 2003). This increase might be because of the polyphenolic oxidation and concomitant generation of new antioxidant compounds (Michalska et al. 2016). In the case of convection drying, both enzymatic and non-enzymatic oxidative reactions take place, whereas for lyophilization the exposure to oxygen is low, therefore oxidation takes place at a minor level (Wojdyło et al. 2014; Gumusay et al. 2015). A study by Michalska et al. (2016) observed that different results were found for antioxidant activity using FRAP and ABTS. Higher antioxidant values were reported for plums that were dried using microwave and vacuum methods of drying, and lower values were obtained for plums dried by convection drying.

To ensure the provision of fruits and vegetables to consumers, especially during an off-season, with the same amount of nutritional and sensory profiles, processing technologies in food applications have been evolved to new heights (Stacewicz-Sapuntzakis et al. 2001). Tender fruits are more susceptible to postharvest damage and physiological changes thus they are in particular need of processing for widespread consumption. Harvesting plums at the appropriate stage of maturity is important for desirable sensory characteristics in the processed product. However, lack of a reliable maturity index in plums poses a challenge for consistent harvesting practices. Skin color of the fruit has thus far been considered one of the most important criterion for ripeness and a common indicator for traditional prune-making plums (Usenik et al. 2008). Plums can be consumed fresh, dried (to obtain prunes) or processed into jam, juice, wine and, after distillation, a brandy called slivovitz (Kosmala et al. 2013).

13.5.2 Nonthermal Processing

Though thermal treatments have the ability to inactivate microorganisms and enhance the postharvest shelf life of fruits and vegetables, they also impair the sensory attributes and bioactive compounds. For example, while heating, some physicochemical reactions take place, resulting in altering the organoleptic properties of the respective product as well as a loss in the bioavailability of bioactive compounds (De Carvalho et al. 2013; Garcı´a-Martı´nez et al. 2013; Kamiloglu and Capanoglu 2015; Patras et al. 2010). To overcome all these limitations, a step toward nonthermal processes was taken into consideration. Various nonthermal techniques, like pulsed electric field (PEF), irradiation, dense phase carbon dioxide (DPCD), edible coatings and ozone processing have been used to maintain the various characteristics of fruit and vegetables and also enhance their nutritional value.

13.5.2.1 Freezing

Among the various preservation techniques, freezing is generally used for long-term food storage. Apart from preservation, the process also allows the transportation of products to markets that lack the availability of fresh fruits, thereby making fruit and its products available throughout the year. Approximately 85–90 per cent of the water present in fruits crystallize at freezing temperatures and causes a reduction in water activity, biochemical changes and growth of microbes (De Ancos et al. 2012). The process also decreases the velocity at which most changes take place. The demand for nutraceutical foods is increasing, so the availability of fruit products to consumers has also had to increase. The low temperature used during the freezing process can prevent the loss of nutrients during storage. The literature, however, provides conflicting data on the influence of freezing upon bioactive compounds; these disagreements might result from the processing conditions chosen for the freezing, storing and thawing of fruit and its products in addition to the pretreatment and intrinsic properties of the fruit.

During harvest, the appearance of the fruit is usually crisp, as its water content is high. The water entering into the cells causes the cell wall to stretch and ultimately results in pressure normally referred as "turgor pressure" exhibited by the plant cells. The turgor pressure helps in maintaining the mechanical integrity of the plant and also acts as a force responsible for cell expansion during growth. The quality of the frozen product depends on the technological processing and also the raw material and storage conditions (De Ancos et al. 2012). The freezing temperature chiefly affects the plasma membrane both in the field and after harvesting (Galindo et al. 2004).

During the freezing process, the lower concentration of solute and higher freezing point results in the formation of ice crystals in the extracellular environment

(Breton et al. 2000). The development of crystals begins at temperatures of around –2°C, and after the formation of ice crystals the plant cells are cooled for a period of time. The development of ice crystals into the extracellular environment causes a difference in osmotic pressure and results in the withdrawal of water from the supercooled cell, which ultimately causes tissue dehydration, also called freeze concentration (Fava et al. 2006). In addition, protoplasm shrinkage and the breakdown of the cellular wall also takes place (Yamada et al. 2002). The penetration of ice into the cell is prevented by the undamaged cell wall.

The process of freezing influences the chemical composition, color, flavor and texture of the fruit. Its effect on enzymatic and microbial activity also affects the chemical composition. The process is usually identified as a postharvest technique that preserves the flavor in the best way possible (Skrede 1996). The flavor is greatly produced by the volatile components present inside the fruit as the fruit ripens (Macku and Jennings 1987). Various researchers have suggested that the concentration of compounds present in the final product continues to increase after harvest, even for non-climacteric fruits, as was observed in strawberries (Miszczak et al. 1995).

Further, anthocyanin stability is also affected differently across cultivars. Various studies have demonstrated that immediately after freezing there are either no changes or there is a rise in anthocyanin. A study by Bonat Celli (2015) observed that in the case of plums the anthocyanin content increased upon freezing, but no effect on carotenoids was observed. Similarly, Leong and Oey (2012) reported an increase of anthocyanin in plums, and this increase was found to be dependent on fruit matrix. While some fruits, like raspberries (Syamaladevi et al. 2011; De Ancos et al. 2000), saw no increase in anthocyanin upon freezing, for strawberries and sweet cherries (Poiana et al. 2010a) a small increase was observed.

13.5.2.2 Pulsed Electric Field

The pulsed electric field (PEF) is one of the nonthermal technologies based on short durations of high-intensity electric pulses applied to the materials present between two electrodes (Raso and Heinz 2006). The purpose of this technology is to enhance the shelf life of food by inactivating the microorganisms. To enhance the stability of secondary metabolites of fruits and vegetables, moderate-intensity pulsed electric fields (MIPEFs) have been studied as a possible treatment. The technology is usually desired for liquid foods because the flow of the electrical current is more efficient and because transferring an electric current from one point to another is easy due to the presence of charged molecules. As consumers always prefer fresh, nutritious foods, preserving food commodities using new preservation techniques is a major concern for maintaining freshness, increasing shelf life and ensuring the stability of bioactive compounds.

The use of PEF for food processing has gained much attention, as it is not only useful for food preservation but also for enhancing the extraction of intracellular metabolite (Ade-Omowaye et al. 2001; Fincan et al. 2004), increasing the drying efficiency (Ade-Omowaye et al. 2001; Taiwo et al. 2002), changing enzyme activity (Yeom et al. 2000) and producing secondary metabolites by creating stress inside the plant.

In summary, various reports have noted that the application of PEF results in higher juice yield, higher juice purity, enhanced antioxidant activity, faster extraction and enhanced rate of drying. However, these effects are difficult to compare because of the difference in electric systems and processes. Parameters including the field strength of the applied electric field, the number of generated pulses and the geometry of pulses, temperature fluctuation, chamber design and energy input should be taken into consideration.

13.5.2.3 Irradiation

Irradiation of food utilizes energy released from ionizing radiation (Urbain 1986), which is the proliferation of energy through space, and has the capacity to infiltrate the food particles. This further intermingles with the atoms and molecules of food and food contaminants, resulting in the formation of ions and the removal of one or more electrons from an atom, and ultimately modifies chemical and biological properties. Basically, this ionization process requires radiation energy, and this energy is being captivated by electrons and increases the level of electrons from ground state to excited state. After the electrons absorb the energy, they leave the atom and are free from nuclear control. This results in the formation of free electrons with a negative charge and ions with a positive charge. Radiations having short wavelengths (>4 eV), including ultraviolet, X-rays and gamma rays, have the ability to cause ionization. For food irradiation, gamma rays are preferred. Irradiation of food has a wide range of applications, including sprout inhibition in potatoes, onions and garlic; elimination of parasites and insects in dried and fresh fruits, meat and seafood and beans; delayed ripening of fruits and vegetables and enhanced shelf life of perishable foods (Crawford and Ruff 1996). The dose of radiation indicates the energy transferred and absorbed. The standard international unit for radiation is gray (Gy).

Radiation technology is gaining importance in food processing, as consumer demand for fresh food material is increasing. In the case of fresh product, firmness is the most important parameter affecting consumer acceptance. During storage, firmness is lost because of the breakdown of pectin present in the middle lamella and cell wall (Silva et al. 2012). The enzymes responsible for the degradation of pectin include polygalacturonase and pectinametil esterase (Singh and Dwivedi 2008). An irradiation dosage of 0.03–0.5 kGy causes inactivation of these enzymes, hence the same dosage has been used to slow down ripening in food products (Uthairatanakij et al. 2006). The process does not only cause delay in ripening but also results in increasing health benefits, concentration of bioactive compounds like that of phenolics and anthocyanins and also antioxidant activity. This increase of antioxidant activity might be due to an increase in the enzymatic activity of PAL and the release of phenolic compounds by radiolytic products.

13.5.2.4 Dense Phase Carbon Dioxide

Dense phase carbon dioxide (DPCD) refers to supercritical CO_2 and liquid CO_2. The technique is an alternative to pasteurization and uses pressure in coalescence with CO_2 so as to preserve the quality of the food. The pressure used is less than 90 MPa, which is far lower than that used in high-pressure processing (400–600 MPa)

(Damar and Balaban 2006). The technique has gained much interest in a variety of food industries. The unique nature of CO_2 makes it pleasing for use in the preservation of food. The zero surface tension and low viscosity of carbon in its supercritical state helps it to penetrate the pores of food.

Carbon dioxide and its highly reactive species has an intense effect on the antioxidants and enzymes present in food. There is very little data available on the impact of DPCD on bioactive compounds of fruits. Different authors have carried out separate research comparing DPCD to other thermal processing and have concluded that HTST (90°C for 60 seconds) also inactivates microorganisms, but it results in a loss of ascorbic acid, whereas DPCD prevents the loss of antioxidant compounds like that of anthocyanin and beta-carotene, thus resulting in its stability. This stability might be because of the concentration gradient of carbon dioxide and also of controlled pressure. In addition, stability is also dependent on the inactivation activity of PPO enzymes (Gui et al. 2006; Pozo-Insfran et al. 2007).

Until now, it has been reported in the literature that antioxidant impacts from DPCD mainly occur in ascorbic acid, carotenoids and total antioxidant activity. The information about the effect on other phytochemicals or individual polyphenols is unknown. Furthermore, it has also been reported that DPCD shows a better impact on antioxidant activity than other emerging techniques (Ortuño et al. 2013).

13.6 CONCLUSION

Plums have high nutritive value, increased dietary fiber content and are also an excellent source of polyphenols, antioxidants, ascorbic acid, carotenoids, minerals, carbohydrates and organic acids. Moreover, plums also contain volatile compounds. The composition and concentration of all these compounds in a fruit depend on genetic makeup as well as environmental factors. The various pre- and postharvest treatments influence the bioactive compounds of plum fruit. But the preharvest use of AVG was found to decrease phenolic content in the case of Black Amber and Fortune plums. On the other hand, oxalic acid causes an increase in the activity of enzymes associated with antioxidant activity, like SOD, POD, CAT and APX. In the case of postharvest treatments, AVG delays the anthocyanin loss and also reduces the occurrence of chilling injuries in fruit. Salicylic acid results in an increase in phenolic concentration with an increase in PAL activity, an enzyme responsible for biosynthesis of the phenolic pathway. Furthermore, MCP at a concentration of 0.5 ml/L reduces the activity of cell wall–degrading enzymes, including polylactouranase, galactosidase and glucosidase.

In addition to chemical treatments, environmental parameters also influence the bioactive compounds. The superiority of light alters the photo-oxidative properties of plants by adjusting the antioxidant defense system. The impact of sunlight has been found to have a progressive impact on bioactive compounds, as it enhances the activity of ascorbic acid and anthocyanins. Exposure of plums to ultraviolet rays causes anthocyanin accumulation. Besides ultraviolet rays, temperature difference during the day and at night also affects the accumulation of anthocyanin. Changes in temperature adjusts phytochemical composition, which ultimately causes variations in antioxidant activity and the eating quality of the plum. At temperatures of

0°C and 25°C, no buildup of anthocyanin occurs. Since the plums are highly rich in antioxidants and other polyphenolic compounds, it can be concluded that the fruit can be used as a source for the development of nutraceutical foods.

13.7 REFERENCES

Ade-Omowaye, B.I.O., Angersbach, A., Eshtiaghi, N.M., and Knorr, D. 2001. Impact of High Intensity Electric Field Pulses on Cell Permeabilisation and as Pre-Processing Step in Coconut Processing. *Innovative Food Science and Emerging Technologies* 1: 203–209.

Ahmad, B., Zaid, A., Sadiq, A., Bashir, S., and Wani, S.H. 2019. Role of Selective Exogenous Elicitors in Plant Response to Abiotic Stress Tolerance. *Plant Abiotic Stress Tolerance* 273–290.

Ahmed, M.J., Singh, Z., and Khan, A.S. 2009. Postharvest Aloe Vera Gel-Coating Modulates Fruit Ripening and Quality of 'Arctic Snow' Nectarine Kept in Ambient and Cold Storage. *International Journal of Food Science & Technology* 44 (5): 1024–1033.

Ali, S., Khan, A.S., Malik, A.U., and Shahid, M. 2016. Effect of Controlled Atmosphere Storage on Pericarp Browning, Bioactive Compounds and Antioxidant Enzymes of Litchi Fruits. *Food Chemistry* 206: 18–29.

Ames, B.N., Shigenaga, M.K., and Hagen, T.M. 1993. Oxidants, Antioxidants and the Degenerative Diseases of Aging. *Proceedings of the National Academy of Sciences of the United States of America* 90:7915–7922.

Arakawa, O. 1993. Effect of Ultraviolet Light on Anthocyanin Síntesis in Light Colored Sweet Cherry cv. Nato Nishiki. *Journal of the Japanese Society for Horticultural Science* 62: 543–546.

Auger, C., Najim, A.A., Bornet, A., Rouanet, J.M., Gasc, F., Cros, G., and Teissedre, P.L. 2004. Catechins and Procyanidins in Mediterranean Diets. *Food Research International* 37: 233–245.

Awad, M.A., De Jager, A., and Van Westing, L.M. 2000. Flavonoid and chlorogenic Acid Levels in Apple Fruit: Characterization of Variation. *Scientia Horticulturae* 83: 249–263.

Bal, E. 2019. Postharvest AminoethoxyVinylGlycine (AVG) Treatment affects Maturity and Storage Life of Plum. *Journal of Agricultural Science and Technology* 21 (6): 1569–1579.

Bantis, F., Ouzounis, T., and Radoglou, K. 2016. Artificial LED Lighting Enhances Growth Characteristics and Total Phenolic Content of *Ocimum basilicum*, but Variably Affects Transplant Success. *Scientia Horticulturae* 198: 277–283.

Barkoh, A., Schlechte, J.W., Hamby, S., and Kurten, G. 2005. Effects of Rice Bran, Cottonseed Meal, and Alfalfa Meal on pH and Zooplankton. *North American Journal of Aquaculture* 67: 237–243.

Behboudian, M.H., Marsal, J., Girona, J., and López, G. 2011. Quality and Yield Responses of Deciduous Fruits to Reduced Irrigation. *Horticultural Review* 38: 149–189.

Bhaswant, M., Fanning, K., Netzel, M., Mathai, M.L., Panchal, S.K., and Lindsay, B. 2015. Cyanidin 3-Glucoside Improves Diet-Induced Metabolic Syndrome in Rats. *Pharmacology Research* 102: 208–217.

Bonat Celli, G., Ghanem, A., and Su-Ling Brooks, M. 2015. Influence of Freezing Process and Frozen Storage on the Quality of Fruits and Fruit Products. *Food reviews international* 32: 280–304.

Bouayed, J., Rammal, H., Dick, A., Younos, C., and Soulimani, R. 2007. Chlorogenic Acid, a Polyphenol from *Prunus domestica* (Mirabelle), with Coupled Anxiolytic and Antioxidant Effects. *Journal of the Neurological Sciences* 262: 77–84.

Boulton, R. 2001. The Co Pigmentation of Anthocyanins and Its Role in the Color of Red Wine: A Critical Review. *American Journal of Enology and Viticulture* 52: 67–87.

Brambilla, A., Maffi, D., and Rizzolo, A. 2011. Study of the Influence of Berry-Blanching on Syneresis in Blueberry Purées. *Procedia Food Science* 1: 1502–1508.

Breton, G., Danyluk, J., Ouellet, F., and Sarhan, F. 2000. Biotechnological Applications of Plant Freezing Associated Proteins. *Biotechnology Annual Review* 6: 57–99.

Buendía, B., Allende, A., Nicolás, Alarcón, J.J., and Gil, I. 2008. Effect of Regulated Deficit Irrigation and Crop Load on the Antioxidant Compounds of Peaches. *Journal of Agricultural and Food Chemistry* 56: 3601–3608.

Candan, A.P., Graell, J., and Larrigaudiere, C. 2007. Roles of Climacteric Ethylene in the Development of Chilling Injury in Plums. *Postharvest Biology and Technology* 47: 107–112.

Cantin, C.M., Crisosto, C.H., and Day, K.R. 2008. Evaluation of the Effect of Different MAP Box Liners on the Quality and Shelf Life of 'Friar' Plums. *Horticulture Technology* 18.

Cantos, E., Espín, J.C., Fernández, M.J., Oliva, J., and Tomás-Barberán, A. 2003. Postharvest UV- C-Irradiated Grapes as a Potential Source for Producing Stilbene-Enriched Red Wines. *Journal of Agricultural and Food Chemistry* 51: 1208–1214.

Cao, A.S., Zheng, Y., Yang, Z., Wanga, K., and Ruia, H. 2009. Effect of Methyl Jasmonate on Quality and Antioxidant Activity of Postharvest Loquat Fruit. *Journal of Science of Food and Agriculture* 89: 2064–2070.

Cao, G., Sofic, E., and Prior, L.R. 1997. Antioxidant and Pro Oxidant Behavior of Flavonoids: Structure-Activity Relationships. *Free Radical Biology and Medicine* 22: 749–760.

Cassidy, A., Hanley, B., and Lamuela-Raventos, R.M. 2000. Isoflavones, Lignans and Stilbenes-Origins, Metabolism and Potential Importance to Human Health. *Journal of Science of Food and Agriculture* 80: 1044–1062.

Chalk, P.M., Craswell, E.T., Polidoro, J.C., and Chen, D. 2015. Fate and Efficiency of 15N-Labelled Slow- and Controlled-Release Fertilizers. *Nutrient Cycling and Agroecosystem* 102: 167–178.

Chatzitheodorou, I.T., Sotiropoulos, T.E., and Mouhtaridou, G.I. 2004. Effect of Nitrogen, Phosphorus, Potassium Fertilization and Manure on Fruit Yield and Fruit Quality of the Peach Cultivars 'Spring Time' and 'Red Haven'. *Agronomy Research* 2: 135–143.

Chen, L., Vigneault, C., Raghavan, G.S.V., and Kubow, S. 2007. Importance of the Phytochemical Content of Fruits and Vegetables to Human Health. *Stewart Postharvest Review* 3:2.

Chun, O.K., Kim, D.O., Moon, H.Y., Kang, H.G., and Lee, C.Y. 2003. Contribution of Individual Polyphenolics to Total Antioxidant Capacity of Plums. *Journal of Agriculture and Food Chemistry* 51: 7240–7245.

Colle, I., Lemmens, L., Van Buggenhout, S., Van Loey, A., and Hendrickx, M. 2010. Effect of Thermal Processing on the Degradation, Isomerization, and Bio Accessibility of Lycopene in Tomato Pulp. *Journal of Food Science* 75: 753–759.

Crawford, L.M., and Ruff, E.H. 1996. A Review of the Safety of Cold Pasteurization Through Irradiation. *Food Control* 7: 87–97.

Crisosto, C.H., Garner, D., Crisosto, G.M., and Bowerman, E. 2004. Increasing 'Blackamber' Plum (Prunus salicina Lindell) Consumer Acceptance. *Postharvest Biology and Technology* 34: 237–44.

Crisosto, C.H., Johnson, R.S., Dejong, T., and Day, K.R. 1997. Orchard Factors Affecting Postharvest Stone Fruit Quality. *Journal of American Society for Horticultural Science Alexandria* 32 (5): 820–823.

Crisosto, C.H., and Kader, A.A. 2000. *Plum and Fresh Prune Postharvest Quality Maintenance Guidelines*. Department of Pomology, University of California Communication Services.

Crisosto, C.H., Mitchell, F.G., and Ju, Z.G. 1999. Susceptibility to Chilling Injury of Peach, Nectarine, and Plum Cultivars Grown in California. *Journal of American Society for Horticultural Science* 34: 1116–1118.

Cuevas, F.J., Pradas, I., Ruiz-Moreno, M.J., Arroyo, F.T., Perez-Romero, L.F., and Montenegro, J.C., Moreno-Rojas, J.M. 2015. Effect of Organic and Conventional Management on Bio-Functional Quality of Thirteen Plum Cultivars (*Prunus salicina Lindl.*). *Plos One* 10 (8): 0136596.

Daane, K.M., Johnson, R.S., Michailides, T.J., Crisosto, C.H., Dlott, J.W., Ramirez, H.T., Yokota, G.T., and Morgan, D.P. 1995. Excess Nitrogen Susceptibility to Raises Nectarine Disease and Insects. *California Agriculture* 49 (4): 13–17.

Damar, S., and Balaban, M.O. 2006. Review of Dense Phase CO2 Technology: Microbial and Enzyme Inactivation, and Effects on Food Quality. *Journal of Food* 71 (1): 1–11.

Das, S.K., and Jana, B.B. 2003. Pond Fertilization Regimen: State-of-the-Art. *Journal of Applied Aquaculture* 13: 35–66.

Davarynejad, G.H., Zarei, M., Nasrabadi, M.E., and Ardakani, E. 2015. Effects of Salicylic Acid and Putrescine on Storability, Quality Attributes and Antioxidant Activity of Plum cv. "Santa Rosa". *Journal of Food Science and Technology* 52: 2053–2062.

De Ancos, B., Sánchez-Moreno, C., Pascual-Teresa, S., and Cano, M.P. 2012. Freezing Preservation of Fruits. In *Handbook of Fruits and Fruit Processing*, edited by N. Sinha, J.S. Sidhu, J. Barta, J.S.B. Wu, and M.P. Cano, 2nd ed., 103–119. John Wiley & Sons.

De Ancos, B., Ibanez, E., Reglero, G and Cano, M.P. 2000. Frozen Storage Effects on Anthocyanins and Volatile Compounds of Raspberry Fruit. *Journal of Agricultural and Food Chemistry* 48 (3): 873–879.

De, Carvalho, J.M., Maia, G.A., da Fonseca, A.V.V., de Sousa, P. H.M., and Rodrigues, S. 2013. Effect of Processing on Physicochemical Composition, Bioactive Compounds and Enzymatic Activity of Yellow Mombin (Spondias mombin L.) Tropical Juice. *Journal of Food Science and Technology* 52 (2): 1182–1187.

DeJong, T.M., and Goudriaan, J. 1989. Modeling Peach Fruit Growth and Carbohydrate Requirements: Reevaluation of the Double-Sigmoid Pattern. *Journal of American Society for Horticulture Science* 114: 800–804.

Deluc, L.G., Quilici, D.R., Decendit, A., Grimplet, J., Wheatley, M.D., Schlauch, K.A., Mérillon, J.M., Cushman, J.C., and Cramer, G.R. 2009. Water Deficit Alters Differentially Metabolic Pathways Affecting Important Flavor and Quality Traits in Grape Berries of Cabernet Sauvignon and Chardonnay. *BMC Genomics* 10:212.

Díaz-Mula, H.M., Serrano, M., and Valero, D. 2012. Alginate Coatings Preserve Fruit Quality and Bioactive Compounds During Storage of Sweet Cherry Fruit. *Food and Bioprocess. Technology* 5: 2990–2997.

Dong, C., Fu, Y., Liu, G., and Liu, H. 2014. Growth, Photosynthetic Characteristics, Antioxidant Capacity and Biomass Yield and Quality of Wheat (Triticum aestivum L.) Exposed to LED Light Sources with Different Spectra Combinations. *Journal of Agronomy and Crop Science* 200: 219–230.

Dong, L., Zhou, H.W., Sonego, L., Lers, A., and Lurie, S. 2001. Ripening of 'Red Rosa' Plums: Effect of Ethylene and 1-Methylcyclopropene. *Australian Journal of Plant Physiology* 28: 1039–1045.

Donovan, J.L., Meyer, A.S., and Waterhouse, A.L. 1998. Phenolic Composition and Antioxidant Activity of Prunes and Prune Juice (*Prunus domestica*). *Journal of Agricultural and Food Chemistry* 46: 1247–1252.

Elmegeed, G.A., Ahmed, H.A., and Hussain, J.S. 2005. Novel Synthesized Amino Steroidal Heterocycles Intervention for Inhibiting Iron-Induced Oxidative Stress. *European Journal of Medical Chemistry* 40: 1283–1294.

Erkan, M., and Eski, H. 2012. Combined Treatment of Modified Atmosphere Packaging and 1- Methylcyclopropene Improves Postharvest Quality of Japanese Plums. *Turkish Journal of Agriculture and Forestry* 36: 563–575.

Ezinne, O.I., and Karen, E.C. 2016. Review: A Systematic Review on the Health Effects of Plums (Prunus domestica and Prunus salicina). *Phytotherapy Research* 30: 701–731.

Fang, N., Yu, S., and Prior, R.L. 2002. LC/MS/MS, Characterization of Phenolic Constituents in Dried Plums. *Journal of Agricultural and Food Chemistry* 50: 3579–3585.

Fanning, K., Edwards, D., Netzel, M., Stanley, R., Netzel, G., and Russel, D. 2013. Increasing Anthocyanin Content in Queen Garnet Plum and Correlations with in-Field Measures. *Acta Horticulturae* 985: 97–104.

Fatma, K.M., Shaaban, G.A.M., El-Hadidy, T.S.H., and Mahmoud, M. 2020. Effects of Salicylic Acid, Putrescine and Moringa Leaf Extract Application on Storability, Quality Attributes and Bioactive Compounds of Plum cv. 'Golden Japan'. *Future of Food: Journal on Food, Agriculture and Society* 8 (2).

Fava, J., Alzamora, S.M., and Castro, M.A. 2006. Structure and Nanostructure of the Outer Tangential Epidermal Cell Wall in Vaccinium Corymbosum L. (blueberry) Fruits by Blanching, Freezing Thawing and Ultrasound. *Food Science Technology International Journal* 12: 241–251.

Fereres, E., and Goldhamer, D.A. 1990. Deciduous Fruit and Nut Trees. In *Irrigation of Agricultural Crops*, edited by B.A. Stewart, and D.R. Nielsen. Monograph, 30, 987–1017. A.S.A. Madison.

Ferguson, I.B., and Boyd, L.M. 2002. Inorganic Nutrients and Fruit Quality. In *Fruit Quality and Its Biological Basis*, edited by M. Knee, 17–45. Blackwell Publishing Ltd.

Fincan, M., DeVito, F., and Dejmek, P. 2004. Pulsed Electric Field Treatment for Solid-Liquid Extraction of Red Beetroot Pigment. *Journal of Food Engineering* 64: 381–388.

Fridovich I. 1978. The Biology of Oxygen Radicals. *Science*. 201: 875–880.

Galindo, F.G., Herppich, W., Gekas, V., and Sjöholm, I. 2004. Factors Affecting Quality and Postharvest Properties of Vegetables: Integration of Water Relations and Metabolism. *Critical Reviews in Food Science and Nutrition* 44: 39–154.

Garcı´a-Martı´nez, E., Igual, M., Martı´n-Esparza, M., and Martı´nez-Navarrete, N. 2013. Assessment of the Bioactive Compounds, Color, and Mechanical Properties of Apricots as Affected by Drying Treatment. *Food and Bioprocess Technology* 6 (11): 3247–3255.

Ghasemnezhad, M., Shiri, M.A., and Sanavi, M. 2010. Effect of Chitosan Coatings on Some Quality Indices of Apricot (Prunus armeniaca L.) During Cold Storage. *Caspian Journal of Environmental Sciences* 8: 25–33.

Girona, J., Mata, M., Arbones, A., Alegre, S., Rufat, J., and Marsal, J. 2003. Peach Tree Response to Single and Combined Regulated Deficit Irrigation Regimes Under Shallow Soils. *Journal of American Society for Horticulture Science* 128: 432–440.

Gol, N.B., Patel, P.R., and Ramana Rao, T.V. 2013. Improvement of Quality and Shelf-Life of Strawberries with Edible Coatings Enriched with Chitosan. *Postharvest Biology and Technology* 85: 185–195.

Goodrich, R.M., Anderson, J.L., and Stoewsand, G.S. 1989. Glucosinolate Changes in Blanched Broccoli and Brussels Sprouts. *Journal of Food processing and Preservation* 13: 275–280.

Gui, F., Chen, F., Wu, J., Wang, Z., Liao, X., and Hu, X. 2006. Inactivation and Structural Change of Horseradish Peroxidase Treated with Supercritical Carbon Dioxide. *Food Chemistry* 97 (3): 480–489.

Gumusay, O.A., Borazan, A.A., Ercal, N., and Demirkol, O. 2015. Drying Effects on the Antioxidant Properties of Tomatoes and Ginger. *Food Chemistry* 173: 156–162.

Gutteridge, J.M. 1993. Free Radicals in Disease Processes: A Compilation of Cause and Consequence. *Free Radical Research Communications* 19: 141–158.

Handley, D.F., and Johnson, R.S. 2000. Late Summer Irrigation of Water Stressed Peach Tress Reduces Fruit Double and Deep Sutures. *Journal of Horticultural Science* 35:771.

Harris, M.L., Shiller, H.J., Reilly, P.M., Donowitz, M., Grisham, M.B., and Bulkley, G.B. 1992. Free Radicals and Other Reactive Oxygen Metabolites in Inflammatory Bowel Disease: Cause, Consequence or Epiphenomenon *Pharmacology & Therapeutics* 53: 375–408.

Hasan, M., Ghosh, R., and Bae, H. 2017. An Overview of LEDs Effects on the Production of Bioactive Compounds and Crop Quality. *Molecules* 22: 1420.

Heo, J.W., Kang, D.H., Bang, H.S., Hong, S.G., Chun, C., Kang, K.K. 2012. Early Growth, Pigmentation, Protein Content, and Phenylalanine Ammonia-Lyase Activity of Red Curled Lettuces Grown Under Different Lighting Conditions. *Korean Journal of Horticultural Science and Technology* 30: 6–12.

Hertog, M.G.L., Hollman, P.C.H., and Venema, D.P. 1992. Optimization of a Quantitative HPLC Determination of Potentially Anti Carcinogenic Flavonoids in Vegetables and Fruits. *Journal of Agricultural and Food Chemistry* 40: 1591–1598.

Huan, C., Xu, Q., Shuling, S., Dong, J., and Zheng, X. 2020. Effect of Benzothiadiazole Treatment on Quality and Anthocyanin Biosynthesis in Plum Fruit During Storage at Ambient Temperature. *Journal of the Science of Food and Agriculture*. DOI:10.1002/jsfa.10946

Iglesias-Fernandez, R., Matilla, A.J., Rodriguez-Gacio, M.C., Fernandez-Otero, C., and de la Torre, F. 2007. The Polygalacturonase Gene PdPG1 Is Developmentally Regulated in Reproductive Organs of Prunus Domestica L. Subsp Insititia. *Plant Science* 172: 763–772.

Johkan, M., Shoji, K., Goto, F., Hashida, S., and Yoshihara, T. 2010. Blue Light-Emitting Diode Light Irradiation of Seedlings Improves Seedling Quality and Growth After Transplanting in Red Leaf Lettuce. *American Journal of Horticultural Science* 45: 1809–1814.

Johnson, R.S., and Handley, D.F. 2000. Using Water Stress to Control Vegetative Growth and Productivity of Temperate Fruit Trees. *American Journal of Horticultural Science* 35: 1048–1050.

Kader, A.A. 2003. A Perspective on Postharvest Horticulture (1978–2003). *American Journal of Horticultural Science* 38: 1004–1008.

Kader, F., Rovel, B., Girardin, M., and Metche, M. 1997. Mechanism of Browning in Fresh Highbush Blueberry Fruit (Vaccinium corymbosum L.). Role of Blueberry Polyphenol Oxidase, Chlorogenic Acid and Anthocyanins. *Journal of Science of Food and Agriculture* 74: 31–34.

Kamiloglu, S., and Capanoglu, E. 2015. Polyphenol Content in Figs (Ficus carica L.): Effect of Sun-Drying. *International Journal of Food Properties* 18 (3): 521–535.

Karaman, S., Ozturk, B., Aksit, H., and Erdogdu, T. 2013. The Effects of Pre-Harvest Application of Aminoethoxyvinylglycine (AVG) on the Bioactive Compounds and Fruit Quality of "Fortune" Plum Variety During Cold Storage. *Food Science and Technology International* 19: 567–576.

Kayano, S., Kikuzaki, H., Fukutsaka, N., Mitani, T., and Nakatani, N. 2002. Antioxidant Activity of Prune (*Prunus domestica* L.) Constituents and a New Synergist. *Journal of Agricultural and Food Chemistry* 50: 3708–3712.

Khan, A.S. 2016. Differences in Fruit Growth and Ripening of Early-, Mid- and Late-Season Maturing Japanese Plum Cultivars. *Fruits* 71: 329–338.

Khan, A.S., and Singh, Z. 2007. Methyl Jasmonate Promotes Fruit Ripening and Improves Fruit Quality in Japanese Plum. *Journal of Horticultural Science and Biotechnology* 82 (5): 695–670.

Khan, A.S., and Singh, Z. 2008.1-MCP Application Affects Ethylene Production, Storage Life and Quality of 'Tegan Blue' Plum. *Acta Horticulturae* 774: 143–150.

Kim, D.O., Chun, O.K., Kim, Y.J., Moon, H., and Lee, C.Y. 2003a. Quantification of Polyphenolics and Their Antioxidant Capacity in Fresh Plums. *Journal of Agricultural and Food Chemistry*. 51: 6509–6515.

Kim, D.O., Jeong, S.W., and Lee, C.Y. 2003b. Antioxidant Capacity of Phenolic Phytochemicals from Various Cultivars of Plums. *Food Chemistry* 81: 321–326.

Kinsella, J.E., Frankel, E., German, B., and Kanner, J. 1993. Possible Mechanisms for the Protective Role of Antioxidants in Wine and Plant Foods. *Food Technology* 47: 85–89.

Kosmala, M., Milala, J., Kołodziejczyk, K., Markowski, J., Zbrzeźniak, M., and Renard, M.G.C.C. 2013. Dietary Fiber and Cell Wall Polysaccharides from Plum (*Prunus domestica L.*) Fruit, Juice and Pomace: Comparison of Composition and Functional Properties for Three Plum Varieties. *Food Research International* 54: 1787–1794.

Kotepong, P., Ketsa, S., and VanDoom, W.G. 2011. A White Mutant of Malay Apple Fruit (*Syzygium malaccense*) Lacks Transcript Expression and Activity for the Last Enzyme of Anthocyanin Synthesis, and the Normal Expression of a MYB Transcription Factor. *Functional Plant Biology* 38: 75–86.

Kozai, T., Fujiwara, K., and Runkle, E.S. 2016. *LED Lighting for Urban Agriculture*. Springer Science Business Media.

Kris-Etherton, P.M., Hecker, K.D., Bonanome, A., Coval, S.M., Binkoski, A.E., Hilpert, K.F., and Etherton, T.D. 2002. Bioactive Compounds in Foods: Their Role in the Prevention of Cardiovascular Disease and Cancer. *The American Journal of Medicine* 113 (9): 71–88.

Kumar, P., Sethi, S., Sharma, R.R., Srivastav, M., and Varghese, E. 2017. Effect of Chitosan Coating on Postharvest Life and Quality of Plum During Storage at Low Temperature. *Scientia Horticulturae* 226: 104–109.

Kumar, M.S., Burgess, S.N., and Luu, L.T. 2004. Review of Nutrient Management in Freshwater Polyculture. *Journal of Applied Aquaculture* 16: 17–44.

Lawrence, B.T., and Melgar, J.C. 2020. Variable Fall Climate Conditions on Carbon Assimilation and Spring Phenology of Young Peach Trees. *Plants* 9: 1353.

Lee, M.K., Arasu, M.V., Park, S., Byeon, D.H., Chung, S.O., Park, S.U., Lim, Y.P., and Kim, S.J. 2016. LED Lights Enhance Metabolites and Antioxidants in Chinese Cabbage and Kale. *Brazilian Archieves of Biology and Technology* 59: 16150546.

Lee, S.K., and Kader, A.A. 2000. Pre Harvest and Postharvest Factors Influencing Vitamin C Content of Horticultural Crops. *Postharvest Biology and Technology* 20: 207–220.

Lekkham, P., Srilaong, V., Pongprasert, N., and Kondo, S. 2016. Anthocyanin Concentration and Antioxidant Activity in Light-Emitting Diode (LED)-Treated Apples in a Greenhouse Environmental Control System. *Fruits* 71: 269–274.

Leong, S.Y., and Oey, I. 2012. Effects of Processing on Anthocyanins, Carotenoids and Vitamin C in Summer Fruits and Vegetables. *Food Chemistry* 133: 1577–1587.

Lin, K.H., Huang, M.Y., Huang, W.D., Hsu, M.H., Yang, Z.W., and Yang, C.M. 2013. The Effects of Red, Blue, and White Light-Emitting Diodes on the Growth, Development, and Edible Quality of Hydroponically Grown Lettuce (*Lactuca sativa* L. var. *capitata*). *Scientia Horticulturae* 150: 86–91.

Lombardi-Boccia, G., Lucarini, M., Lanzi, S., Aguzzi, A., and Cappelloni, M. 2004. Nutrients and Antioxidant Molecules in Yellow Plums (Prunus domestica L.) from Conventional and Organic Productions: A Comparative Study. *Journal of Agricultural and Food Chemistry* 52: 90–94.

Lopez, C.A., Botia, M., and Alcaraz, C.F. 2003. Effects of Foliar Sprays Containing Calcium, Magnesium and Titanium on Plum (Prunus domesticaL.) Fruit Quality. *Journal of Plant Physiology* 160 (12): 1441–1446.

Maatallah, S., Guizani, M., Hjlaoui, H., Boughattas, N.E.H., Lopez-Lauri, F., and Ennajeh, M. 2015. Improvement of Fruit Quality by Moderate Water Deficit in Three Plum Cultivars (Prunus salicina L.) Cultivated in a Semi-Arid Region. *Fruits* 70: 325–332.

Macku, C., and Jennings, W.G. 1987. Production of Volatiles by Ripening Bananas. *Journal of Agricultural Food Chemistry* 35: 845–848.

Maftoonazad, N., Ramaswamy, H.S., and Marcotte, M. 2008. Shelf-Life Extension of Peaches Through Sodium Alginate and Methyl Cellulose Edible Coatings. *International Journal of Food Science and Technology* 43: 951–957.

Manganaris, G.A., Vicente, A.R., and Crisosto, C.H. 2008. Effect of Pre-Harvest and Post-Harvest Conditions and Treatments on Plum Fruit Quality: CAB Reviews Perspectives in Agriculture Veterinary Science. *Nutrition and Natural Resources* 3.

Marschner, H. 1995. *Mineral Nutrition of Higher Plants*, 2nd ed., 889. Academic Press.

Martínez-Esplá, A., Serrano, M., Valero, D., Martínez-Romero, D., Castillo, S., and Zapata, P.J. 2017. Enhancement of Antioxidant Systems and Storability of Two Plum Cultivars by Pre Harvest Treatments with Salicylates. *International Journal of Molecular Science* 18: 1–14.

Martínez-Espláa, A., Serranob, M., Martínez-Romeroa, D., Valeroa, D., and Zapataa, P.J. 2018. Oxalic Acid Pre Harvest Treatment Increases Antioxidant Systems and Improves Plum Quality at Harvest and During Postharvest Storage. *Journal of Science of Food and Agriculture* 9 (1): 235–243.

Martinez-Romero, D., Dupille, E., Guillen, F., Valverde, J.M., Serrano, M., and Valero, D. 2003. 1-Methylcyclopropene Increases Storability and Shelf Life in Climacteric and Nonclimacteric Plums. *Journal of Agricultural and Food Chemistry* 51: 4680–4686.

Martínez-Romero, D., Zapata, P.J., Guillén, F., Paladines, D., Castillo, S., Valero, D., and Serrano, M. 2017. The Addition of Rosehip Oil to Aloe Gels Improves Their Properties as Postharvest Coatings for Maintaining Quality in Plum. *Food Chemistry* 217: 585–592.

Marty, I., Bureau, S., Sarkissian, G., Gouble, B., Audergon, J.M., and Albagnac, G. 2005. Ethylene Regulation of Carotenoid Accumulation and Carotenogenic Gene Expression in Colour-Contrasted Apricot Varieties (Prunus armeniaca). *Journal of Experimental Botany* 56: 1877–1886.

Menniti, A.M., Donati, I., and Gregori, R. 2006. Responses of 1-MCP Application in Plums Stored Under Air and Controlled Atmospheres. *Postharvest Biology and Technology* 39: 243–246.

Menniti, A.M., Gregori, R., and Donati, I. 2004. 1-Methylcyclopropene Retards Postharvest Softening of Plums. *Postharvest Biology and Technology* 31: 269–275.

Michalska, A., Wojdyło, A., Lech, K., Łysiak, G.P., and Figiel, A. 2016. Physicochemical Properties of Whole Fruit Plum Powders Obtained Using Different Drying Technologies. *Food Chemistry* 207: 223–232.

Miletić, N., Popović, B., Mitrović, O., and Kandić, M. 2012. Phenolic Content and Antioxidant Capacity of Fruits of Plum cv. 'Stanley' ('Prunus domestica'L.) as Influenced by Maturity Stage and On-Tree Ripening. *Australian Journal of Crop Science* 6 (4): 681–687.

Miletić, N., Mitrović, O., Popović, B., Mašković, P., Mitić, M., and Petković, M. 2019. Chemical Changes Caused by Air Drying of Fresh Plum Fruits. *International Food Research Journal* 26.

Miszczak, A., Forney, C.F., and Prange, R.K. 1995. Development of Aroma Volatiles and Color During Postharvest Ripening of 'Kent' Strawberries. *Journal of American Society of Horticultural Science* 120: 650–655.

Mubarak, A., Swinny, E.E., Ching, S., Jacob, S.R., Lacey, K., Hodgson, J., Croft, K.D., and Considine, M.J. 2012. Polyphenol Composition of Plum Selections in Relation to Total Antioxidant Capacity. *Journal of Agriculture and Food Chemistry* 60: 10256–10262.

Murathan, Z.T., Arslan, M., and Erbil, N. 2020. Analyzing Biological Properties of Some Plum Genotypes Grown in Turkey. *International Journal of Fruit Science* 20: 1729–1740.

Murray, X.J., Holcroft, D.M., Cook, N.C., and Wand, S.J.E. 2005. Postharvest Quality of 'Laetitia' and 'Songold' (Prunus salicina Lindell) Plums as Affected by Preharvest Shading Treatments. *Postharvest Biology and Technology* 37: 81–92.

Muneer, S., Kim, E.J., Park, J.S., and Lee, J.H. 2014. Influence of Green, Red and Blue Light Emitting Diodes on Multiprotein Complex Proteins and Photosynthetic Activity Under Different Light Intensities in Lettuce Leaves (Lactuca sativa L.). *International Journal of Molecular Science* 15: 4657–4670.

Nakatani, N., Kayano, S., Kikuzaki, H., Sumino, K., Katagiri, K., and Mitani, T. 2000. Identification, Quantitative Determination, and Antioxidative Activities of Chlorogenic Acid Isomers in Prune (*Prunus domestica* L.). *Journal of Agriculture and Food Chemistry* 48: 5512–5516.

Naor, A. 2006a. Irrigation Scheduling and Evaluation of Tree Water Status in Deciduous Orchards. *Horticultural Reviews* 32: 111–166.

Naor, A. 2006b. Irrigation Scheduling of Peach—Deficit Irrigation at Different Phenological Stages and Water Stress Assessment. *Acta Horticulturae* 713: 339–349.

Naor, A., Stern, R., Peres, M., Greenblat, Y., Gal, Y., and Flaishman, M.A. 2005. Timing and Severity of Postharvest Water Stress Affect Following Year Productivity and Fruit Quality of Field-Grown 'Snow Queen' Nectarine. *Journal of American Society of Horticultural Science* 130: 806–812.

Nayak, B., Berrios, J.E.J., Powers, J.R., and Tang, J. 2011. Thermal Degradation of Anthocyanins from Purple Potato (cv. Purple Majesty) and Impact on Antioxidant Capacity. *Journal of Agricultural and Food Chemistry* 59: 11040–11049.

Nguyen, M., Francis, D., and Schwartz, S. 2001. Thermal Isomerization Susceptibility of Carotenoids in Different Tomato Varieties. *Journal of Science of Food and Agriculture* 81: 910–917.

Nile, S.H., and Park, S.W. 2014. Edible Berries: Bioactive Components and Their Effect on Human Health. *Nutrition* 30: 134–144.

Nwanna, E.E., and Oboh, G. 2007. Antioxidant and Hepato Protective Properties of Polyphenol Extracts from *Telfairia occidentalis* (fluted pumpkin) Leaves on Acetaminophen-Induced Liver Damage. *Pakistan Journal of Biological Sciences* 10: 2682–2687.

Oboh, G., Raddatz, H., and Henle, T. 2008. Antioxidant Properties of Polar and Non-Polar Extracts of Some Tropical Green Leafy Vegetables. *Journal of the Science of Food and Agriculture* 88: 2486–2492.

Ortuño, C., Duong, T., Balaban, M., and Benedito, J. 2013. Combined High Hydrostatic Pressure and Carbon Dioxide Inactivation of Pectin Methylesterase, Polyphenol Oxidase and Peroxidase in Feijoa Puree. *The Journal of Supercritical Fluids* 82: 56–62.

Ozturk, B., Kucuker, E., Karaman, S., and Ozkan, Y. 2012. The Effect of Cold Storage and Aminoethoxyvinylglycine (AVG) on Bioactive Compounds of Plum (Prunus salicina L. cv. "Black Amber"). *Postharvest Biology and Technology* 72: 35–41.

Ozturk, B., Ozkan, Y., Altuntas, E., Yıldız, K., and Saracoglu, O. 2013. Effect of Aminoethoxyvinylglycine on Biochemical, Physicomechanical, and Colour Properties of cv. "Braeburn" Apple (Malus domestica Borkh). *Semina: Ciencias Agrarias* 34: 1111–1120.

Ozturk, B., Yildiz, K., and Kucuker, E. 2015. Effect of Pre-Harvest Methyl Jasmonate Treatments on Ethylene Production, Water-Soluble Phenolic Compounds and Fruit Quality of Japanese Plums. *Journal of Science of Food and Agriculture* 95: 583–591.

Paliyath, G., Murr, D.P., Handa, A.K., and Lurie, S. eds. 2008. *Postharvest Biology and Technology of Fruits, Vegetables, and Flowers*. Wiley-Blackwell Publishing.

Panahirad, S., Naghshiband-Hassani, R., Bergin, S., Katam, R., and Mahna, N. 2020. Improvement of Postharvest Quality of Plum (Prunus domestica L.) Using Polysaccharide-Based Edible Coatings. *Plants* 9: 1148.

Park, S.U., Ahn, D.J., Jeon, H.J., Kwon, T.R., Lim, H.S., Choi, B.S., Baek, K.H., and Bae, H. 2012. Increase in the Contents of Ginsenosides in Raw Ginseng Roots in Response to Exposure to 450 and 470 nm Light from Light-Emitting Diodes. *Journal of Ginseng Research* 36: 198–204.

Parry, M., Rosenzweig, C., and Livermore, M. 2005. Climate change, global food supply and risk of hunger. Philosophical Transactions of the Royal Society of London B. *Biological Sciences* 360: 2125–2138.

Patras, A., Brunton, N.P., O'Donnell, C., and Tiwari, B. 2010. Effect of Thermal Processing on Anthocyanin Stability in Foods; Mechanisms and Kinetics of Degradation. *Trends in Food Science and Technology* 21 (1): 3–11.

Piga, A., Del Caro, A., and Corda, G. 2003. From Plums to Prunes: Influence of Drying Parameters on Polyphenols and Antioxidant Activity. *Journal of Agricultural and Food Chemistry* 51: 3675–3681.

Poiana, M.A., Moigradean, D., and Alexa, E. 2010a. Influence of Home Scale Freezing and Storage of Antioxidant Properties and Color Quality of Different Garden Fruits. *Bulgarian Journal of Agricultural Science* 16 (2): 163–171.

Pozo-Insfran, D., Follo-Martinez, D., Talcott, S., and Brenes, C. 2007. Stability of Co Pigmented Anthocyanins and Ascorbic Acid in Muscadine Grape Juice Processed by High Hydrostatic Pressure. *Journal of Food Science* 72 (4): 247–253.

Prior, R.L., and Cao, G. 2000. Antioxidant Phytochemicals in Fruits and Vegetables: Dietary and Health Implications. *Horticultural Sciences* 35: 588–592.

Proteggente, A.R., Pannala, A.S., Paganga, G., van Buren, L., Wagner, E., Wiseman, S., van de Put, F., Dacombe, C., and Rice- Evans, C.A. 2002. The Antioxidant Activity of Regularly Consumed Fruit and Vegetables Reflects Their Phenolic and Vitamin Composition. *Free Radical Research* 36: 217–233.

Qin, J., Culver, D.A., and Yu, N. 1995. Effect of Organic Fertilizer on Heterotrophs and Autotrophs: Implications for Water Quality Management. *Aquaculture Research* 26: 911–920.

Quail, P.H., Boylan, M.T., Parks, B.M., Short, T.W., Xu, Y., and Wagner, D. 1995 Phytochromes: Photosensory Perception and Signal Transduction. *Science* 268: 675–680.

Rahman, I., Biswas, S.K., and Kirkham, P.A. 2006. Regulation of Inflammation and Redox Signalling by Dietary Polyphenols. *Biochemical Pharmacology* 72: 1439–1452.

Raso, J., and Heinz, V. 2006. Pulsed Electric Field for the Food Industry. In *Fundamental and Applications*, 3–26. Springer.

Rein, M.J. 2005. *Co Pigmentation Reactions and Color Stability of Berry Anthocyanins*. PhD thesis, Department of Applied Chemistry and Microbiology, University of Helsinki.

Rettke, M.A., Pitt, T.R., Maier, N.A., and Jones, J.A. 2006. Quality of Fresh and Dried Fruit of Apricot (cv. Mooprark) in Response to Soil Applied Nitrogen. *Australian Journal of Experimental Agriculture Melbourne* 46 (1): 123–129.

Rossi, M., Giussani, E., Morelli, R., Lo Scalzo, R., Nani, R.C., and Torreggiani, D. 2003. Effect of Fruit Blanching on Phenolics and Radical Scavenging Activity of Highbush Blueberry Juice. *Food Research International* 36: 999–1005.

Rudell, D.R., Mattheis, J.P., Fan, X., and Fellman, J.K. 2002. Methyl Jasmonate Enhances Anthocyanin Accumulation and Modifies Production of Phenolics and Pigments in 'Fuji' Apples. *Journal of the American Society for Horticultural Science* 127: 435–441.

Ruiz-Garcia, Y., Romero Cascales, I., Gil-Muñoz, R., Fernandez-Fernandez, J.I., Lopez-Roca, J.M., and G6mez-Plaza, E. 2012. Improving Grape Phenolic Content and Wine Chromatic Characteristics Through the Use of Two Different Elicitors: Methyl Jasmonate Versus Benzothiadiazole. *Journal of Agricultural Food Chemistry* 60: 1283–1290.

Ruiz, R. 2006. Effects of Different Potassium Fertilizers on Yield, Fruit Quality and Nutritional Status of 'Fairlane' Nectarine Trees and on Soil Fertility. *Acta Horticulturae* 721: 185–190.

Rungapamestry, V., Duncan, A.J., Fuller, Z., and Ratcliffe, B. 2007. Effect of Cooking Brassica Vegetables on the Subsequent Hydrolysis and Metabolic Fate of Glucosinolates. *Proceedings of Nutrition Society* 66: 69–81.

Sahamishirazi, S., Moehring, J., Claupein, W., and Graeff, H.S. 2017. Quality Assessment of 178 Cultivars of Plum Regarding Phenolic, Anthocyanin and Sugar Content. *Food Chemistry* 214: 694–701.

Saltveit, M.E. 2003. Is it Possible to Find an Optimal Controlled Atmosphere? *Postharvest Biology and Technology* 27: 3–13.

Santhakumar, A.B., Kundur, A.R., Fanning, K., Netzel, M., Stanley, R., and Singh, I. 2015a. Consumption of Anthocyanin-Rich Queen Garnet Plum Juice Reduces Platelet Activation Related Thrombogenesis in Healthy Volunteers. *Journal of Functional Foods* 12: 11–22.

Sayyari, M., Babalar, M., Kalantari, S., Martinez-Romero, D., Guillen, F., Serrano, M., Valero, D. 2011.Vapour Treatments with Methyl Salicylate or Methyl Jasmonate Alleviated Chilling Injury and Enhanced Antioxidant Potential During Postharvest Storage Of Pomegranates. *Food Chemistry* 124: 964–970.

Schroeder, G.L. 1978. Autotrophic and Heterotrophic Production of Microorganisms in Intensely- Manured Fish Ponds and Related Fish Yields. *Aquaculture* 14: 303–325.

Shukitt-Hale, B., Kalt, W., Carey, A.N., Vinqvist-Tymchuk, M., Mcdonald, J., and Joseph, J.A. 2009. Plum Juice, but not Dried Plum Powder Is Effective in Mitigating Cognitive Deficits in Aged Rats. *Nutrition* 25: 567–573.

Silva, J.M., Villar, H.P., and Pimentel, R.M.M. 2012. Structure of the Cell Wall of Mango After Application of Ionizing Radiation. *Radiation Physics and Chemistry* 81: 1770–1775.

Sima, P.R., Rahim, N.H., Nasser, M. 2020. Pectin-based edible coating preserves antioxidative capacity of plum fruit during shelf life. *Food Science and Technology International* 1–10.

Singh, S.P., and Singh, Z. 2013. Controlled and modified atmospheres influence chilling injury, fruit quality and ant oxidative system of Japanese plums (*Prunus salicina* Lindell). *International Journal of Food Science and Technology* 48: 363–374.

Singh, P., and Dwivedi, U.N. 2008. Purification and Characterization of Multiple Forms of Polygalacturonase from Mango (*Mangifera indica* cv. Dashehari) Fruit. *Food Chemistry* 111: 345–349.

Singh, Z., and Khan, A.S. 2010. Physiology of Plum Fruit Ripening. *Stewart Postharvest Review* 2: 3.

Skrede, G. 1996. Fruits. In *Freezing Effects on Food Quality*, edited by L.E. Jeremiah, 183–245. Marcel Dekker Inc.

Srilaong, V., and Tatsumi, Y. 2003. Changes in respiratory and antioxidative parameters in cucumber fruit (*Cucumis sativus* L.) stored under high and low oxygen concentrations. *Journal of the Japanese Society for Horticultural Science* 72: 525–532.

Stacewicz-Sapuntzakis, M., Bowen, P.E., Hussain, E.A., Damayanti-Wood, B.I., and Farnsworth, N.R. 2001. Chemical composition and potential health effects of prunes- a functional food. *Critical Reviews in Food Science and Nutrition* 41: 251–287.

Stanger, M.C., Steffens, C.A., Amarante, C.V.T.D., Brackmann, A., and Anese, R.O. 2017. Quality preservation of 'Laetitia' plums in active modified atmosphere storage. *Revista Brasileira de Fruticultura* 39: 1–7.

Stern, R.A., Flaishman, M., Ben-Arie, R. 2007. Effect of synthetic auxins on fruit size of five cultivars of Japanese plum (Prunus salicina Lindl.). *Scientia Horticulturae* 112: 304–309.

Stintzing, F.C., Stintzing, A.S., Carle, R., Frei, B., and Wrolstad, R.E. 2002. Color and antioxidant properties of cyanidin-based anthocyanin pigments. *Journal of Agricultural and Food Chemistry* 50: 6172–6181.

Svelander, C.A., Tibäck, E.A., Ahrné, L.M., Langton, M.I.B.C., Svanberg, U.S.O., and Alminger, M.A.G. 2010. Processing of tomato impact on in vitro bioaccessibility of lycopene and textural properties. *Journal of Science of Food and Agriculture* 90: 1665–1672.

Syamaladevi, R.M., Sablani, S.S., Tang, J., Powers, J., and Swanson, B.G. 2011. Stability of anthocyanins in frozen and freeze-dried raspberries during long-term storage: In relation to glass transition. *Journal of food science* 76: 414–421.

Taiwo, K.A., Angersbach, A., and Knorr, D. 2002. Influence of high intensity electric field pulses and osmotic dehydration on the rehydration characteristics of apple slices at different temperatures. *Journal of Food Engineering* 52: 185–192.

Thakur, R., Pristijono, P., Golding, J.B., Stathopoulos, C.E., Scarlett, C.J., Bowyer, M., Singh, S.P., and Vuong, Q.V. 2018. Development and Application of Rice Starch Based Edible Coating to Improve the Postharvest Storage Potential and Quality of Plum Fruit (Prunus salicina). *Scientia Horticulturae* 237: 59–66.

Tibäck, E.A., Svelander, C.A., Colle, I.J.P., Altskär, A.I., Alminger, M.A.G., Hendrickx, M.E.G., Ahrné, L.M., and Langton, M.I.B.C. 2009. Mechanical and thermal Pretreatments of Crushed Tomatoes: Effects on Consistency and in Vitro Accessibility of Lycopene. *Journal of Food Science* 74: 386–395.

Timilsena, Y.P., Adhikari, R., Casey, P., Muster, T., Gill, H., and Adhikari, B. 2015. Enhanced efficiency fertilisers: A review of formulation and nutrient release patterns. *Journal of Science of Food and Agriculture* 95: 1131–1142.

Tomás-Barberán, F.A., Espin, J.C. 2001. Phenolic compounds and related enzymes as determinants of quality in fruits and vegetables. *Journal of the Science of Food and Agriculture* 81: 853–876.

Tsuji, M., Harakawa, M., and Komiyama, Y. 1983. Inhibition of increase of pulp color and phenylalanine ammonia-lyase activity in plum fruit at high temperature (30°C). *Journal of the Japanese Society for Food Science and Technology* 30: 688–692.

Turturica, M., Nicoleta, S., Bahrim, G., and Rapeanu, G. 2015. Effect of thermal treatment on phenolic compounds from plum (prunus domestica) extracts e A kinetic study. *Journal of Food Engineering* 171: 200–207.

Urbain, W.M. 1986. *Food Irradiation*. Academic Press, Inc., U.S. Department of Agriculture.

Usenik, V., Kastelec, D., Veberič, R., and Štampar, F. 2008. Quality changes during ripening of plums (*Prunus domestica* L.). *Food Chemistry* 111 (4): 830–836.

Usenik, V., Štampar, F., and Veberič, R. 2009. Anthocyanins and fruit color in plums (*Prunus domestica* L.) during ripening. *Food Chemistry* 114 (2): 529–534.

Uthairatanakij, A., Jitareerat, P., and Kanlavanarat, S. 2006. Effects of Irradiation on Quality Attributes of Two Cultivars of Mango. *Acta Horticulture* 712: 885–892.

Valero, D., Díaz-Mula, H.M., Zapata, P.J., Guillén, F., Martínez-Romero, D., Castillo, S., Serrano, M. 2013. Effects of Alginate Edible Coating on Preserving Fruit Quality in Four Plum Cultivars During Postharvest Storage. *Postharvest Biology and Technology* 77: 1–6.

Vinson, J.A., Su, X.H., Zubik, L., and Bose, P. 2001. Phenol antioxidant quantity and quality in foods: Fruits. *Journal of Agricultural and Food Chgoncalvesemistry* 49 (11): 5315–5321.

Volden, J., Borge, G.I.A., Hansen, M., Wicklund, T., and Bengtsson, G.B. 2009. Processing (blanching, boiling, steaming) effects on the content of glucosinolates and antioxidant-related parameters in cauliflower (Brassica oleracea L. ssp. botrytis). *LWT Food Science and Technology* 42: 63–73.

Wang L, Chen S, Kong W, Li S, Archbold D. 2006. Salicylic acid pretreatment alleviates chilling injury and affects the antioxidant system and heat shock proteins of peaches during cold storage. *Postharvest Biology and Technology* 41: 244–251.

Wang, S.Y., and Zheng, W. 2005. Pre harvest application of methyl Jasmonate increases fruit quality and antioxidant capacity in raspberries. *International Journal of Food Science and Technology* 40: 187–195.

Wang, L.M., Sang, W.N., Xu, R.R., and Cao, J.K. 2020. Alteration of Flesh Color and Enhancement of Bioactive Substances via the Stimulation of Anthocyanin Biosynthesis in 'Friar' Plum Fruit by Low Temperature and the Removal. *Food Chemistry* 310.

Wang, R., Wang, L., Yuan, S., Li, Q., Pan, H., and Cao, J.K. 2018. Compositional modifications of bioactive compounds and changes in the edible quality and antioxidant activity of 'Friar' plum fruit during flesh reddening at intermediate temperatures *Food Chemistry* 254: 26–35.

Wang, H., Cao, G., and Prior, R.L. 1996. Total antioxidant capacity of fruits. *Journal of Agricultural and Food Chemistry* 44: 701–705.

White, P.J., and Broadley, M.R. 2003. Calcium in plants. Annals of Botany. 92: 487–511.

Wohlfarth, G.W., and Schroeder, G.L. 1979. Use of manure in fish farming a review. *Agricultural Wastes* 1: 279–299.

Wojdylo, A., Figiel, A., Lech, K., Nowicka, P., and Oszmian´ ski, J. 2014. Effect of convective and vacuum—microwave drying on the bioactive compounds, color, and antioxidant capacity of sour cherries. *Food and Bioprocess Technology* 7: 829–841.

Wu, M.C., Hou, C.Y., Jiang, C.M., Wang, Y.T., Wang, C.Y., Chen, H.H., and Chang, H.M. 2007. A novel approach of LED light radiation improves the antioxidant activity of pea seedlings. *Food Chemistry* 101: 1753–1758.

Wu, X., and Prior, R.L. 2005. Systematic identification and characterization of anthocyanins by HPLC-ESI-MS/MS in common foods in the United States: Fruits and berries. *Journal of Agricultural and Food Chemistry* 53: 2589–2599.

Xu, X., and Tian, S. 2008. Salicylic acid alleviated pathogen—induced oxidative stress in harvested sweet cherry fruit. *Postharvest Biology and Technology* 49: 379–385.

Yamada, T., Kuroda, K., Jitsuyama, Y., Takezawa, D., Arakawa, K., and Fujikawa, S. 2002. Roles of the plasma membrane and the cell wall in the responses of plant cells to freezing. *Planta* 215: 770–778.

Yang, Z.F., Zheng, Y.H., and Cao, S.F. 2013. Influence of Harvest Maturity on Fruit Quality, Color Development and Phenylalanine Ammonia-Lyase (PAL) Activities in Chinese Bayberry During Storage. *Acta Horticulturae* 1012: 171–175.

Yeom, H.W., Streaker, C.B., Zhang, Q.H., and Min, D.B. 2000. Effects of Pulsed Electric Fields On the Quality of Orange Juice and Comparison with Heat Pasteurization. *Journal of Agricultural and Food Chemistry* 48 (10): 4597–4605.

Zaidi, S.F., Muhammad, J.S., Shahryar, S., Usmanghani, K., Gilani, A.H., and Jafri, W. 2012. Anti-Inflammatory and Cytoprotective Effects of Selected Pakistani Medicinal Plants in Helicobacter Pylori-Infected Gastric Epithelial Cells. *Journal of Ethnopharmacology* 141: 403–410.

14 Effect of Preharvest and Postharvest Factors on Quality of Plum

Muhammad Kamran Khan, Muhammad Imran, Muhammad Haseeb Ahmad, Aliza Zulifqar and Nimra Saeed
Department of Food Science, Faculty of Life Sciences, Government College University, Faisalabad-38000, Pakistan

CONTENTS

14.1 PREHARVEST FACTORS

Through technological advancements in preharvesting techniques, the required quality of fruit can be attained to satisfy consumers. There are numerous preharvesting practices that may define the quality of fruits in an orchard. At the same time, biochemical disorders must be kept in mind to limit harm to the fruit in terms of

DOI: 10.1201/9781003205449-14

its value and ripening conditions. These disorders may lead to the development of unwanted conditions of fruit, such as unpleasant aroma, softness and the accumulation of anthocyanin leading to changes in colored and reduction in acidity. Changes in fruit-ripening methodologies and circumstances contribute to value-addition and increasing flavor. For instance, exposure to light on cultivated areas is important. Access to direct and indirect sunlight makes a difference in fruit quality, as the upper parts of trees receive more light, contributing to the retention of plum fruit quality (Manganaris et al. 2008). To maintain quality, plums must be dealt with carefully from the orchard to the consumer's hands.

14.1.1 Orchard Health

Orchard soil management and secondary metabolites have direct interaction with each other in terms of syndromes and other related issues, such as the watering of fields and the physical management of the orchard. A successful plum orchard depends on cultivars, growing conditions and a training system. Lombardi-Boccia et al. (2004) found that higher levels of tocopherol, beta-carotene and total phenolic compounds are present if the fruits grew on natural land. Further, Chocano et al. (2016) found that managing soil with compost resulted in better yield and quality of plum fruits. Adding sufficient amounts of compost increases the carbon content and microbial pool of the soil, making the soil capable of developing better-quality products.

Plum and peach orchards in temperate regions are usually planted on sloppy terrain, and there are several difficulties from prehasvesting to postharvesting. In such orchards, weeds must be demolished through manual weeding or by putting grass mulching (15 cm thick) in the basin area. In addition, lawn grass must be planted to fill in free spaces around the fruit plants and basin area. For proper and continuous supply of nutrients to plants, long and deep rooted growth of roots must be avoided among intercrops, as they use nutrients more rapidly than fruit plants and can cause a decrease in nutrients in orchards. For peach and plum orchards, intercrops such as cowpeas, peas, sbeans, cabbage, cauliflower, garlic, colocasia and ginger are advised. Over the time, plum yield can decrease if the orchard is not well maintained, resulting in an inability to meet production criteria during cultivation season (Milosevic et al. 2007).

With the passage of time, orchard soil health can deteriorate. Biological, physical and chemical aspects determine the ability of soil to grow fruitful plants and compete with plant pathogens. Proper functioning of soil can be estimated by its ability to absorb and release the nutrients, to absorb water and its movement. Provision of necessary oxygen supply to plants and soil microbes and may correspond to environmental changes. On the other hand, a variety of chemicals is also necessary for soil health, such as pH, cation exchange capacity (CEC), salinity and the presence of any contaminants, like heavy metals or persistent pesticide residues. Stronger roots would be less affected by fluent water flow and absorbed fertilizers and raise the output of orchards by better understanding of the biological, physical and chemical aspects of soil.

Prerequisites must be followed before planting, such as choosing an appropriate area for growing fruit, topsoil and best possible seed. It is important to obtain the best productivity of fruits via multiple preharvesting orchard management activities, including training and pruning, manure, fertilizer application, irrigation, weed control and insect pest and disease management. The shelf life of plum fruits is very short, so they require immediate pre- and postharvest management practices to maintain fruit characteristics.

14.1.2 Pruning

Many researchers have stated that pruning intensity directly affects offshoot development (Demirtas et al. 2010). Due to the important amount of carotenes and fibrous and anthocyanin content of plums, this fruit plays a crucial role in the human diet (Sommano et al. 2013). Black plum cultivars in particular are popular among consumers. Recently, great confusion has been observed in pruning and the planting density of plums in the Mediterranean region. Desired production has not been reached yet because of erroneous planting and pruning applications. Generally, classical pruning methods have been used in these orchards. Due to the recent increase in the production of Japanese plum cultivars in the Mediterranean, it is now necessary to find suitable pruning methods for the ecological demands of those cultivars in that region. Hrotkó et al. (2001) stated that plum orchards were increasing, but there was lack of information on pruning and other technical methods of orchard maintenance. Researchers emphasized the necessity for the plantation of plum orchards with dwarf species to increase fruit quality, facilitate the harvest and decrease the labor costs. Constructed shape in pruning allows air to enter the crown and increases the success rate of disease and pest control (Simon et al. 2007). Similarly, there is a positive relation between the amount of light entering the crown and the coloring of the fruit (Dennis Jr 2003). Success in pruning systems may be changed according to cultivar, labor costs, fruit price and location. However, it was observed that pruning systems might change certain biological characteristics of cultivars, such as fertilization time, flowering density and fertilization percentage (Stephan et al. 2008). Pruning plays a crucial role in both decreasing the juvenile period and increasing the productivity and quality of fruit trees (Naira and Moieza 2014). It minimizes vegetative buds and promotes growth of shoots, owing to changes in hormonal circumstances to the supply of required nutrients to not only newly formed roots but also to top areas. Fruit are dominant sinks in the fight between vegetative and reproductive growth, and reproductive sinks have a greater priority for water and nutrients than other plant parts.

Numerous implementations affect the flavor of fruits including sugars, acids, and aroma compounds. The concentration of soluble solids in fruits is negatively affected by low photosynthetic activity caused by incorrect canopy management, high-density plantation, usage of black hail netting, and forceful pruning. Pruning is not only important for the quality of the fruit in the current season but also for the growth of new branches for increased productivity the next season. Thus, the main consideration of pruning is to produce fruit of the required quality by getting rid of unwanted parts. By following some guidelines while pruning, plum growth can be increased in different areas on an annual basis. For example, in tropical environments, 30–70 cm and 25–30 cm under mid-hills is considered suitable for obtaining good production; 40–50% thinning out and 75% heading back of shoots is recommended for good quality fruit yield. Non-productive and damaged branches must be pruned. On two-year-old wood, the majority of plum types have spurs. These spurs have a lifespan of five to six years. Each season, pruning is required to allow for some spur regrowth. The pruning is done to the point when yearly shoot growth of 25–50 cm is induced. For proper fruiting, 25–30% thinning of shoots and 50–75% heading back of shoots is recommended in fruit-bearing plum trees.

14.1.3 Source of Nutrients

Utilizing minerals in the best possible way forces plants to perform at their best level (Hewett 2006). Orchard preparation is mainly dependent on fertilization as a basic need. There is a lot of contribution of these nutrients that permits plants to work more efficiently in terms of production. Fertilizers can also be harmful to plants and orchards, and inappropriate fertilization (in terms of too much or too little) may lead to economic losses as well as plant and human health impacts. Fertilizers can also have harmful effects on the environment (Singh et al. 2012). Almost half of the supplied nutrients are not used by plants and hence lost. Pesakovic et al. (2012) mentioned that the use of chemical nutrients in high ratios (above 600 kg ha-1 of NPK) in plum-growing technology causes a reduction in microorganisms in the plum rhizosphere, which further decreases the production capability of the plum. Thus, it is quite necessary to develop strategies to optimize production and increase nutritional values so that help in advanced plum-growing technology along with environmental protection will be gained.

Fertilization is one of the main agrotechnical events that increase the yield-per-unit area (Georgiev et al. 2019). To determine the optimal fertilizer rates/doses, it is important to identify first the nutritional needs of the trees through plant and soil analysis. Nitrogen (N) and potassium (K) are the primary nutrients required by plants. In stone fruits, increments of nitrogen supply increase vegetative growth but slow down the maturity process of the fruits. Nitrogen deficiency also caused under-sized fruits, limited yield, and tasteless fruit. By boosting the photosynthetic procedure in leaves along with organic acids, sucrose and optimal potassium (K), fruit quality increases. Calcium (Ca) is especially important for fruit trees since it plays a role in a variety of biochemical and morphological processes. Its participation plays a vital role in delaying fruit senescence. Furthermore, the importance of calcium cannot be denied, as it plays a beneficial role in fruit development and has a healthy impact on cell wall structure (Kader 2002). Calcium can also be administered to the fruit by foliar sprays before harvest or during postharvest dips.

It is well known that fertilizers are needed to improve vegetative activity and increase the productivity of fruit trees. Leaves are the main and most commonly used plant organ, so analyzing and determining their chemical composition gives an idea of the degree of nutrients to supply to the plants (Wang et al. 2015). According to Panayotova et al. (2018), foliar nourishment gives good results in a number of vegetable and fruit crops. To affect plants, leaf fertilizers must be easily absorbed, transported quickly and must easily release their ions. Iron (Fe) complexes with important protein-forming enzymes in plants are linked to chloroplasts, which play an essential role in chlorophyll synthesis. Chlorophyll is the main green photosynthetic pigment found in plants through which they absorb the light energy (Kiang et al. 2007). The macronutrients, such as potassium (K) and magnesium (Mg), are involved in the photosynthesis process and in a number of plant functions (Tränkner et al. 2018). This requires monitoring the nutrient content of the foliage of fruit species.

The uptake of heavy metals is reduced by the influence of living microbial cells from biofertilizers. Mosa et al. (2014) noted that it is better to supply plants with biofertilizer elements rather than chemical fertilizers for optimal growth. Von-Bennewitz et al. (2008) found that in stone and pome fruits biofertilization acts as a simulating technique for proper growth. Yeast manipulates the plant progression, nutritional strength and physical and chemical properties of plum tree fruits (Mansour et al. 2011). Other than the major components already discussed, there are small amounts of other nutrients too that actively contribute to the health of fruits and are applied in different forms, including as sprays, through the natural supply of water, added to the soil or a mixture of these. The decomposition of waste, in the form of fallen leaves, fruits and pruning debris, as well as cover crop clippings, recycles micronutrients inside orchards. The amount of micronutrients lost in fruits and pruning debris carried out of the orchard, as well as leaching, are all permanent losses of micronutrients.

14.1.4 Quantity and Time of Irrigation

Properly irrigated fruit and nut trees grow up in a proper manner. On the other hand, improper amounts of irrigation cause many defects, such as wilting, curling and sunburnt leaves and small, shriveled and sunburnt fruit. Solid soluble parts in the soil can improve with less supply of water, but this may result in smaller fruit. Low fruit load enhances carbon partitioning into fruits and increases soluble solid content. Irrigation affects the plant's water and nutrient supply, as well as the nutritional and antioxidant potential of the fruit.

Personal observations with plum growers indicated that under the Egyptian conditions, fruit have shown poor peel color, low sugar and high acidity. This might be due to the hot and dry growing season, which requires sufficient rates of irrigation during fruit ripening to avoid heat stress and preharvest fruit abscission (Farag 2015). Plant–water interaction has been of particular consideration in current years; it is usually known that moderate water stress definitely improves plum fruit quality. Deficit irrigation is known as a method that increases a plant's water-use efficiency and improves fruit quality (Intrigliolo et al. 2013). For the growth of stone fruit, a three-step procedure is involved. During stage 1, a classic double-sigmoid pattern with rapid exponential growth during cell division occurs. Stage 2 involves slow growth due to pit hardening and embryo development and is comparatively shorter than stage 1. The third and final stage follows rapid growth due to cell enlargement prior to harvest. Each of the described stages has its own time of completion under provided circumstances. The response of stone fruit trees to deficit irrigation relies on the stage of fruit growth at which it was applied (Maatallah et al. 2015). Therefore, deficit irrigation applied to plum trees during the period after pit hardening until the end of harvest (stage 3) resulted in a reduction in fruit size and total yield, but improved fruit composition, color and quality (Steduto et al. 2012). Fruit sensitivity to water stress during stage 2 is slight because of the slow growth rate during this period (Lawrence and Melgar 2020).

Water acts as an essential component of plant structure as well as a medium for biological activities. The need for water remains the same for all kinds of plum trees, whether they are young or old. As the season becomes hotter from April to June, this demand for water increases due to extensive vaporization. In November, temperate weather allows for a delay in watering the plants. There are some particular time periods when plum plants need water. In the hilly areas, plum orchards need to be watered every 10–12 days. On the other hand, from March to July, plums need to be watered on a weekly basis using the basin methodology. In water-scarce areas, watering methods can be performed by using the drip method, which is more efficient and saves more water. Under the drip method, plants are watered every other day for hour, using three drippers per plant. To maintain soil moisture in orchards in rainy areas, the basin area should be covered with thick grass, usually up to 15 cm thick, or with black polythene during the month of March.

There is an average of 8–10 irrigations in one season when watering every 10–15 days. There is no need for irrigation in a rainy climate. In September, October and November, the interval may be extended to 20 days. A gap of one week is required for watering after fruit set, with (6–8) irrigation settlements advised for increased productivity of Santa Rosa plum fruits. Agricultural and horticultural products can change their shape under the effects of chemicals being used (e.g., CaCO3 sprays on plum) along with extra water treatments. Residual surface chemicals should be avoided for crops that are quickly harmed or cannot be exposed to free water (like bramble fruits and strawberries).

14.1.5 Sprays of Fungicides and Other Chemicals

Copper-based fungicides are widely used to act against different varieties of mycoses. In plum, one of the most dangerous is klyasterosporiosis, which is caused by *Stigmina carpophila* (Lév.) M.B. Ellis. Leaves, shoots, flower buds, ovaries and fruits get infected by this pathogenic disease, which demolishes the overall productivity of the trees and also fruit quality (Yakuba et al. 2021). The spread of klyasterosporiosis on shoots has increased by up to 40% and on leaves by up to 90% under the conditions of the region. Environmental health is negatively affected by copper too. But at same time, there is no best option available to control fungicides other than this. Fungicides with copper elements are more known than others because there is less risk involved against pathogen resistance in their utilization and they are cheap to purchase (Yakuba et al. 2021).

There are many copper compounds that are best known for plum rescue, such as copper sulfate, copper oxychloride and copper hydroxide (Kuehne et al. 2017). Only two copper-containing fungicides have been registered for industrial plum plants, which poses the issue of investigating their ultimate effects during changes occurring in the surroundings. Plants are largely made up of leaves, which are used for absorption. These also help trees to grow, as numerous elements are generated within them. Titanium containing two bio activators are mixed with calcium and sprayed during preharvesting, and further studies were conducted to investigate the resulting fruit quality (Alcaraz-Lopez et al. 2004). The effect of developing fruit fungal areas shown during postharvesting, although this development occurred in the growing season. To control illnesses before and after harvest, fungicides are often used during preharvest.

The effectiveness of various preharvest chemicals sprayed, varies with the variety, environmental conditions and stage of fruit development at which it is applied. Plum fruits treated with 4% CaCl2 maintained higher values of TSS, total sugars, reducing sugars and acidity during the entire storage period under ambient conditions (Hayat et al. 2013). The optimum formulation, whether as a fungicide or a natural component, for controlling postharvest rot of fruits and vegetables at the preharvest stage should meet the following requirements (Smilanick 2012):

- Fruit must remain unharmed.
- For surroundings and for humans its harmful effects must be negotiable.
- The food storage process must be affected by residuals.
- It must be compatible with normal processes, inexpensive and simple to apply.

14.1.6 Tillage Operation (Weed Control)

The presence of foreign materials on or in the product might significantly impair its appeal. Extraneous matter includes things like growth medium, which is a frequent mushroom ailment. Similarly, the presence of organic materials not demanded by fields in excessive amounts is especially not recommended for leafy vegetables. The presence of these items decreases the commodity's desirability. Other foreign objects, such as insect webbing, excreta and dead bodies, are likewise unwelcome (Nassar et al. 2001).

Weeds are one of the major components causing a decrease in water supply in growing fields, as they absorb large amounts of water and demand nitrogen and other relevant elements. Monocot weeds, such as *Cyperus rotundus*, *Commelina nudiflora*, *Digitaria sp* and *Setaria gluuca*, and dicot weeds, such as amaranthus, viridis, chenopodium and medicago, were found in plum orchards, consuming the majority of the nutrients in the soil. Cultivated areas should always be kept weed free, and this can be fulfilled by using manual methods or mechanical processes, especially during leaf fall season. Weedicides should be used to keep the weeds under control. Weedicides such as simazine and atrazine, applied at 3.0 kg/ha before weed emergence in March, are followed by two sprays of the post-emergence weedicide glyphosate, administered at 4.5 kg/ha in July and August. Mulching also aids in the management of weed growth in the tree basin area. Apart from intercropping and cover cropping, one of the procedures used in orchards is manual weed removal (Chandel 2021). Weeds in plum orchards are managed by post-emergence glyphosate sprays at a rate of 800 ml/ha during rainy seasons (Shamrao 2020).

It is not necessary to maintain a weed-free area under the trees during the dormant season, so cover crops or ornamentals can be cultivated to enhance the soil or simply to look pleasant. Because mature trees have an established root system, are full-sized and do not need to grow as much, they can tolerate more weeds, turf, or cover crops growing inside their dripline. There are many annual and perennial weeds in plum orchards. These weeds fight for water, nutrients, and light with plum trees (in newly developed orchards). Weeds can limit the development, vigor, and performing production of plants; therefore, to obtain these necessities is more of a worry. Weeds continue to be an issue in older orchards because they can raise the danger of frost damage early in the season, host plum tree pests and pathogens, obstruct irrigation systems, compete for water and nutrients, and cause problems during harvest. A successful orchard weed management program has numerous components (Bentley 2002).

14.1.7 Others

Genotypic variation: Before crop plantation, cultivators must select cultivars, as biotechnological adaptations and horticulture breeding contributes significantly not only to increasing the crop but also keeping newly developed product secure (Kader 2002). Healthy selection among multiple races of fruit is one of the most important factors needed for cultivating. Further, genetics also plays a significant role in cultivating a fruit to unleash outcome during its storage (Crisosto and Costa 2008).

Canopy light condition: Because of the light environment, plum fruits ripen in descending order on trees. Thus, fruits on the lower part of the tree remain small in size and mature 10–14 days later than fruits on the upper part of trees. In addition, fruits on all plum plant postures need enough light for timely ripening and quality growth (Manganaris et al. 2008).

Titanium treatment: Increased fruit size, fruit firmness retention, reduced weight loss during storage and superior fruit surface color has been attained by titanium treatments. All these healthier attributes have been gained due to improved calcium, iron, zinc and copper uptake incited by titanium (Manganaris et al. 2008). Titanium has also been assessed in combination with calcium, magnesium and algae extract and has shown good results (Alcaraz-Lopez et al. 2004).

Growth regulators: Synthetic auxins can be applied at the start of pit hardening in order to stimulate fruit cell extension, fruit size and production enhancement (Stern et al. 2007). The application of gibberellic acid (GA3) has shown enhancement of tree growth and fruit yield (Servili et al. 2017).

14.2 POSTHARVEST FACTORS

Postharvest quality maintenance of plums begins from the field and continues until the consumer holds it in their hands. After harvesting, the fruit remains alive and performs all the functions of a living tissue. As plums are highly perishable fruits, postharvest quality loss is rapid, contributing to the quality management of the fruit. These factors include maturity stage, methods of harvesting, time of harvesting, precooling (after harvesting and before storage), sorting and grading, coatings, packaging and packaging materials, storage and temperature of storage, transportation (duration and road conditions) and method of loading or unloading fruits. By optimizing the postharvest conditions and applying the best handling practices, the postharvest quality of the fruit can be maintained.

14.2.1 Maturity Stage

A fruit's maturity stage is the only factor that will guarantee its eating quality for the consumer. Both premature and overripened fruit will affect the product's overall storage and quality (Kaur et al. 2018). Early harvesting before plums are fully ripened will extend their storage life but reduce their consumption by consumers; premature harvesting also results in a rapid loss of moisture, which will result in the physiological deterioration of the fruit. In contrast, if plums are harvested later in the maturity stage, they will be rich in taste and flavor but their storage life will decrease; therefore, harvesting of properly matured fruit is necessary (Majeed and Jawandha 2016). Maturity stages are measured based on the physiological and horticulture characteristics of plums. However, dependence of different harvest conditions on maturity stage can vary, e.g., transportation time and duration, storage duration and the market where it is being delivered. For stone fruits, it is best to let them ripen on the plant to attain maximum quality attributes (Prasad et al. 2018). To determine the correct maturity stage for harvesting, there are different perimeters; this calculation is known as the maturity index (Topp et al. 2012). First, there are two important factors, i.e., physiological maturity (when a fruit reaches a sufficient stage of development while attached to a plant) and horticulture maturity (when a fruit possesses the characteristics as desired by the consumer) (Sekse et al. 2013). Before moving to the methods used for measuring the stage of maturity, certain points are focused on:

1. Plums should be harvested during their peak growth conditions.
2. The harvesting of plum should coincide with acceptable flavor and appearance.
3. Market requirements should be kept in mind while choosing mature fruits for harvesting.

There are several different maturity index parameters, including plum color, size, firmness, soluble solids, titratable acidity and more. In stone fruits, skin color is considered one of the most critical factors for harvesting as well as for consumer acceptance (Miletic et al. 2012). But skin color alone cannot be used because plums can sometimes achieve their dark red or violet color before the fruit is fully ripened. So, along with skin color, fruit firmness is another measurement that can determine the market life of harvested plum (C. H. Crisosto and Kader 2000). The most relevant parameters of plum maturity are soluble solid contents and titratable acidity. Generally, plums with high soluble solid contents, i.e., more than 12%, have a high level of consumer acceptance (Nunes et al. 2009). Different stages of maturity in plum are shown in Figure 14.1.

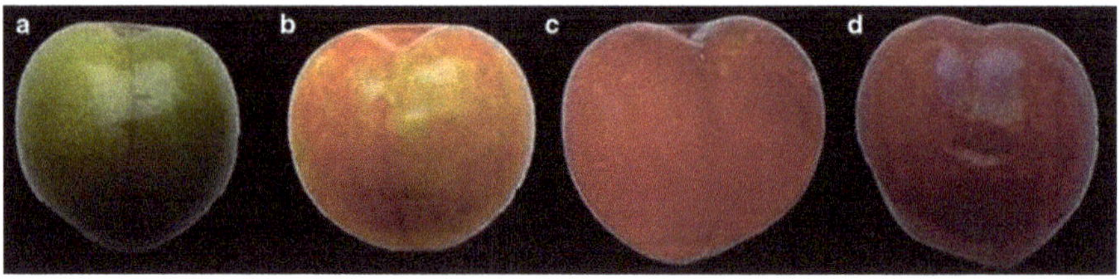

FIGURE 14.1 Maturity stages: a) green mature, b) color break, c) full color development, d) over-ripe.

FIGURE 14.2 Manual harvesting of plums.

14.2.2 Method of Harvesting

Harvesting method selection is an important factor that plays a role in determining the shelf life of the fruit as well as the physical appearance of that fruit, i.e., consumers' first preference (Manganaris et al. 2008). Harvesting via an improper method will result in damage, bruises or dark spots, which will make the fruit unattractive. There are different methods used for harvesting according to fruit type, such as hand harvesting and mechanical harvesting (Sharma 2010). For plums, hand harvesting is typically used. Fruits are picked by hand and put into bags and then dumped into baskets for further storage and transportation. The baskets in which harvested plums are placed should be covered with a soft lining to prevent any damage to the fruit (Zuzunaga et al. 2001). Harvesting by hand (Figure 14.2) is time consuming and demands more manpower, but it causes less damage to the fruit and results in good produce. Harvesting rates can be increased by increasing manpower, and though it takes longer, hand harvesting is less expensive than mechanical harvesting (Vangdal et al. 2008).

14.2.3 Time of Harvesting

Plum does not ripen uniformly at once, so harvesting should be done in two to four pickings. As plums are highly perishable fruits, increases in temperature or sunburn can cause significant quality defects due to a loss of moisture (Crisosto et al. 2004). The optimal time considered for picking or harvesting the plum is before 10:00 a.m., or before temperatures get too warm, and the fruits should be placed under some shade or precooled and immediately transferred to a packing house for further steps, such as sorting and grading. Precooling is essential to remove the field heat, and it should be done as soon as possible after harvesting (Pineiro and Diaz 2006). Precooling will slow down the deterioration reactions initiated by heat. Harvesting plums in the early morning and delivering them to the packing house before 10:00 a.m. increases the storage life of the fruit, as well as its quality, by preserving the fruit's native characteristics (Indiarto et al. 2020). However, it also depends on the targeted market; for example, fruits harvested for the fresh fruit market are harvested in the morning and handled with great care, but plums harvested for canning are typically handled with less care.

14.2.4 Sorting and Grading

Sorting and grading are also important steps after harvesting and are mainly performed in the packing house or at the precooling area. Sorting is done to separate out the plums that have any type of bruises or have been subjected to any diseases. Sorting is done manually by workers and is necessary because if any diseased fruit is packed with healthy produce it will result in the loss of the whole package (Käthner 2016).

Grading is done based on different physical parameters, like size, shape, diameter, color and weight. The extent of the defects to the fruit can also be graded. Different types of graders, from simple to complex, are used. Depending on the targeted market or consumer, plum grading is done either manually or by some mechanical means. Sometimes mechanical graders are fixed up for one character, and some graders simultaneously grade fruit on the bases of size and color (Hussein et al. 2020; Prasad et al. 2018).

One of the mechanical graders used for plum grading is UNITEC's Plum Vision 3, which is the exclusive technology for the selection of the external and internal quality of plums and totally preserves the integrity of the product. At the same time, these technologies offer considerable cost savings in the quality-selection phase (Sharif et al. 2020). Another one is the Fstsort roller grader that is designed for sorting and grading plums, prunes and apricots, which are similar in size and shape. It includes a roller inspection table for picking out bad fruits and separating them from the line; a water tank, brushes and sprays can also be added for cleaning if there is a need for a washing and sorting line. Plums are graded in different classes according to specific country standards on the bases of size, color, integrity and variety, i.e., extra class, class I and class II.

14.2.5 Edible Coatings

The coating of fruits or vegetables started in China in the 12th century when citrus fruits were coated with wax to prevent moisture loss and maintain overall quality. As time and technology advanced, different coatings were developed, one of the most used being edible coatings (Panahirad et al. 2019). In edible coatings, food-grade material is used, and a thin layer of coating is applied to the fruit layer, which provides a type of modified atmosphere by acting as a barrier to gases and water movements. Plums are highly perishable fruits, and their rate of quality loss increases as moisture migrates or by the transfusion of gases, which adds in postharvest losses. In Pakistan (Swat District), a study claimed a total of 21.5% postharvest losses at different levels from farm to fork (Olivas et al. 2008).

For plums, applying edible coatings at certain points is critical. Coating should not cause any physical damage to the fruit; it should be nontoxic, as it will be eaten along with the fruit, and it should provide a barrier to gaseous exchange but not completely stop the fruit respiration process, which would result in off flavor due to the production of undesirable volatile compounds (Riva et al. 2020). Plum coatings are generally composed of proteins, lipids and polysaccharides. Proteins include wheat protein, whey protein, gelatin and corn protein, but they are limited due to the potential for allergic reactions and because these types of coatings are prone to cracking. In lipids, different waxes and acyl glycerols are used. Lipid coatings provide the best moisture barrier to plum fruit but a poor gas barrier (Campos et al. 2011). Polysaccharide edible coatings are usually made of chitosan, alginate, pectin, cellulose and gum arabic due to their excellent gas barrier and mechanical properties (Shahzad et al. 2013; Mahajan et al. 2018).

A study reported that edible coatings made of carboxymethyl celluloses and pectin either alone or in combination exert the best effects, not only in maintaining the quality but also in improving the vitamin C, anthocyanin and phenolic compounds content by preserving the nutritional value of the fruit (Panahirad et al. 2020).

14.2.6 Packaging and Packaging Materials

Packaging and packaging materials play an important role in preserving fruit quality; they do not improve the quality, but maintain it by providing the required environment or conditions. Plums are soft, fragile and highly perishable fruits; any mishandling during packaging or in selecting packaging material will result in losses (Sharma 2010). Plums need soft packaging material to protect them from mechanical damage or bruising; packaging should also be ventilated, as plants' respiration process continues after fruits are picked. Mostly, plums are packed in CFB (corrugated fiber boxes) or boxes made from a combination of CFB and plywood; these are easy to handle, hygienic, low in weight and recyclable (Kader and Rolle 2004). Plums are packed in layers with a sheet of paper between each layer to avoid cross-contamination. The inner surface of the box is also covered with soft material to protect the fruit. Anew technology in this regard is a plastic film that acts as a barrier for moisture and preserves the nutritional quality of the fruit (Hofman et al. 2002). Table 14.1 shows the special features of some newly used packaging materials.

TABLE 14.1
Different Types of Packaging Materials Used for Plum Fruit during Postharvest Flow

Packaging material	Special features
Starch wrapping	Used for retail marketing of fresh produce in the form of cling plastic films from stretch wrapping.
Plastic woven sacks	These bags are made of high-density polyethylene or polypropylene, are light in weight and can be reused. They are used for packaging stone fruits to transport them over short distances.
Plastic trays/crates	These are hygienic, light in weight, sturdy and recyclable and are used in multistrip packaging.
Transformed plastic trays	These trays have cavities to hold individual fruits, which prevents the fruits from rubbing against each other, which often leads to bruising or surface cracks.
Corrugated polypropylene corrugated boxes	These are light in weight, hygienic, water resistant, sturdy and have a light bursting strength. They are useful in multistrip packaging.
Modified atmosphere packaging	The internal atmosphere can be modified with a combination of certain gases and the selection of suitable packaging material.

14.2.7 Storage

After harvesting, plum fruit is still alive and performing all its respiratory functions. The fruit's ethylene level, which ripens the fruit, will also trigger overripening. The main objectives of storage are to place the fruit in optimum conditions to attain the maximum storage life. If plum is picked at full maturity, not under mature or over-ripened, then storage conditions for plum are 1–0°C at 85–95% relative humidity (Manganaris et al. 2008). Plum stored in these conditions will have a shelf life of up to four weeks, but it also depends on the variety, species and maturity stage when the fruits were picked (Manganaris et al. 2008). Two types of storage are used for plums: low temperature storage and control and modified atmosphere storage.

Low temperature storage: Temperature is the most important and promising factor that limits the shelf life of fresh produce. According to several studies, storage at low temperature after harvesting is more effective than heat treatment. Plums are sensitive to chilling temperatures, so they cannot be stored at low temperatures for a long time, but during storage at 1–0°C, extensive care is taken to control the thermostatic conditions, along with relative humidity, to prevent freeze damage to the fruit (Wangchu et al. 2021). Storage life can be increased from four to eight weeks at low temperatures by dipping the plums in calcium chloride for two minutes.

Control and modified atmosphere storage: Controlled and modified atmosphere also used for plum storage by maintaining 2–3% O_2 and 2–8% CO_2 where fruits can be retained for two to three months (Díaz-Mula et al. 2011). Applying 1-MCP methylene inhibitor to plums before storage has been reported to be the most effective in enhancing the storage time. It has been tested in a range of concentrations (0.05–1 ml/l), resulting in marked delay of fruit softening, as well as a reduction in the activity of several cell wall–degrading enzymes, such as polygalacturonase, galactosidase and endo-glucanase/glucosidase (Díaz-Mula et al. 2011). However, the beneficial effects observed during storage get reduced when the fruit is treated at a more advanced ripening stage (Eksteen et al. 1986). The modified atmosphere packaging technique was useful to delay the ripening process of plum cultivars through a delay in the changes in color and the losses of firmness and acidity, and in turn an extension of shelf life could be achieved.

14.3 CONCLUSION

Good quality fruit plays a vital role in the nutritional security of the community and generates high income for farmers. Preharvest and postharvest factors have great effect on the final quality of plum fruits. These factors directly and indirectly influence the growth and nutrient composition of plum fruits. Lack of control of preharvest factors may lead to non-uniform maturation, firmness and level of different bioactive compounds. On the other hand, postharvest factors, like timely harvesting, appropriate harvesting method, sorting and grading, may help the plum grower gain maximum financial benefit and provide fine quality plum fruit to the market. New approaches, like high-tech horticultural practices and the management of natural resources, can be effective strategies to obtain good quality plum fruits.

14.4 REFERENCES

Alcaraz-Lopez, C., Botía, M., Alcaraz, C.F., and Riquelme, F. 2004. Effects of Calcium-Containing Foliar Sprays Combined with Titanium and Algae Extract on Plum Fruit Quality. *Journal of Plant Nutrition* 27 (4): 713–729.

Bentley, W. (2002). *UC IPM Pest Management Guidelines, Plum*, vol. 3462. University of California, ANR/Communications Services.

Campos, C.A., Gerschenson, L.N., and Flores, S.K. 2011. Development of Edible Films and Coatings with Antimicrobial Activity. *Food and Bioprocess Technology* 4 (6): 849–875.

Chandel, J. 2021. *Block-1 Temperate Fruit*. Indira Gandhi National Open University.

Chocano, C., García, C., González, D., de Aguilar, J.M., and Hernández, T. 2016. Organic plum Cultivation in the Mediterranean Region: The Medium-Term Effect of Five Different Organic Soil Management Practices on Crop Production and Microbiological Soil Quality. *Agriculture, Ecosystems & Environment* 221: 60–70.

Crisosto, C., and Costa, G. 2008. Preharvest Factors Affecting Peach Quality. In *The Peach: Botany, Production and Uses*, edited by D. Layne, D. Bassi. CAB eBooks 536–550.

Crisosto, C.H., Garner, D., Crisosto, G.M., and Bowerman, E. 2004. Increasing 'Blackamber' Plum (*Prunus salicina Lindell*) Consumer Acceptance. *Postharvest Biology and Technology* 34 (3): 237–244.

Crisosto, C.H., and Kader, A.A. 2000. Plum and Fresh Prune Postharvest Quality Maintenance Guidelines. *Department of Plant Sciences, University of California, Davis CA* 95616: 1–8.

Demirtas, M.N., Bolat, I., Ercisli, S., Ikinci, A., Olmez, H.A., Sahin, M., Celik, B. et al. 2010. The Effects of Different Pruning Treatments on the Growth, Fruit Quality and Yield of 'Hacihaliloglu' Apricot. *Acta Sci. Pol., Hortorum Cultus* 9 (4): 183–192.

Dennis Jr, F. 2003. Flowering, Pollination and Fruit Set and Development. *Apples: Botany, Production and Uses* 153–166.

Díaz-Mula, H., Martínez-Romero, D., Castillo, S., Serrano, M., and Valero, D. 2011. Modified Atmosphere Packaging of Yellow and Purple Plum Cultivars. 1. Effect on Organoleptic Quality. *Postharvest Biology and Technology* 61 (2–3): 103–109.

Eksteen, G., Visagie, T., and Laszlo, J. 1986. Controlled Atmosphere Storage of South African Grown Nectarines and Plums. *Deciduous Fruit Grower* 36 (4): 128–132.

Farag, K. 2015. Performance of Adopted Fruit Species and Cultivars to Egyptian-Desert Agriculture and Their Major Production Problems. *Advances in Plants & Agriculture Research* 2 (2): 00041.

Georgiev, D., Mihova, T., and Georgieva, M. 2019. Effect of Fertilization on Biochemical Composition of Fruits of Black Currant and Red Currant. *Journal of Mountain Agriculture on the Balkans* 22 (1): 228–237.

Hayat, S., Mir, M., Singh, D., Hayat, N., and Khan, F. 2013. Effect of Post-Harvest Treatment and Storage Conditions on Quality of Plum cv. *Santa Rosa. Progressive Horticulture* 45 (1): 89–94.

Hewett, E.W. 2006. An Overview of Preharvest Factors Influencing Postharvest Quality of Horticultural Products. *International Journal of Postharvest Technology and Innovation* 1 (1): 4–15.

Hofman, P., Fuchs, Y., and Milne, D. 2002. Harvesting, Packing, Postharvest Technology, Transport and Processing. *The Avocado: Botany, Production and Uses* 363–401.

Hrotkó, K., Magyar, L., Klenyán, T., and Simon, G. 2001. *Effect of Rootstocks on Growth and Yield Efficiency of Plum Cultivars.* Paper presented at the VII International Symposium on Plum and Prune Genetics, Breeding and Pomology 577.

Hussein, Z., Fawole, O.A., and Opara, U.L. 2020. Harvest and Postharvest Factors Affecting Bruise Damage of Fresh Fruits. *Horticultural Plant Journal* 6 (1): 1–13.

Indiarto, R., Izzati, A.N., and Djali, M. 2020. Post-Harvest Handling Technologies of Tropical Fruits: A Review. *International Journal* 8 (7).

Intrigliolo, D.S., Ballester, C., and Castel, J.R. 2013. Carry-Over Effects of Deficit Irrigation Applied Over Seven Seasons in a Developing Japanese Plum Orchard. *Agricultural Water Management* 128: 13–18.

Kader, A.A. 2002. *Pre-and Postharvest Factors Affecting Fresh Produce Quality, Nutritional Value, and Implications for Human Health.* Paper presented at the Proceedings of the International Congress Food Production and the Quality of Life.

Kader, A.A., and Rolle, R.S. 2004. *The Role of Post-Harvest Management in Assuring the Quality and Safety of Horticultural Produce*, vol. 152. Food & Agriculture Org.

Käthner, J. 2016. *Interaction of Spatial Variability Characterized by Soil Electrical Conductivity and Plant Water Status Related to Generative Growth of Fruit Trees.* Universität Potsdam.

Kaur, H., Sawhney, B., and Jawandha, S. 2018. Evaluation of Plum Fruit Maturity by Image Processing Techniques. *Journal of Food Science and Technology* 55 (8): 3008–3015.

Kiang, N.Y., Siefert, J., and Blankenship, R.E. 2007. Spectral Signatures of Photosynthesis. I. Review of Earth Organisms. *Astrobiology* 7 (1): 222–251.

Kuehne, S., Roßberg, D., Röhrig, P., von Mehring, F., Weihrauch, F., Kanthak, S., Gitzel, J. et al. 2017. The Use of Copper Pesticides in Germany and the Search for Minimization and Replacement Strategies. *Organic Farming* 3 (1): 66–75.

Lawrence, B.T., and Melgar, J.C. 2020. Variable Fall Climate Conditions on Carbon Assimilation and Spring Phenology of Young Peach Trees. *Plants* 9 (10): 1353.

Lombardi-Boccia, G., Lucarini, M., Lanzi, S., Aguzzi, A., and Cappelloni, M. 2004. Nutrients and Antioxidant Molecules in Yellow Plums (*Prunus domestica L.*) from Conventional and Organic Productions: A Comparative Study. *Journal of Agricultural and Food Chemistry* 52 (1): 90–94.

Maatallah, S., Guizani, M., Hjlaoui, H., Boughattas, N.E.H., Lopez-Lauri, F., and Ennajeh, M. 2015. Improvement of Fruit Quality by Moderate Water Deficit in Three Plum Cultivars (*Prunus salicina L.*) Cultivated in a Semi-Arid Region. *Fruits* 70 (6): 325–332.

Mahajan, B.C., Tandon, R., Kapoor, S., and Sidhu, M.K. 2018. Natural Coatings for Shelf-Life Enhancement and Quality Maintenance of Fresh Fruits and Vegetables—A Review. *Journal Postharvest Technology* 6 (1): 12–26.

Majeed, R., and Jawandha, S. 2016. Enzymatic Changes in Plum (Prunus salicina Lindl.) Subjected to Some Chemical Treatments and Cold Storage. *Journal of Food Science and Technology* 53 (5): 2372–2379.

Manganaris, G., Vicente, A., and Crisosto, C. 2008. Effect of Pre-Harvest and Post-Harvest Conditions and Treatments on Plum Fruit Quality. *CAB Reviews: Perspectives in Agriculture, Veterinary Science, Nutrition and Natural Resources* 3 (9): 1–9.

Mansour, A., Ahmed, F., Abdelaal, A., Eissa, R., and Amira, A. 2011. Selecting the Best Method and Dose of Yeast for Kelsey Plum Trees. *Journal of Applied* 7 (7): 1218–1221.

Miletic, N., Popovic, B., Mitrovic, O., and Kandic, M. 2012. Phenolic Content and Antioxidant Capacity of Fruits of Plum cv. 'Stanley'(*'Prunus domestica'L.*) as Influenced by Maturity Stage and On-Tree Ripening. *Australian Journal of Crop Science* 6 (4): 681–687.

Milosevic, T., Glisic, I., and Milosevic, N. 2007. *Dense Planting Effect on the Productive Capacity of Some Plum Cultivars*. Paper presented at the I Balkan Symposium on Fruit Growing 825.

Mosa, W.F.A.E.-G., Paszt, L.S., and Abd EL-Megeed, N.A. 2014. The Role of Bio-Fertilization in Improving Fruits Productivity—A Review. *Advances in Microbiology* 4 (15): 1057.

Naira, A., and Moieza, A. 2014. Summer Pruning in Fruit Trees. *African Journal of Agricultural Research* 9 (2): 206–210.

Nassar, H.F., Stearns, D.T., and Beattie, D.J. 2001. Monday 23 July Evening Tuesday 24 July·Morning. *HortScience* 36 (3): 514.

Nunes, C., Rato, A.E., Barros, A.S., Saraiva, J.A., and Coimbra, M.A. 2009. Search for Suitable Maturation Parameters to Define the Harvest Maturity of Plums (Prunus domestica L.): A Case Study of Candied Plums. *Food Chemistry* 112 (3): 570–574.

Olivas, G., Dávila-Aviña, J., Salas-Salazar, N., and Molina, F. 2008. Use of Edible Coatings to Preserve the Quality of Fruits and Vegetables During Storage. *Stewart Postharvest Review* 3 (6): 1–10.

Panahirad, S., Naghshiband-Hassani, R., Bergin, S., Katam, R., and Mahna, N. 2020. Improvement of Postharvest Quality of Plum (Prunus domestica L.) Using Polysaccharide-Based Edible Coatings. *Plants* 9 (9): 1148.

Panahirad, S., Naghshiband-Hassani, R., Ghanbarzadeh, B., Zaare-Nahandi, F., and Mahna, N. 2019. Shelf Life Quality of Plum Fruits (*Prunus domestica L.*) Improves with Carbox ymethylcellulose-Based Edible Coating. *HortScience* 54 (3): 505–510.

Panayotova, G., Kostadinova, S., Aleksieva, S., Slavova, N., and Aladzhova, C. 2018. Nitrogen and Phosphorus Balances as Dependent on Durum Wheat Fertilization. *Bulgarian Journal of Agricultural Science* 24 (1): 9–17.

Pesakovic, M., Djukic, D., Mandic, L., and Miletic, R. 2012. Microbiological Activity and Productivity of Soil in Plum Orchard. *Journal of Environmental Protection and Ecology* 13: 951.

Pineiro, M., and Diaz, L.B. (2006). *Improving the Safety and Quality of Fresh Fruit and Vegetables (FFV): A Practical Approach.* Paper presented at the I International Symposium on Fresh Food Quality Standards: Better Food by Quality and Assurance 741.

Prasad, K., Jacob, S., and Siddiqui, M.W. 2018. Fruit Maturity, Harvesting, and Quality Standards. In *Preharvest Modulation of Postharvest Fruit and Vegetable Quality*, 41–69. Elsevier.

Riva, S.C., Opara, U.O., and Fawole, O.A. 2020. Recent Developments on Postharvest Application of Edible Coatings on Stone Fruit: A Review. *Scientia Horticulturae* 262: 109074.

Sekse, L., Simčič, M., and Vidrih, R. 2013. Fruit Surface Colour as Related to Quality Attributes in Two Plum (*Prunus Domestica L.*) Cultivars at Different Maturity Stages. *Europ Journal Horti Science* 78 (1): 13–21.

Servili, A., Feliziani, E., and Romanazzi, G. 2017. Exposure to Volatiles of Essential Oils Alone or Under Hypobaric Treatment to Control Postharvest Gray Mold of Table Grapes. *Postharvest Biology and Technology* 133: 36–40.

Shahzad, M., Akhter, A., Qureshi, A.H., Jehan, N., Ullah, I., and Khan, M. 2013. Assessment of Post-Harvest Losses of Plum in Swat, Pakistan. *Pakistan Journal of Agricultural Research* 26 (3).

Shamrao, B.S. 2020. *Production Technology of Peach, Plum and Apricot in India Prunus*. IntechOpen.

Sharif, N., Sajid, B., Munir, N., and Naz, S. 2020. Sensors for Sorting and Grading of Fruits and Vegetables. In *Sensor-Based Quality Assessment Systems for Fruits and Vegetables*, 57–77. Apple Academic Press.

Sharma, S.K. 2010. *Postharvest Management and Processing of Fruits and Vegetables: Instant Notes*. New India Pub. Agency.

Simon, S., Sauphanor, B., and Lauri, P.-E. 2007. Control of Fruit Tree Pests Through Manipulation of Tree Architecture. *Pest Technology* 1 (1): 33–37.

Singh, S.K., Nidhika, T., and Yamini, S. 2012. Effective Nutrient Management in Fruit Crops. *Asian Journal of Horticulture* 7 (2): 606–609.

Smilanick, J. 2012. Recent Advances on the Use of Natural and Safe Alternatives to Conventional Methods to Control Postharvest Gray. *Postharvest Biology and Technology* 63: 141–147.

Sommano, S., Caffin, N., and Kerven, G. 2013. Screening for Antioxidant Activity, Phenolic Content, and Flavonoids from Australian Native Food Plants. *International Journal of Food Properties* 16 (6): 1394–1406.

Steduto, P., Hsiao, T.C., Fereres, E., and Raes, D. 2012. *Crop Yield Response to Water*, vol. 1028. Food and Agriculture Organization of the United Nations.

Stephan, J., Sinoquet, H., Donès, N., Haddad, N., Talhouk, S., and Lauri, P.-E. 2008. Light Interception and Partitioning Between Shoots in Apple Cultivars Influenced by Training. *Tree Physiology* 28 (3): 331–342.

Stern, R.A., Flaishman, M., and Ben-Arie, R. 2007. Effect of Synthetic Auxins on Fruit Size of Five Cultivars of Japanese Plum (*Prunus salicina Lindl.*). *Scientia Horticulturae* 112 (3): 304–309.

Topp, B.L., Russell, D.M., Neumüller, M., Dalbó, M.A., and Liu, W. 2012. Plum. In *Fruit Breeding*, 571–621. Springer.

Tränkner, M., Tavakol, E., and Jákli, B. 2018. Functioning of Potassium and Magnesium in Photosynthesis, Photosynthate Translocation and Photoprotection. *Physiologia Plantarum* 163 (3): 414–431.

Vangdal, E., Vanoli, M., Eccher Zerbini, P., Jacob, S., Torricelli, A., and Spinelli, L. (2008). *TRS-Measurements as a Nondestructive Method Assessing Stage of Maturity and Ripening in Plum (Prunus domestica L.)*. Paper presented at the III International Conference Postharvest Unlimited 2008 858.

Von-Bennewitz, E., Hlušek, J., and Lošák, T. 2008. Nutritional Status, Vegetative and Generative Behaviour of Apple Trees After the Application of Two Biopreparations. Acta Universitatis *Agriculturae et Silviculturae* Mendelianae *Brunensis* 56: 13–18.

Wang, G.Y., Zhang, X.Z., Yi, W., Xu, X.F., and Han, Z.H. 2015. Key Minerals Influencing Apple Quality in Chinese Orchard Identified by Nutritional Diagnosis of Leaf and Soil Analysis. *Journal of Integrative Agriculture* 14 (5): 864–874.

Wangchu, L., Angami, T., and Mandal, D. (2021). Plum. In *Temperate Fruits*, 297–331. Apple Academic Press.

Yakuba, G., Podgornaya, M., Mishchenko, I., Didenko, N., and Chernov, V. (2021). *Evaluation of the Application of Fungicides of Inorganic Copper Compounds in Apple and Plum Agroecosystems of the Krasnodar Territory*. Paper presented at the E3S Web of Conferences.

Zuzunaga, M., Serrano, M., Martinez-Romero, D., Valero, D., and Riquelme, F. 2001. Comparative Study of Two Plum (Prunus salicina Lindl.) Cultivars During Growth and Ripening. *Food Science and Technology International* 7 (2): 123–130.

15 Novel Extraction Methods of Anthocyanins from Plums, Plum Products, and By-Products

Giorgiana M. Cătunescu[1], Ioana M. Bodea[2], Ruth Hornedo-Ortega[3], M. Carmen Garcia-Parrilla[3], Ana M. Troncoso[3], and Ana B. Cerezo[3]

[1] Department of Technical and Soil Sciences, Faculty of Agriculture, University of Agricultural Sciences and Veterinary Medicine Cluj-Napoca, Calea Manaştur 3–5, 400372 Cluj-Napoca, Romania

[2] Department of Paraclinical and Clinical Sciences, Faculty of Veterinary Medicine, University of Agricultural Sciences and Veterinary Medicine Cluj-Napoca, Calea Mănăştur 3–5, 400372 Cluj-Napoca, Romania

[3] Departamento de Nutrición y Bromatología, Toxicología y Medicina Legal, Facultad de Farmacia, Universidad de Sevilla, C/Profesor García González 2, 41012 Sevilla, Spain

CONTENTS

15.1 INTRODUCTION

Anthocyanins are a subclass of flavonoids found in dark-colored fruits (red, purple, or black) and vegetables, thus in plums as well (Popescu et al. 2018; Gościnna et al.

DOI: 10.1201/9781003205449-15

2021; Sahamishirazi et al. 2017; Navarro et al. 2018; Cabrera-Bañegil et al. 2020). The main anthocyanins reported in plums are cyanidin 3-glucoside, cyanidin 3-rutino side, peonidin-3-O-glucoside, and cyanidin 3-xylosylglucosides (Popescu et al. 2018; Panahirad et al. 2020; Tomic et al. 2019; Garcia-Parra et al. 2018; Sahamishirazi et al. 2017; Hernandez-Herrero and Frutos 2014; Cendres et al. 2014, 2012; Chun et al. 2003).

Anthocyanins are important compounds because they were shown to contribute to the in vitro and in vivo bioactivity of plums, plum products, and by-products (Constantin et al. 2018; Netzel et al. 2006; Popescu et al. 2018). An array of health benefits are attributed to the consumption of plums and plum products, mainly related to their antioxidant and anti-inflammatory activities (Netzel et al. 2006; Slimestad et al. 2009; Hernandez-Herrero and Frutos 2014; Cuevas et al. 2015; Constantin et al. 2018; Tomic et al. 2019; Panahirad et al. 2020; Walkowiak-Tomczak et al. 2008) and their adjuvant effect in the prevention of metabolic disorders (Gościnna et al. 2021; Basanta et al. 2016). It was shown that plum intake was linked to a lowering of cholesterol levels, including a cardioprotective activity and a positive effect on preventing obesity and neurodegenerative diseases (Kuo et al. 2015; Noratto et al. 2015; Wu et al. 2019). Some authors have also reported that plum consumption may have an inhibitory effect on the proliferation of some cancer cells (Liang and Liu 2019; Alsolmei et al. 2019). Thus, consumers are very interested in adding antho cyanin-rich foods to their diet, including plums and prunes (Prasain et al. 2018). Additionally, anthocyanins are currently being utilized as natural food-coloring agents by the food industry (Yao et al. 2016; Aliano-Gonzalez et al. 2020; Zhu et al. 2019; Casedas et al. 2018; Pereira et al. 2015; Barnes et al. 2009; Cătunescu et al. 2021) because they are increasingly preferred by the consumers (Yuan et al. 2020; Tennant and Klingenberg 2016).

In this context, the anthocyanin content is considered an important biomarker for plums, plum products, and by-products (González-Cebrino et al. 2013; Fanning et al. 2014; Johnson, Collins, Walsh et al. 2020). However, they are not only indicators of the nutritional value of plums but also of interest for horticulturists that are starting to breed cultivars based on these quality parameters. The plum variety Queen Garnet is a good example in this sense, as it was obtained and is marketed for its very high anthocyanin content (Johnson, Collins, Walsh et al. 2020; Fanning et al. 2014).

Anthocyanins are mostly found in plum peels, concretely in the vacuoles in their epidermis (Hernandez-Herrero and Frutos 2014; Vlaic et al. 2018; Medina-Meza and Barbosa-Cánovas 2015; Navarro et al. 2018; Cendres et al. 2014; Olawuyi et al. 2020; Chun et al. 2003; Cosmulescu et al. 2015; Basanta et al. 2016), while procyanidins are distributed in both the epicarp and mesocarp, but present in slightly higher concentration in the peels (Cendres et al. 2012; Cosmulescu et al. 2015). Thus, the pomace resulted from juicing still contains a significant amount of anthocyanins (Olawuyi et al. 2020; Milala et al. 2013).

The content and profile of anthocyanins naturally differs with many factors dependent on their biological and agronomic profile, such as variety and cultivar, pedo-climatic conditions, and growing stage and fruit maturation at harvest (Netzel et al. 2006; Slimestad et al. 2009; Cuevas et al. 2015; Sahamishirazi et al. 2017; Tomic et al. 2019; Walkowiak-Tomczak et al. 2008; Michalska et al. 2017). The treatments plums are subjected to prior to anthocyanin extraction have an important effect as well. Thus, when plums are exposed to light, heat treatment, or crushing and juicing, anthocyanins degrade substantially because they are oxidized by the native polyphenol oxidases (PPO) (Versari et al. 1997; Khandare et al. 2011; Olawuyi et al. 2020). Additionally, PPO is able to oxidize the simple phenols generating o-diphenols together with the very reactive o-quinones, which in turn oxidize anthocyanins (Buckow et al. 2010; Kader et al. 1997). This chain reaction leads to the formation of brown pigments. Meanwhile, the sugar moiety of the anthocyanins can be hydrolyzed by β-glucosidase (Skrede et al. 2000). The obtained aglycones anthocyanidins can be easier oxidized by PPO, speeding up the oxidation reactions. These chain reactions can be observed in fruit juices when the anthocyanin-rich peels are crushed during pressing (Buckow et al. 2010).

In addition to these factors, the extraction procedure employed to obtain and purify the anthocyanins from plums has arguably one of the most important impacts. It is a necessary step before the characterization of phytochemical compounds present in foods, including plums and plums by-products. The characterization of plum anthocyanins is required, among others, to discriminate among different varieties (Johnson, Collins, Mani et al. 2020; Slimestad et al. 2009; Lozano et al. 2008), to identify cultivars with high anthocyanin content suitable to obtain genetically improved varieties (Sahamishirazi et al. 2017), to detect the natural source of bioactive compounds (Netzel et al. 2006; Constantin et al. 2018; Chaiyasut et al. 2016), to discover potential sources of natural food coloring (Hernandez-Herrero and Frutos 2014), or even to select the most convenient technology (drying, high-pressure processing) for the development of plum by-products, such as dry plums (prunes) and purees (Vangdal et al. 2017; González-Cebrino et al. 2013). Therefore, the method and the conditions elected are crucial to obtain a high-anthocyanin extraction yield.

Two main types of extraction procedures are usually employed nowadays for anthocyanins: conventional solvent extraction and new non-conventional methods, which can sometimes be paired with the conventional techniques (Hidalgo and Almajano 2017). Solvent extraction is the typical procedure employed to extract anthocyanins from plums and their products (Slimestad et al. 2009; Lozano et al. 2008; Hernandez-Herrero and Frutos 2014; Joshi and Devi 2015; Chaiyasut et al. 2016; Sahamishirazi et al. 2017; Johnson, Collins, Walsh et al. 2020); it generally uses acidified organic solvents or mixtures of organic solvents (e.g., ethanol, methanol) in various concentrations and proportions to the sample weight. The solvent is intended to disrupt the cell walls in the epicarp and extract the anthocyanins in the acidified environment (Rodriguez-Saona and Wrolstad 2001). However, many conventionally used solvents are not generally recognized as safe (GRAS) (i.e., methanol), and even for the innocuous solvents (i.e., ethanol) high concentration and large volumes are needed to obtain an industry-relevant yield (Ferreira et al. 2020; Wang et al. 2016). This shows the importance of developing better extraction techniques that have minor effects upon the anthocyanin profile and are less time consuming but also environmentally friendly. Additionally, many pre-treatments were proposed prior to the extraction of anthocyanins from plums, plum products, and by-products intended to affect the peel tissue and increase extraction rates and yields (Cendres et al. 2014). In this context, non-conventional methods were developed and gained interest both for laboratory and industrial use. Among these methods, the most promising are ultrasound-assisted extraction (Usenik et al. 2009; Cuevas et al. 2015; Medina-Meza and Barbosa-Cánovas 2015; Popescu et al. 2018; Manzano Durán et al. 2019; Gościnna et al. 2021) and microwave-assisted extraction (Cendres et al. 2014; Gościnna et al. 2021; Haddadi-Guemghar et al. 2014; Cendres et al. 2012; Wang et al. 2014; Michalska et al. 2019; Popescu et al. 2018).

In general, there appears to be a lack of consensus regarding the optimum pre-extraction processing and extraction methods for anthocyanins. Additionally, the current extraction protocols are not yet standardized, thus the currently available data on the quantity and profile of anthocyanins in plums is neither accurate nor precise (Gościnna et al. 2021). Thus, the results reported in the current literature are not directly comparable (Johnson, Collins, Walsh et al. 2020). Moreover, the reports seem to be scarce for plums and their products. Thus, the aim of this chapter is to review some of the newer extraction methods employed for plums, plum products, and by-products to obtain anthocyanins both in a laboratory and industrial settings.

15.2 CONVENTIONAL EXTRACTION METHODS OF PLUM ANTHOCYANINS

The anthocyanins are polar compounds, with a positive charge conferred by the oxygen on the pyran ring (Barnes et al. 2009). Thus, they need a slightly acidified aqueous solution of organic solvents to ensure maximum extraction efficiency (Nicoue et al. 2007; Silva et al. 2017; Barnes et al. 2009). Conventionally, the anthocyanins from plums are extracted by using solvents such as acidified methanol or ethanol solution, although acetone and acetonitrile have also been used (Table 15.1). However, besides the large array of solvents that can be employed, the conventional extraction process is characterized by a range of parameters that can vary widely, such as the solvent-to-sample ratio, the acid used for acidification, temperature, and solvent contact period with the sample (Gościnna et al. 2021; Medina-Meza and Barbosa-Cánovas 2015; Johnson, Collins, Mani et al. 2020; Constantin et al. 2018; Chaiyasut et al. 2016).

Few studies have evaluated the effect of solvent extraction conditions on plum anthocyanins (Joshi and Devi 2015; Johnson, Collins, Mani et al. 2020). Most of the literature available focuses on the characterization of total and individual anthocyanins present in plum fruit, puree, and peel, some of them as colorants (Table 15.1). However, there is not a consensus regarding the best extraction conditions for plum anthocyanins. As can be seen in Table 15.1, methanol is the most used solvent (6 out of the 10 studies) with concentrations varying between 90% and 100%, followed by ethanol. Acetone was used only to extract anthocyanins from plum peels as natural food colorants (Hernandez-Herrero and Frutos 2014).

Additionally, the extraction solvent is usually acidified to stabilize the flavylium ion (Cătunescu et al. 2021; Silva et al. 2017; Barnes et al. 2009; Nicoue et al. 2007). Solvent acidification was shown to play a crucial role (Truong et al. 2012) because the denaturation of the cell membranes led to increased anthocyanin extraction efficiency in other foods. The use of hydrochloric acid is the most common acidifier used to extract anthocyanins from plums (Lozano et al. 2008; González-Cebrino et al. 2013;

TABLE 15.1
Summary of Optimum Extraction Conditions in Different Studies

Extraction method	Samples	Species, variety	Origin	Conditions	Anthocyanin concentration	References
liquid-solid extraction	plums	Illawarra plum (*Podocarpus elatus* Endl), Burdekin plum (*Pleiogynium timorense* (DC.) Leenh)	Australia	deionized water (5% formic acid); mixed for 2 minutes; centrifuged at 4000 rpm for 10 minutes	cyanidin 3-glucoside 6.072–19.296 μmol of CE/g of FW; pelargonidin 3-glucoside, 0.094 ± 0.004 μmol of CE/g of FW	Netzel et al. (2006)
	plums	*P. domestica* L. varieties: Avalon, Jubileum, Valor	Norway	methanol (0.5% trifluoroacetic acid) for 24 h	cyanidin 3-glucoside: 0.50–4.14 mg 100/g FW; cyanidin 3-rutinoside: 1.15–6.38 mg 100/g FW; peonidin 3-glucoside: 0.01–0.39 mg 100/g FW; peonidin 3-rutinoside: 0.09–0.16 mg 100/g FW	Slimestad et al. (2009)
	plums	*P. salicina* Lindl., six varieties (Black Amber, Suplumeleven, Fortune, Larry Ann, Suplumsix, Songold)	Spain	methanol (+1% HCl) for 24 hours under refrigeration	10.93–28.93 mg/100g FW	Lozano et al. (2008)
	plum puree	*P. salicina* Lindl.	Spain	methanol solution (1% HCl)	3.2–4.1 mg/100g FW	González-Cebrino et al. (2013)
	plum peels	–	Spain	two-phase extraction in the dark: an extraction with acetone acidified (15% HCl) using a 4:1 solvent:sample ratio for four hours; a repeated extraction using acidified (0.15% HCl) 70% aqueous acetone	–	Hernandez-Herrero and Frutos (2014)

(Continued)

TABLE 15.1 *Continued*

Extraction method	Samples	Species, variety	Origin	Conditions	Anthocyanin concentration	References
	plum pomace (skin and stone)	*Prunus salicinia* L. var. Santa Rosa	India	ethanol proportion (50% and 100%); citric acid (0.1% and 0.2%); solid-solvent ratio (8:1 and 10:1)	highest anthocyanin concentration (325 mg/100 mL): 50% ethanol + 0.2% citric acid, 10:1 solid-solvent ratio;minimum concentration (169 mg/100 mL): 50% ethanol + 0.1% citric acid, 8:1 solid-solvent ratio	Joshi and Devi (2015)
	plums	*P. domestica*	Thailand	methanol: 0.1 N HCl; solid-solvent ratio of 85:15		Chaiyasut et al. (2016)
	plums	*Prunus domesticus* L (178 different cultivars)	Germany	acidified methanol (0.1% HCl), mixed and centrifuged (4000 rpm, 20°C)	cyanidin-3-rutinoside: 1.10–216.83 mg CGE 100/g FW; cyanidin-3-glucoside: 6.38–73.26 mg CGE 100/g FW	Sahamishirazi et al. (2017)
	dried plums	*Prunus domestica* L. varieties: Jubileum, Victoria, Reeve	Western Norway	1:1 of 13.7 mol/L ethanol and 0.06 mol/L HCl; vortexed for 30 seconds and shaken for 60 minutes at room temperature; centrifuged at 25,000 g at 4°C	0.9–131.6 mg CG/kg DW	Vangdal et al. (2017)
	plum jams and marmalades	–	Romania	aqueous ethanol (70%, v/v), two hours shaking	14.53–0.40 mg/kg DM	Constantin et al. (2018)
	plums	*Prunus domestica*	Australia	protocol 1: 90% aqueous methanol; mixing (50 rpm for 60 minutes); centrifugation (1000 g for 10 minutes) repeated twice protocol 2: 50% aqueous ethanol; mixing (150 rpm for 15 min); centrifugation (1800 g for 10 minutes), repeated twice	total anthocyanin content for methanolic samples: 1.91–11.89 mg malv-3-glu/100g FW; for ethanolic samples: 0.67–4.41 mg malv-3-glu/100g FW	Johnson, Collins, Walsh et al. (2020)

	plums	*Prunus domestica* and *P. salicina*, five varieties (black plums, Croc Eggs plums, Dapple Dandy, red plums, sugar plums)	Australia	90% methanol, mixed (50 rpm for 60 minutes), centrifuged (1000 g for 10 minutes), repeated twice	total anthocyanin content: 0.3–21.5 mg/100g FW cyanidin-3-galactoside: 0.49 mg/100g FW; cyanidin-3-glucoside: 0.03–8.23 mg/100g FW; cyanidin-3-rutinoside: 0.06–3.66 mg/100g FW; peonidin-3-glucoside: 0.03–0.1 mg/100g FW	Johnson, Collins, Mani et al. (2020)
ultrasound- assisted extraction	plums	*Prunus domestica*, four cultivars: Jojo, Valor, Cacanska rodna, Cacanska najbolja	Slovenia	methanol (1% HCl)	cyanidin-3-rutinoside (4.1 to 23.4 mg/100g FW), peonidin 3-rutinoside (2.2–6.1 mg cyanidin 3-glucoside equivalents/100g FW), cyanidin 3-glucoside (0.2–5.6 mg/100g FW), cyanidin 3-xyloside (0.6–1.5 mg cyanidin 3-glucoside equivalents/100g FW) and peonidin 3-glucoside (traces-0.19 mg/100g FW)	Usenik et al. (2009)
	plums	*Prunus salicinia* Lindl.	Spain	50% aqueous acetone; mixed for 1 minute; sonicated for 15 minutes; centrifugation at 5000 rpm for 15 minutes at 4°C	total anthocyanin content: 0.55–29.82 mg GAE/100g FW	Cuevas et al. (2015)
	plum peels	*Prunus domestica* var. Casselman	USA	water as solvent; ultrasound conditions: power of 400 W, frequency of 24 kHz, maximum amplitude of 120 μm; temperatures: 25°C and 50°C	–	Medina-Meza and Barbosa-Cánovas (2015)
	dried plums	*Prunus domestica* L.	Romania	solvent: distilled water, extraction at 40 kHz frequency and 200 W power for 30 minutes at room temperature (20°C)	40–90 mg/100 mg FW	Popescu et al. (2018)

(Continued)

TABLE 15.1 *Continued*

Extraction method	Samples	Species, variety	Origin	Conditions	Anthocyanin concentration	References
	plums	*Prunus salicina* Lindl.	Spain	84% aqueous methanol; extraction at 40 MHz frequency for 30 minutes	–	Manzano Durán et al. (2019)
	plums	*Prunus domestica* L. (18 plum cultivars)	Serbia	methanol, 1:5 (w/v), solid:solvent ratio, extraction for 1 hour using a cooled ultrasonic bath; centrifugation of supernatants (10,000 rpm for 10 minutes)	cyanidin-3-rutinoside: nd—6.29 ± 1.29 mg/100g FW; cyanidin-3-glucoside: nd—8.54 ± 1.34 mg/100g FW; peonidin-3-glucoside: nd—4.46 ± 0.80 mg/100g FW; peonidin-3- rutinoside: nd—0.77 ± 0.16 mg/100g FW, with Nada variety richest in anthocyanins and Zlatka the poorest	Tomic et al. (2019)
	plum mash	*Prunus salicina* L.	South Korea	15, 30, 45, 60, and 75 minutes, extraction at a frequency of 38 kHz, a power of 600 W, a temperature of 45°C	total anthocyanins 0.13 ± 0.01 mg/100 mL	Olawuyi et al. (2020)
	plums	*Prunus sp*. L. (Bluefree, Stanley, and Sweet Common Prune (two varieties: Węgierka and Zwykła)	Poland	aqueous methanol solution (80%, v/v) in an ultrasound bath at 80 W and 12 kHz at 22°C for 20 minutes	total anthocyanin content (1419.7 mg/kg DM); cyanidin-3-glucoside (501.7 mg/ kg DM); cyanidyn-3-rutinoside (822.6 mg/kg DM); peonidin-3-glucoside (83.4 mg/kg DM); peonidin-3-rutinoside (12 mg/ kg DM)	Gościnna et al. (2021)
microwave-assisted extraction	fresh and powdered plums	*Prunus sp*. L. (Bluefree, Stanley, and Sweet Common Prune (two varieties: Węgierka and Zwykła)	Poland	80% methanol aqueous solution, 10:1 solvent:sample ratio, at 600 W and 60°C for 5 minutes, centrifugation for 15 minutes at 4°C at 4500 rpm	total anthocyanin content (1475.4 mg/kg DM); cyanidin-3-glucoside (516.7 mg/kg DM); cyanidyn-3-rutinoside (859.3 mg/kg DM); peonidin-3-glucoside (85.5 mg/kg DM); peonidin-3-rutinoside (14.0 mg/kg DM)	Gościnna et al. (2021)

dried plums	*Prunus domestica* L. (Grase Româneşti variety)	Romania	400 W for 80 seconds, then extracted with ethanol solution (85:15%, v:v) acidified with 0.1M HCl,	–	Popescu et al. (2018)
fresh and frozen plums	*Prunus domestica* L. (President variety)	France	no solvent, microwave oven, a frequency of 450 MHz, a maximum power of 1000 W (with adjustments of 10 W increments)	–	Cendres et al. (2011)
frozen plums	*Prunus domestica* L. (Najbolia variety)	France	waters, 0.5, 1, and 1.5 W per gram of frozen plum (corresponding to 250, 500, and 750 W).	total anthocyanin content: 90 mg/L for 0.5 W/g; 98 mg/L for 1 W/g; 71 for 1.5 W/g; 6 mg/kg cyanidin-3-O-glucoside; 36.2 mg/kg cyanidin-3-O-rutinoside; 6.9 mg/kg peonidin-3-Orutinoside	Cendres et al. (2012)
plums	"Sanhua" plum *Prunus salicina* Lindl. (Dami, Xiaomi, Jima, and Zaoshi varieties)	China	microwaved for 45, 90, and 150 seconds at 700 W, cooled to room temperature; extracted three times with 90% methanol acidified with 0.5% formic acid; 1:2 (w/v) samples:solvent ratio, sonicated for 5 minutes, the combined supernatants were evaporated at 40°C and recovery of the remaining extract in 3 mL pure methanol	highest extraction yields for a 150-second treatment for all varieties; total anthocyanin content 173 mg Cyn-Glu/L for Dami; 169 mg Cyn-Glu/L for Xiaomi, 179 mg Cyn-Glu/L for Jima, 187 mg Cyn-Glu/L for Zaoshi; cyanidin 3-O-glucoside 46 mg/L for Dami; 46 mg/L for Xiaomi, 47 mg/L for Jima, 47 mg/L for Zaoshi; cyanidin 3-O-rutinoside: 127 mg/L for Dami; 123 mg/L for Xiaomi, 132 mg/L for Jima, 140 mg/L for Zaoshi;	Wang et al. (2014)

(Continued)

TABLE 15.1 *Continued*

Extraction method	Samples	Species, variety	Origin	Conditions	Anthocyanin concentration	References
combined methods, including novel methods	plum mash	*Prunus salicina* L.	South Korea	15, 30, and 60 minutes at a temperature of 45°C, a power of 600 W, and a frequency of 38 kHz combined with enzyme treatment (pectinase 0.2% [v/v])	total anthocyanins: 0.66 ± 0.23 mg/100 mL for 15 minutes; 0.76 ± 0.05 mg/100 mL for 30 minutes; 0.89 ± 0.05 mg/100 mL for 60 minutes	Olawuyi et al. (2020)
	lyophilized dried plum powder	*Prunus domestica* (European and Japanese plums)	–	step 1: methanol and HCl 99:1 (v/v), a ratio of 1:4 sample:solvent; then microwaved at 450 W for 15 seconds, sonicated at 35 kHz, for 15 minutes at 25°C, let set for 1 hour at room temperature (20°C), centrifuged (3500 rpm for 20 minutes) step 2 repeated three times: the solid was resuspended in 3 mL of methanol and HCl 99:1 (v/v), microwaved, sonicated for 30 minutes, centrifuged for 15 minutes	–	Issaad et al. (2017)
	plums	*Prunus domestica* L. (cv. President and Najbolia)	–	frozen 18°C, microwave heated at power density 1 W/g		Cendres et al. (2014)
	dried plum pomace	*Prunus domestica* L.	Poland	convective drying (50–90°C, with 10°C increments) and vacuum microwave treatment at 120–480 W, with 120 W increments	cyanidin 3-O-glucoside: 2.2–3.2 mg/100g DM for MW treatment; 1.2–4.5 for combined MW and CD treatment; cyanidin 3-O-rutinoside: 2.2–3.2 mg/100g DM for MW treatment; 4.1–5.5 for combined MW and CD treatment;	Michalska et al. (2019)

					peonidin 3-O-rutinoside 0.3–0.5 mg/100g DM for MW treatment; 0.5–0.2 for combined MW and CD treatment;	
liquid-solid extraction/ column chromatography	plum pomace	*Prunus salicinia* L. (Santa Rosa variety)	India	solid/water 1:1, boiled for 30 minutes; Amberlite XAD-16 (100 cm x 40 mm Φ); elution solvent 20–40% ethanol	–	Dwivedi et al. (2014)
	plum puree	*Prunus salicina* Lindl	Spain	methanol (+1% HCl); centrifuged at 22,300 × g at 4°C for 10 minutes; repeated twice. purification with a Sep-Pak C18 preactivated with water and methanol	2.35–3.06 mg/100g FW	García-Parra et al. (2014)
	plum juice	*Prunus domestica* L.	Germany	ultrasonic water bath at 4°C; centrifuged twice for 30 minutes, with 10,000 g at 4°C, SPE cartridge prepacked with 1 mL C18; 0.5 mL of water for cleaning and 1 mL methanol for compounds elution	0.1–0.6 mg/ 100g fruit; cyanidin-3-rutinoside: < 2–8 mL/L; cyanidin-3-glucoside: < 1–2 mg/L; peonidin-3-rutinoside: < 1–2 mg/L; peonidin-3-glucoside: < 0.1–0.2 mg/L	Goldner et al. (2015)
	plum puree	*Prunus salicina* Lindl. cv. Crimson Globe	Spain	methanol (+1% HCl); centrifugation (22,300 g for 10 minutes at 4°C; repeated twice; purification with a Sep-Pak C18 preactivated with water and methanol	total anthocyanin content: 65–110.5 mg/100g	Garcia-Parra et al. (2018)

(*Continued*)

TABLE 15.1 *Continued*

Extraction method	Samples	Species, variety	Origin	Conditions	Anthocyanin concentration	References
	plum husk	*Prunus domestica*		methanol: water: formic acid solution (80:19:1 [v/v/v]), 20-minute ultrasound treatment, extraction with hexane and ethyl acetate; Amberlite XAD-7HP resin in a ratio of 1:10 of sample in weight and resin; washings with water; methanol in a ratio of 1:5 for anthocyanin elution; repeated three times; purification: sephadex LH-20 column and an aqueous solution of 0.1% trifluoroacetic acid used as mobile phase	total anthocyanin content: 12.46 ± 1.58 mg/g DW cyanidin-3-rutinoside 16 mg	Fernandez-Aulis et al. (2020)

Note: GAE—gallic acid equivalents; CE—catechin equivalents; DW—dried weight; FW—fresh weight; CG—cyanidin 3-glucoside equivalents; DM—dry matter; malv-3-glu—malvidin-3-O-glucoside; DW—dry weight; Cyn-Glu—cyanidin 3-O-glucoside; Cyn-Rut—cyanidin 3-O-rutinoside; MW—microwave; CD—conventional drying; nd—not detected.

Hernandez-Herrero and Frutos 2014; Chaiyasut et al. 2016; Sahamishirazi et al. 2017). Trifluoroacetic acid (TFA) and formic acid have been also used (Netzel et al. 2006; Slimestad et al. 2009). It must be taken into consideration, however, that strong acids, such as hydrochloric acid, can hydrolyze the acetylated anthocyanins and thus affect the natural profile of anthocyanins in the samples.

Therefore, the effect of different acids in combination with different solvents in the anthocyanin extraction yield of plums (including individual compounds) should be further tested.

Anthocyanin concentration in plums and plums by-products are expressed per fresh weight (FW) (Netzel et al. 2006; Slimestad et al. 2009; Lozano et al. 2008; González-Cebrino et al. 2013; Sahamishirazi et al. 2017; Johnson, Collins, Mani et al. 2020; Usenik et al. 2009; Cuevas et al. 2015; Popescu et al. 2018), per dry matter (DM) (Constantin et al. 2018; Gościnna et al. 2021), per dry weight (DW) (Vangdal et al. 2017), and per extract (Olawuyi et al. 2020; Goldner et al. 2015; Wang et al. 2014) (Table 15.1). Thus, it needs to be pointed out that there is no consensus in the expression of neither total anthocyanin content nor particular anthocyanins. This is relevant for comparison purposes, i.e., among the different varieties, products, and processing methods. In these terms, the most recommended and usual approach is to express the anthocyanin concentration in mg/100g FW.

To the best of our knowledge, only two studies compare different solvent conditions for the extraction of plum anthocyanins (Joshi and Devi 2015; Johnson, Collins, Walsh et al. 2020). Johnson, Collins, Walsh, et al. (2020) compared the effect of ethanol/water and methanol/water extraction solvents on the yield of anthocyanins from six commercial plum varieties (*Prunus domestica*) grown in Australia (Table 15.1). Briefly, in the first extraction step, a 90% (v/v) of aqueous methanol solution was used as solvent, then the sample-solvent mixture was centrifuged for 60 minutes at 50 rpm, followed by centrifugation at 1000 g for 10 minutes. A re-extraction step was included. On the other hand, the second extraction used ethanol and followed the procedure intended for the quantification of anthocyanins in grapes proposed by the Australian Wine Research Institute. It utilized a 50% (v/v) aqueous ethanol solvent, then the sample was placed in a shaker (at 150 rpm) for 15 minutes at 20°C, followed by centrifugation at 1800 g for 10 minutes. A re-extraction step was also included. Acidification of extraction solvent was not employed to prevent the denaturation of the natural anthocyanin composition. Total anthocyanin content varied between 1.91 mg and 11.89 mg malvidin-3-O-glucoside (malv-3-glu)/100g FW for methanol solvent and between 0.67 mg and 4.41 mg malv-3-glu/100g FW for ethanol extraction. The results showed significantly higher total and individual anthocyanin yield for the methanolic extracts, except for the black plum variety, for which no significant differences were found. The authors explained that methanol presented a higher polarity compared to ethanol, which is more convenient to extract polar compounds like anthocyanins. Additionally, the low boiling point of methanol makes it suitable to concentrate the anthocyanin extract.

Joshi and Devi (2015), on the other hand, evaluated the effect of ethanol concentration (50% and 100%), citric acid (0.1% and 0.2%), and solid-solvent ratio (8:1 and 10:1) on the extraction yield of anthocyanins in plum (*Prunus salicinia* L. var. Santa Rosa) pomace (skin and stones), which was intended as a natural coloring agent for foodstuffs. The citric acid was used because the anthocyanin extract was intended as a food ingredient. The combination that showed the highest anthocyanin extraction yield (325 mg/100 mL) was 50% ethanol and 0.2% citric acid with a solvent-solid ratio of 10:1, while the minimum concentration (169 mg/100 mL) was reported for the mixture of 50% ethanol and 0.1% citric acid at a ratio of 8:1. Although citric acid is a weaker acid than HCl, it seems to still stabilize the anthocyanin cationic form (Joshi and Devi 2015).

15.3 ULTRASOUND-ASSISTED EXTRACTION METHODS OF PLUM ANTHOCYANINS

Ultrasound-assisted extraction is increasingly being used to extract anthocyanins from plums, plum products, and by-products because it can increase the yield and extraction rate while reducing the volume of solvent and extraction duration (Gościnna et al. 2021; Manzano Durán et al. 2019; Medina-Meza and Barbosa-Cánovas 2015; Popescu et al. 2018). Anthocyanins are better extracted because of cavitation produced by the energy derived from ultrasounds. This phenomenon consists of the development of microbubbles that collapse on the surface of the extraction matrix (He et al. 2016; Mazza et al. 2019). In this way, the used solvent penetrates the cells of the plum peels more easily, which facilitates the extraction of anthocyanins. The collapse of the bubbles also creates a mechanical effect that results in the supplementary enhancement of anthocyanin release.

It was shown that anthocyanin stability could be improved by using lower frequencies for the ultrasound treatment. However, the time of extraction increases significantly when choosing lower frequencies for the extraction (Yuan et al. 2020). Ultrasound-assisted extraction can be paired with heat treatments to increase the number of microbubbles and improve extraction efficiency (Aliano-Gonzalez et al. 2020). On the other hand, temperatures above 70°C have been shown to affect the anthocyanins by producing a quick degradation (Ju and Howard 2003).

Thus, significant increases in total anthocyanin content may be achieved by using ultrasound-assisted extraction. Several studies have shown a significantly increased extraction efficiency when using aqueous methanol or acidified methanol as solvents and ultrasounds compared with other methods employing high temperatures, which impact the pigment stability (Gościnna et al. 2021). Hence, the use of methanol paired with ultrasonic extraction can employ a lower extraction temperature and is one of the most commonly used methods. Additionally, it is a cheap, simple, sustainable, and efficient approach that still yields optimum results.

Cuevas et al. (2015) recommended a 50% acetone aqueous solution at a solvent:sample ratio of 20:1 to extract anthocyanins from freeze-dried plums. The solvent-sample mixture was then sonicated for 15 minutes at 4°C. Additionally, the use of acidified methanol (1% HCl) coupled with sonication was proposed to extract anthocyanins from an array of Slovenian plum varieties (Table 15.1) (Usenik et al. 2009).

Manzano Durán et al. (2019), on the other hand, used mathematical modeling (response surface methodology) to study the combined effects of extraction solvent (water, pure methanol, and aqueous methanolic solutions) and duration of ultrasound treatment (5–30 minutes) at 4°C. The authors proposed an extraction time of 30 minutes and a methanol concentration of 84% to optimally extract phenolic compounds from different Japanese plum (*Prunus Salicina* Lind.) varieties.

Gościnna et al. (2021) employed an ultrasound treatment on fresh and powdered plums extracted with an aqueous methanol solution (80%). They obtained an increased yield when extracting four main anthocyanins as follows: cyanidin-3-rutinoside, cyaniding-3-glucoside, peonidin-3-glucoside, and peonidin-3-rutinoside. The overall increase in anthocyanin content reached up to 29%.

Tomic et al. (2019) used cooled methanol and ultrasonic extraction for one hour to extract phenolic compounds, anthocyanins among them. They reported five main anthocyanins in different plum varieties from Serbia, with cyanidin-3-glucoside and cyanidin-3-rutinoside being the most abundant.

Plum peels are a by-product that results mainly from the juice industry. However, they can be used in a circular bioeconomy to extract phytochemicals because the exocarp is rich in bioactive compounds, specifically anthocyanins (Hernandez-Herrero and Frutos 2014; Olawuyi et al. 2020; Milala et al. 2013). Ultrasonication at 25°C and 50°C was evaluated to assess this extraction method as an environmentally friendly alternative to hot water extraction at 70°C. The treatment at 50°C (Table 15.1) gave better extraction yields for total anthocyanins from this substrate (Medina-Meza and Barbosa-Cánovas 2015).

Lastly, ultrasound treatment may be used at the industrial level to stabilize pigments in plums before convective drying treatment. In a study by Popescu et al. (2018), a frequency of 40 kHz and a power of 200 W was used to sonicate plums submerged in distilled water at room temperature (20°C) for a period of 30 minutes. They concluded that this ultrasound treatment had a smaller impact on the plum tissue compared to osmotic dehydration and improved the total anthocyanin retention in the obtained prunes.

The fruit juice industry also uses ultrasound treatments to improve not only the yield but also the phenolic composition and antioxidant activity of fruits (Olawuyi et al. 2020; Soltani Firouz et al. 2019). Olawuyi et al. (2020) compared various durations of sonication on plum mash to assess the efficiency in extracting phenolic compounds, anthocyanins included, but also its effect on their antioxidant activity. They employed a frequency of 38 kHz and a power of 600 W at a temperature of 45°C for varying periods of 15–75 minutes. An ultrasound treatment of 30 minutes was considered optimal for a single treatment.

However, a slight degradation of anthocyanins may occur subsequent to ultrasonic treatment (Popescu et al. 2018; Dubrović et al. 2011; Soltani Firouz et al. 2019). This could be explained by the oxidation caused by free radicals, such as the hydroxyl radical, which is formed during sonication (Soltani Firouz et al. 2019), or by polyphenol oxidase (PPO), which degrades the anthocyanin glucosides (Versari et al. 1997; Khandare et al. 2011; Olawuyi et al. 2020). In this sense, Olawuyi et al. (2020) showed a negative correlation (r= -0.912) between total anthocyanin content and PPO activity.

15.4 MICROWAVE-ASSISTED EXTRACTION METHODS OF PLUM ANTHOCYANINS

Microwave-assisted extraction is used to extract anthocyanins because it is rapid, has high extraction efficiency, and requires low volumes of solvent (Liu et al. 2019;

Cătunescu et al. 2021; Mollica et al. 2016; Sparr Eskilsson and Björklund 2000). The microwaves generate frictions among the polar and charged ionic molecules in the sample exposed to microwaves (Yu et al. 2016). This produces selective and localized heating mainly in the water contained in the sample (Haddadi-Guemghar et al. 2014). Thus, the effect is primarily thermal, and the increase in temperature increases the pressure inside the tissue cells. The process causes ruptures in the cell walls and a subsequent release of anthocyanins in the extraction solvent (Teo et al. 2008; Haddadi-Guemghar et al. 2014). The duration of the sample to reach a designated temperature can be reduced with the increase of power (Cendres et al. 2014). In conclusion, the most important parameters that influence the extraction of anthocyanins from plums, plum products, and by-products when employing microwaves, are microwave power and solvent type, as seen in Table 15.1.

Cendres et al. (2011) extracted juice from plums with no added water using microwaves. The extraction was done by a process called hydrodiffusion caused by the heating of the water contained in the fruits. The experimental equipment included a multimode 2450 MHz microwave able to deliver a power of 1000 W with an incremental step of 10 W. The juice was continuously extracted and removed from the extraction cavity until the flow halted or the sample began to overheat.

Cendres et al. (2012) showed that microwave treatments could be used before pressing in the plum juice industry to increase juice yield and the extraction of compounds. Later, Cendres et al. (2014) proposed a microwave hydrodiffusion protocol to obtain juice from fresh and frozen plums. They additionally studied the diffusion of different molecules during extraction, including anthocyanins. It was concluded that this procedure extracted compounds inconsistently dependent on the localization and solubility of molecules but also on the size of the cells and the thickness and resistance of their cell walls. Anthocyanins, which are contained in the epidermis, were extracted in the middle and end phases of the process and at low employed powers.

Gościnna et al. (2021) used microwaves to extract bioactive compounds, anthocyanins included, from fresh and powdered plums. A power level of 600 W employed for 5 minutes was used to extract 10 g in 100 mL of 80% methanol aqueous solution. This assured a temperature of 60°C in the sample. This extraction method produced a 16–41% increase in the total anthocyanin content dependent on the drying procedure applied and the variety of plum. An additional increase was observed for the four main anthocyanins (cyanidin-3-rutinoside, cyanidin-3-glucoside, peonidin-3-rutinoside, and peonidin-3-glucoside) when comparing fresh plums to vacuum-dried prunes extracted using microwaves. But the obtained total anthocyanin content and anthocyanin profile were statistically similar to the ultrasound-treated samples.

However, several studies showed that some of the phenolic compounds determined in conventional-extracted samples were not detected in the extracts obtained by microwave-assisted extraction methods (Milena et al. 2019). This may be explained by the increases in temperature that can appear locally in static extraction flasks.

15.5 COMBINED EXTRACTION PROCEDURES

Several novel extraction methods employ, however, a combination of extraction techniques and several extraction steps (Table 15.1).

A microwave bleaching pre-treatment was shown to increase the content of anthocyanins when obtaining juice from four "Sanhua" plum cultivars (Wang et al. 2014). Several treatment durations were tested for a 700 W microwave oven, and a period of 150 seconds was deemed to be optimal for anthocyanin retention. This increase in anthocyanins was correlated with the inactivation of PPO and peroxidase (POD), which cause the formation of quinones and the further polymerization of phenolics (Kader et al. 1997). The actual extraction procedure consisted of an ultrasound-assisted solvent extraction. A 90% methanol acidified with 0.5% formic acid at a ratio of 1:2 (w/v) sample:solvent was used to repeatedly extract the anthocyanins until tissue discoloration. Additionally, the anthocyanin profile was little altered by this pre-treatment when compared to hot water bleaching and SO_2 treatment (Wang et al. 2014).

Similarly, a 400 W microwave pre-treatment for 80 seconds was reported to improve anthocyanin retention of Romanian plums (Popescu et al. 2018). A standard ultrasound-assisted extraction with acidified (0.1M HCl) ethanol (85:15%, v:v) was used.

Issaad et al. (2017) proposed an ultrasound-microwave-assisted extraction for lyophilized dried plum powders. A 99% aqueous methanol acidified with 1% HCl was used as extraction solvent. A two-step approach was employed. The first step included a microwaving treatment at 450 W for 15 seconds followed by a sonication at 35 kHz for 15 minutes, while for the second, the sonication duration was prolonged to 30 minutes. The second step was repeated three times for optimal extraction. This extraction method was able to produce anthocyanidins (delphinidin, malvidin, peonidin) and anthocyanins (kuromanin chloride).

Cendres et al. (2014) proposed a freezing pre-step for the microwave-assisted extraction of plum juice rich in anthocyanins. This was based on studies that reported up to 30% higher yields when using frozen fruit (Cendres et al. 2012, 2011). In this way, the cell walls in the plum peels are subjected to a dual damaging, one of the freeze–thaw process and another of the pressure caused by the microwave heating. Freezing and thawing break cell membranes because ice crystals have a higher volume. Thus, water can transfer from intra- and extracellular compartments (Petzold and Aguilera 2009). When subjected to microwave heating, the water molecules contained in ice absorb energy, melt, and form hot spots that accelerate melting. In this way, the number of water molecules affected by microwaves increases and the temperature reaches 100°C. This is the first step of extraction, produced by the melting ice. However, when reaching 100°C the second step of the extraction begins as juice is generated by steam. This step is similar to the microwave extraction of juice from fresh fruit and the extraction rate reaches its maximum. The extraction results showed that the anthocyanins were exclusively extracted from the plum peels and that this procedure seemed to enable the extraction of the anthocyanins that are usually hard to extract. While anthocyanins were extracted in the middle to the last phase of the extraction, procyanidins were extracted only at the end. This was seen because of the higher solubility of monomeric polyphenols (Cendres et al. 2014).

Michalska et al. (2019) reported the combined effect of drying and microwave treatments upon plum pomace. The pomace was predried by convention at various temperatures (50–90°C) until the pomace sample reached 0.75 kg of water per kg of pomace (up to 70 minutes). Then the predried samples were microwave dried under a vacuum at an array of powers (120–480 W). The highest retention of anthocyanins was obtained for conventional pomace at 70°C and the combined drying using a microwave power of 120 W. During the convention drying pretreatment, the content of quercetin-3-O-rutinoside and quercetin-3-O-galactoside was reported higher for increased temperatures. From the tested microwave powers, the highest negative effect upon anthocyanins was observed at 240 W. However, regression analysis showed a higher effect of processing duration than of high temperature. The authors proposed the optimum parameters, in terms of bioactive compound retention from plum pomace, anthocyanins included the following: a convective predrying treatment (70°C) and a microwave vacuum drying (120 W). This procedure also shortened the total drying time.

15.6 CONCENTRATION AND PURIFICATION STEP OF PLUM ANTHOCYANINS

However, a subsequent step is required for the concentration and purification of anthocyanins after the liquid-solid extraction. Thus, this preliminary step will enrich the extract and lead to the obtaining of high-purity anthocyanins. Several procedures are being studied to separate plant anthocyanins, such as ion exchange or solvent and high-speed counter-current chromatography (Šmídová et al. 2017; Sheng et al. 2014; Xue et al. 2019). But these procedures usually need prolonged durations, require qualified personnel, and are a good fit only for laboratory or small-scale plants (Xue et al. 2019). In this sense, adsorbent resins were successfully proposed for this purpose because they are cheap and easy to obtain, with good selectivity and high adsorption and desorption capacity (Liu et al. 2004; Cao et al. 2010; Guanlin et al. 2005; Liu et al. 2007; Esatbeyoglu et al. 2016; Chang et al. 2012).

Few studies have described the concentration and purification process of anthocyanins from plums and plum by-products (Table 15.1) with various final objectives. Among these, the following can be mentioned: the effect of different processing techniques for plum puree production (García-Parra et al. 2014; Garcia-Parra et al. 2018), the obtaining of natural food pigments (Dwivedi et al. 2014), and the production of anthocyanin standards (Fernandez-Aulis et al. 2020).

Among the adsorbent resins used to concentrate and purify anthocyanins from plums and plum by-products, it is worth mentioning the Sep-Pak C18 for plum puree and juice, Amberlite XAD-16 for plum pomace, and Amberlite XAD-7HP and Sephadex LH-20 for plum husk (Table 15.1).

To the best of our knowledge, a single study has been conducted to standardize the adsorption conditions to obtain crude anthocyanins from plum-processing wastes (Dwivedi et al. 2014). Different proportions of Amberlite XAD-16 (15%, 20%, 25%, 30%, 35%, and 40%) were tested. Amberlite XAD-16 is a non-ionic acrylic polymer adsorbent. The authors selected this resin because it had been reported to be one of the most suitable adsorbents for anthocyanin purification from fruit juice (Kraemer-Schafhalter et al. 1998; Dwivedi et al. 2014). Additionally, they used different concentrations of ethanol (20%, 40%, 60%, 80%, and 100%) as a desorption solvent. The results showed that a percentage of 35% amberlite XAD-16 was the optimum proportion for a significant adsorption of anthocyanins. A concentration of 60% of ethanol proved to be the best concentration that ensured the highest desorption of anthocyanins (94.96%), while 20% ethanol produced the minimum desorption (36.06%). Higher ethanol proportions, such as 80% and 100%, did not generate a higher anthocyanin recovery. This might be explained by the reduced amount of water because anthocyanins are water-soluble molecules (Dwivedi et al. 2014).

It must be pointed out that, to the best of our knowledge, there is no reference in the scientific literature reporting the optimization conditions for different adsorbent resins used for the extraction of anthocyanins from plums and plums by-products. Thus, further research is required to establish the best procedures and process parameters to obtain higher anthocyanin extraction yield.

15.7 CONCLUSION

Several studies have reported anthocyanins extraction methods from plums, plum products, and by-products, but they are still rather scarce, and they are not in consensus. Additionally, the current protocols are not yet standardized. Thus, their accuracy and precision are not clear, as the reported results cannot be directly comparable. Moreover, there appears to be no consensus in the expression of total anthocyanin content and particular anthocyanins. But the most used expression of anthocyanin seems to be in mg/100g FW.

Solvent extraction methods of plum anthocyanins are the most commonly used and are considered conventional. The usually utilized solvents are acidified methanol or ethanol aqueous solutions, but acetone and acetonitrile are also being used. The most used acid for acidification is HCl, although it was reported that strong acids could affect the natural profile of anthocyanins by hydrolyzing the acetylated anthocyanins.

Thus, conventional extraction methods are characterized by a wide range of solvents that can be employed but also by various solvent-to-sample ratios, acids used for acidification, temperatures, and solvent contact periods. However, no clear recommendation can be made in terms of yield and anthocyanin profile from plums and plum products.

Ultrasound treatment is one of the most used in combination with solvent extraction. Usually, lower frequencies enhance the stability of anthocyanins, but they would require longer treatment durations. Additionally, ultrasound-assisted extraction can be paired with heat treatments that improve extraction efficiency, but they should be kept below 70°C. However, some degradations of anthocyanins were reported for plum products subsequent to ultrasonic processing, especially when paired with heat treatments.

Microwave-assisted extraction is another novel procedure that is considered an efficient and rapid method, with the advantage that it requires lower solvent volumes. Moreover, several procedures were reported for plums, approaches that require no solvent at all, known as hydrodiffusion. Pretreatments at powers between 400 W and 700 W lasting for 80–150 seconds were proposed to obtain anthocyanin-rich extracts and juices. But some studies have reported losses of some phenolic compounds compared to conventional-extracted anthocyanins. These losses could be attributed to the local increases in temperature during microwave extraction.

Additionally, several authors proposed combined extraction procedures, such as ultrasound-microwave-assisted extraction and microwave-assisted extraction with a prefreezing or a convention predrying step.

Regardless of the actual extraction procedure, an additional concentration and purification are needed to obtain anthocyanins and anthocyanin-rich extracts. The most used is an Amberlite XAD-16 adsorbent. The only study tackling the optimization of an Amberlite XAD-16 and ethanol concentrations proposed 35% polymer and 60% ethanol aqueous solution as optimum to concentrate anthocyanins from plums and plums by-products.

In conclusion, more studies are needed to evaluate the extraction efficiency and productivity of anthocyanins from plums, plum products, and by-products. The current literature reports do not allow the proposal of a rapid, optimal extraction method that uses low solvent volumes and employs environmentally friendly procedures. Additionally, standardization is required for a proper comparison of the reporting of extraction results.

15.8 REFERENCES

Aliano-Gonzalez, M.J., Jarillo, J.A., Carrera, C., Ferreiro-Gonzalez, M., Alvarez, J.A., Palma, M., Ayuso, J., Barbero, G.F., and Espada-Bellido, E. 2020. Optimization of a Novel Method Based on Ultrasound-Assisted Extraction for the Quantification of Anthocyanins and Total Phenolic Compounds in Blueberry Samples (Vaccinium corymbosum L.). *Foods* 9.

Alsolmei, F.A., Li, H., Pereira, S.L., Krishnan, P., Johns, P.W., and Siddiqui, R.A. 2019. Polyphenol-enriched Plum Extract Enhances Myotubule Formation and Anabolism While Attenuating Colon Cancer-Induced Cellular Damage in C2C12 Cells. *Nutrients* 11: 1077.

Barnes, J.S., Nguyen, H.P., Shen, S., and Schug, K.A. 2009. General Method for Extraction of Blueberry Anthocyanins and Identification Using High Performance Liquid Chromatography-Electrospray Ionization-Ion Trap-Time of Flight-Mass Spectrometry. *Journal Chromatogr A* 1216: 4728–4735.

Basanta, M.F., Marin, A., De Leo, S.A., Gerschenson, L.N., Erlejman, A.G., Tomás-Barberán, F.A., and Rojas, A.M. 2016. Antioxidant Japanese Plum (*Prunus salicina*) Microparticles with Potential for Food Preservation. *Journal of Functional Foods* 24: 287–296.

Buckow, R., Kastell, A., Terefe, N.S., and Versteeg, C. 2010. Pressure and Temperature Effects on Degradation Kinetics and Storage Stability of Total Anthocyanins in Blueberry Juice. *Journal Agric Food Chem* 58: 10076–10084.

Cabrera-Bañegil, M., Lavado Rodas, N., Prieto Losada, M.N., Cipollone, F.B. et al. 2020. Evolution of Polyphenols Content in Plum Fruits (*Prunus salicina*) with Harvesting Time by Second-Order Excitation-Emission Fluorescence Multivariate Calibration. *Microchemical Journal* 158.

Cao, S.-Q., Pan, S.Y., Yao, X.L., and Fu, H.F. 2010. Isolation and Purification of Anthocyanins from Blood Oranges by Column Chromatography. *Agricultural Sciences in China* 9: 207–215.

Casedas, G., Gonzalez-Burgos, E., Smith, C., Lopez, V., and Gomez-Serranillos, M.P. 2018. Regulation of Redox Status in Neuronal SH-SY5Y Cells by Blueberry (Vaccinium myrtillus L.) Juice, Cranberry (Vaccinium macrocarpon A.) Juice and Cyanidin. *Food and Chemical Toxicology* 118: 572–580.

Cătunescu, G.M., Bodea, I.M., Pop, C.R., Hornedo-Ortega, R., Carmen Garcia-Parrilla, M., Troncoso, A.M., Cerezo, A.B., and Rotar, A.M. 2021. Effect of Processing on the Bioactivity of Blueberries Anthocyanins. In *Blueberries: Nutrition, Consumption and Health*, edited by Laura M. Williams. Nova Science Publisher.

Cendres, A., Chemat, F., Maingonnat, J.F., and Renard, C.M.G.C. 2011. An Innovative Process for Extraction of Fruit Juice Using Microwave Heating. *LWT—Food Science and Technology* 44: 1035–1041.

Cendres, A., Chemat, F., Page, D. et al. 2012. Comparison Between Microwave Hydrodiffusion and Pressing for Plum Juice Extraction. *LWT—Food Science and Technology* 49: 229–237.

Cendres, A., Hoerlé, M., Chemat, F., and Renard, C.M.G.C. 2014. Different Compounds Are Extracted with Different Time Courses from Fruits During Microwave Hydrodiffusion: Examples and Possible Causes. *Food Chemistry* 154: 179–186.

Chaiyasut, C., Sivamaruthi, B.S., Pengkumsri, N., Sirilun, S., Peerajan, S., Chaiyasut, K., and Kesika, P. 2016. Anthocyanin Profile and Its Antioxidant Activity of Widely Used Fruits, Vegetables, and Flowers in Thailand. *Asian Journal of Pharmaceutical and Clinical Research* 9.

Chang, X.-L., Wang, D., Chen, B.Y., Feng, Y.M., Wen, S.H., and Zhan, P.Y. 2012. Adsorption and Desorption Properties of Macroporous Resins for Anthocyanins from the Calyx Extract of Roselle (*Hibiscus sabdariffa* L.). *Journal of Agricultural and Food Chemistry* 60: 2368–2376.

Chun, O.K., Kim, D.O., and Yong Lee, C. 2003. Superoxide Radical Scavenging Activity of the Major Polyphenols in Fresh Plums. *Journal of Agricultural and Food Chemistry* 51: 8067–8072.

Constantin, O.E., Râpeanu, G. et al. 2018. Antioxidative Capacity of and Contaminant Concentrations in Processed Plum Products Consumed in Romania. *Journal of Food Proteins* 81: 1313–1320.

Cosmulescu, S., Trandafir, I., Nour, V., and Botu, M. 2015. Total Phenolic, Flavonoid Distribution and Antioxidant Capacity in Skin, Pulp and Fruit Extracts of Plum Cultivars. *Journal of Food Biochemistry* 39: 64–69.

Cuevas, F.J., Pradas, I., Ruiz-Moreno, M.J., Arroyo, F.T., Perez-Romero, L.F., Montenegro, J.C., and Moreno-Rojas, J.M. 2015. Effect of Organic and Conventional Management on Bio-Functional Quality of Thirteen Plum Cultivars (*Prunus salicina* Lindl.). *PLoS One* 10: e0136596.

Dubrović, I., Herceg, Z., Jambrak, A.R., Badanjak, M., and Dragović-Uzelac, V. 2011. Effect of High Intensity Ultrasound and Pasteurization on Anthocyanin Content in Strawberry Juice. *Food Technology and Biotechnology* 49: 196–204.

Dwivedi, S.K., Joshi, V.K., and Mishra, V. 2014. Extraction of Anthocyanins from Plum Pomace Using XAD-16 and Determination of Their Thermal Stability. *Journal of Scientific & Industrial Research* 73: 57–61.

Esatbeyoglu, T., Rodríguez-Werner, M., Schlösser, A., Liehr, M., Ipharraguerre, I., Winterhalter, P., and Rimbach, G. 2016. Fractionation of Plant Bioactives from Black Carrots (Daucus Carota Subspecies Sativus Varietas Atrorubens Alef.) by Adsorptive Membrane Chromatography and Analysis of Their Potential Anti-Diabetic Activity. *Journal of Agricultural and Food Chemistry* 64: 5901–5908.

Fanning, K.J., Topp, B., Russell, D., Stanley, R., and Netzel, M. 2014. Japanese Plums (*Prunus salicina* Lindl.) and Phytochemicals-Breeding, Horticultural Practice, Postharvest Storage, Processing and Bioactivity. *Journal of Science Food Agriculture* 94: 2137–2147.

Fernandez-Aulis, F., Torres, A., Sanchez-Mendoza, E., Cruz, L., and Navarro-Ocana, A. 2020. New Acylated Cyanidin Glycosides Extracted from Underutilized Potential Sources: Enzymatic Synthesis, Antioxidant Activity and Thermostability. *Food Chemistry* 309: 125796.

Ferreira, L.F. et al. 2020. Citric Acid Water-Based Solution for Blueberry Bagasse Anthocyanins Recovery: Optimization and Comparisons with Microwave-Assisted Extraction (MAE). *LWT* 133.

García-Parra, J., González-Cebrino, F., Cava, R., and Ramírez, R. 2014. Effect of a Different High Pressure Thermal Processing Compared to a Traditional Thermal Treatment on a Red Flesh and Peel Plum Purée. *Innovative Food Science & Emerging Technologies* 26: 26–33.

Garcia-Parra, J., Gonzalez-Cebrino, F., Delgado-Adamez, J., Cava, R., Martin-Belloso, O., Elez-Martinez, P., and Ramirez, R. 2018. Effect of High-Hydrostatic Pressure and Moderate-Intensity Pulsed Electric Field on Plum. *Food Sci Technol Int* 24: 145–160.

Goldner, K., Neumüller, M., and Treutter, D. 2015. Phenolic Contents in Fruit Juices of Plums with Different Skin Colors. *Journal of Applied Botany and Food Quality* 88: 322–326.

González-Cebrino, F., Durán, R., Delgado-Adámez, J., Contador, R., and Ramírez, R. 2013. Changes After High-Pressure Processing on Physicochemical Parameters, Bioactive Compounds, and Polyphenol Oxidase Activity of Red Flesh and Peel Plum Purée. *Innovative Food Science & Emerging Technologies* 20: 34–41.

Gościnna, K., Pobereżny, J., Wszelaczyńska, E., Szulc, W., and Rutkowska, B. 2021. Effects of Drying and Extraction Methods on Bioactive Properties of Plums. *Food Control* 122.

Guanlin, W., Jing, Y., and Hongyan, L. 2005. *Extraction of Anthocyanin from Sweetpotato by Macroporous Resin and Its Bacteriostatic Mechanism.* Scientia Agricultura Sinica.

Haddadi-Guemghar, H., Janel, N., Dairou, J., Remini, H., and Madani, K. 2014. Optimisation of Microwave-Assisted Extraction of Prune (Prunus domestica) Antioxidants by Response Surface Methodology. *International Journal of Food Science & Technology* 49: 2158–2166.

He, B., Zhang, L.L., Yue, X.Y., Liang, J., Jiang, J., Gao, X.L., and Yue, P.X. 2016. Optimization of Ultrasound-Assisted Extraction of Phenolic Compounds and Anthocyanins from Blueberry (Vaccinium ashei) Wine Pomace. *Food Chemistry* 204: 70–76.

Hernandez-Herrero, J.A., and Frutos, M.J. 2014. Colour and Antioxidant Capacity Stability in Grape, Strawberry and Plum Peel Model Juices at Different pHs and Temperatures. *Food Chem* 154: 199–204.

Hidalgo, G.I., and Almajano, M.P. 2017. Red Fruits: Extraction of Antioxidants, Phenolic Content, and Radical Scavenging Determination: A Review. *Antioxidants (Basel)* 6.

Issaad, F.Z. et al. 2017. Flavonoids in Selected Mediterranean Fruits: Extraction, Electrochemical Detection and Total Antioxidant Capacity Evaluation. *Electroanalysis* 29: 358–366.

Johnson, J.B., Collins, T., Mani, J.S., and Naiker, M. 2020. Nutritional Quality and Bioactive Constituents of Six Australian Plum Varieties. *International Journal of Fruit Science* 21: 115–132.

Johnson, J.B., Collins, T., Walsh, K., and Naiker, M. 2020. Solvent Extractions and Spectrophotometric Protocols for Measuring the Total Anthocyanin, Phenols and Antioxidant Content in Plums. *Chemical Papers* 74: 4481–4492.

Joshi, V.K., and Preema Devi, M. 2015. Optimization of Extraction Treatment and Concentration of Extract on Yield and Quality of Anthocyanins from Plum Var. 'Santa Rosa'. *Indian Journal of Natural Products and Resources (IJNPR)[Formerly Natural Product Radiance (NPR)]* 5: 171–175.

Ju, Z.Y., and Howard, L.R. 2003. Effects of Solvent and Temperature on Pressurized Liquid Extraction of Anthocyanins and Total Phenolics from Dried Red Grape Skin. *Journal of Agricultural Food Chemistry* 51: 5207–5213.

Kader, F., Rovel, B., Girardin, M., and Metche, M. 1997. Mechanism of Browning in Fresh Highbush Blueberry Fruit (Vaccinium corymbosum L). Role of Blueberry Polyphenol Oxidase, Chlorogenic Acid and Anthocyanins. *Journal of the Science of Food and Agriculture* 74: 31–34.

Khandare, V., Walia, S., Singh, M., and Kaur, C. 2011. Black Carrot (Daucus carota ssp. sativus) Juice: Processing Effects on Antioxidant Composition and Color. *Food and Bioproducts Processing* 89: 482–486.

Kraemer-Schafhalter, A., Fuchs, H., and Pfannhauser, W. 1998. Solid-Phase Extraction (SPE)—a Comparison of 16 Materials for the Purification of Anthocyanins from Aronia Melanocarpa Var Nero. *Journal of the Science of Food and Agriculture* 78: 435–440.

Kuo, P.-H., Ching, I.L., Chen, Y.H., Chiu, W.C., and Lin, S.H. 2015. A High-Cholesterol Diet Enriched with Polyphenols from Oriental Plums (*Prunus salicina*) Improves Cognitive Function and Lowers Brain Cholesterol Levels and Neurodegenerative-Related Protein Expression in Mice. *British Journal of Nutrition* 113: 1550–1557.

Liang, D., and Liu, Z. 2019. Research on Inhibition Rate of Plum Essence to the Proliferation of Breast Cancer Cells MCF-7 and Liver Cancer Cells HepG2. *Hans Journal of Food and Nutrition Science* 8: 75–79.

Liu, C., Xue, H., Shen, L., Liu, C., Zheng, X., Shi, J., and Xue, S. 2019. Improvement of Anthocyanins Rate of Blueberry Powder Under Variable Power of Microwave Extraction. *Separation and Purification Technology* 226: 286–298.

Liu, X., Xu, Z., Gao, Y., Yang, B., Zhao, J., and Wang, L. 2007. Adsorption Characteristics of Anthocyanins from Purple-fleshed Potato (Solanum. *International Journal of Food Engineering* 5.

Liu, X., Xiao, G., Chen, W., Xu, Y., and Wu, J. 2004. Quantification and Purification of Mulberry Anthocyanins with Macroporous Resins. *Journal of Biomedicine and Biotechnology* 2004: 326–331.

Lozano, M. et al. 2008. Physicochemical and Nutritional Properties and Volatile Constituents of Six Japanese Plum (Prunus salicina Lindl.) Cultivars. *European Food Research and Technology* 228: 403–410.

Manzano Durán, R., Fernández Sánchez, J.E., Velardo-Micharet, B., and José Rodríguez Gómez, M. 2019. Multivariate Optimization of Ultrasound-Assisted Extraction for the Determination of Phenolic Compounds in Plums (Prunus salicina Lindl.) by High-Performance Liquid Chromatography (HPLC). *Instrumentation Science & Technology* 48: 113–127.

Mazza, K.E.L. et al. 2019. Syrah Grape Skin Valorisation Using Ultrasound-Assisted Extraction: Phenolic Compounds Recovery, Antioxidant Capacity and Phenolic Profile. *International Journal of Food Science & Technology* 54: 641–650.

Medina-Meza, I.G., and Barbosa-Cánovas, G.V. 2015. Assisted Extraction of Bioactive Compounds from Plum and Grape Peels by Ultrasonics and Pulsed Electric Fields. *Journal of Food Engineering* 166: 268–275.

Michalska, W., Majerska, L., and Brzezowska, J. 2019. Qualitative and Quantitative Evaluation of Heat-Induced Changes in Polyphenols and Antioxidant Capacity in Prunus domestica L. By-Products. *Molecules* 24: 3008.

Michalska, A., Wojdyło, A., Łysiak, G., and Figiel, A. 2017. Chemical Composition and Antioxidant Properties of Powders Obtained from Different Plum Juice Formulations. *International Journal of Molecular Sciences* 18:176.

Milala, J. et al. 2013. Plum Pomaces as a Potential Source of Dietary Fibre: Composition and Antioxidant Properties. *Journal of Food Science and Technology* 50: 1012–1017.

Milena, V. et al. 2019. Advantages of Contemporary Extraction Techniques for the Extraction of Bioactive Constituents from Black Elderberry (Sambucus nigra L.) Flowers. *Industrial Crops and Products* 136: 93–101.

Mollica, A., Locatelli, M., Macedonio, G., Carradori, S., Sobolev, A.P., De Salvador, R.F., Monti, S.M., Buonanno, M., Zengin, G., and Angeli, A. 2016. Microwave-Assisted Extraction, HPLC analysis, and Inhibitory Effects on Carbonic Anhydrase I, II, VA, and VII Isoforms of 14 Blueberry Italian Cultivars. *Journal of Enzyme Inhibition and Medicinal Chemistry* 31: 1–6.

Navarro, M., Moreira, I., Arnaez, E., Quesada, S., Azofeifa, G., Vargas, F., Alvarado, D., and Chen, P. 2018. Polyphenolic Characterization and Antioxidant Activity of Malus Domestica and Prunus Domestica Cultivars from Costa Rica. *Foods* 7:15.

Netzel, M., Netzel, G., Tian, Q., Schwartz, S., and Konczak, I. 2006. Sources of Antioxidant Activity in Australian Native Fruits: Identification and Quantification of Anthocyanins. *J Agric Food Chem* 54: 9820–9826.

Nicoue, E.E., Savard, S., and Belkacemi, K. 2007. Anthocyanins in Wild Blueberries of Quebec: Extraction and Identification. *Journal of Agricultural and Food Chemistry* 55: 5626–5635.

Noratto, G. et al. 2015. Consumption of Polyphenol-Rich Peach and Plum Juice Prevents Risk Factors for Obesity-Related Metabolic Disorders and Cardiovascular Disease in Zucker Rats. *The Journal of Nutritional Biochemistry* 26: 633–641.

Olawuyi, I.F., Akbarovich, S.A., Kil Kim, C, and Young Lee, W. 2020. Effect of Combined Ultrasound-Enzyme Treatment on Recovery of Phenolic Compounds, Antioxidant Capacity, and Quality of Plum (Prunus salicina L.) Juice. *Journal of Food Processing and Preservation* 45.

Panahirad, S., Naghshiband-Hassani, R., and Mahna, N. 2020. Pectin-Based Edible Coating Preserves Antioxidative Capacity of Plum Fruit During Shelf Life. *Food Sci Technol Int* DOI:1082013220916559.

Pereira, V.A., Natália Queiroz de Arruda, I., and Stefani, R. 2015. Active Chitosan/PVA Films with Anthocyanins from Brassica oleraceae (Red Cabbage) as Time—Temperature Indicators for Application in Intelligent Food Packaging. *Food Hydrocolloids* 43: 180–188.

Petzold, G., and Aguilera, J.M. 2009. Ice Morphology: Fundamentals and Technological Applications in Foods. *Food Biophysics* 4: 378–396.

Popescu, E.C. et al. 2018. Effect of Convective Dryness Combined with Osmotic Dehydration, Blanching, Microwave and Ultrasonic Treatment on Bioactive Compounds and Rehydration Capacity of Dried Plums. *Revista De Chimie* 69: 1949–1953.

Prasain, J.K., Barnes, S., and Michael Wyss, J. 2018. Chapter 24 — Analyzing Ingredients in Dietary Supplements and Their Metabolites. In *Polyphenols: Mechanisms of Action in Human Health and Disease*, edited by Ronald Ross Watson, Victor R. Preedy, and Sherma Zibadi, 2nd ed., 337–346. Academic Press.

Rodriguez-Saona, L.E., and Wrolstad, R.E. 2001. Extraction, Isolation, and Purification of Anthocyanins. *Current Protocols in Food Analytical Chemistry* F1.1.1–F1.1.11.

Sahamishirazi, S., Moehring, J., Claupein, W., and Graeff-Hoenninger, S. 2017. Quality Assessment of 178 Cultivars of Plum Regarding Phenolic, Anthocyanin and Sugar Content. *Food Chem* 214: 694–701.

Sheng, F., Wang, Y., Zhao, X., Tian, N., Hu, H., and Li, P. 2014. Separation and Identification of Anthocyanin Extracted from Mulberry Fruit and the Pigment Binding Properties Toward Human Serum Albumin. *Journal of Agricultural and Food Chemistry* 62: 6813–6819.

Silva, S. et al. 2017. Production of a Food Grade Blueberry Extract Rich in Anthocyanins: Selection of Solvents, Extraction Conditions and Purification Method. *Journal of Food Measurement and Characterization* 11: 1248–1253.

Skrede, G., Wrolstad, R.E., and Durst, R.W. 2000. Changes in Anthocyanins and Polyphenolics During Juice Processing of Highbush Blueberries (Vaccinium corymbosum L.). *Journal of Food Science* 65: 357–364.

Slimestad, R., Vangdal, E., and Brede, C. 2009. Analysis of Phenolic Compounds in Six Norwegian Plum Cultivars (*Prunus domestica* L.). *J Agric Food Chem* 57: 11370–11375.

Šmídová, B. et al. 2017. The Pentafluorophenyl Stationary Phase Shows a Unique Separation Efficiency for Performing Fast Chromatography Determination of Highbush Blueberry Anthocyanins. *Talanta* 166: 249–254.

Soltani Firouz, M., Farahmandi, A., and Hosseinpour, S. 2019. Recent Advances in Ultrasound Application as a Novel Technique in Analysis, Processing and Quality Control of Fruits, Juices and Dairy Products Industries: A Review. *Ultrasonics Sonochemistry* 57: 73–88.

Sparr Eskilsson, C., and Björklund, E. 2000. Analytical-Scale Microwave-Assisted Extraction. *Journal of Chromatography A* 902: 227–250.

Tennant, D.R., and Klingenberg, A. 2016. Consumer EXPOSURES to Anthocyanins from Colour Additives, Colouring Foodstuffs and from Natural Occurrence in Foods. *Food Additives & Contaminants: Part A* 33: 959–967.

Teo, C.C., Tan, S.N., Yong, J.W., Hew, C.S., and Ong, E.S. 2008. Evaluation of the Extraction Efficiency of Thermally Labile Bioactive Compounds in Gastrodia elata Blume by Pressurized Hot Water Extraction and Microwave-Assisted extraction. *J Chromatogr A* 1182: 34–40.

Tomic, J., Stampar, F., Glisic, I., and Jakopic, J. 2019. Phytochemical Assessment of Plum (*Prunus domestica* L.) Cultivars Selected in Serbia. *Food Chem* 299: 125113.

Truong, V.D. et al. 2012. Pressurized Liquid Extraction and Quantification of Anthocyanins in Purple-Fleshed Sweet Potato Genotypes. *Journal of Food Composition and Analysis* 26: 96–103.

Usenik, V., Stampar, F., and Veberic, R. 2009. Anthocyanins and Fruit Colour in Plums (Prunus domestica L.) During Ripening. *Food Chemistry* 114: 529–534.

Vangdal, E., Picchi, V., Fibiani, M., and Lo Scalzo, R. 2017. Effects of the Drying Technique on the Retention of Phytochemicals in Conventional and Organic Plums (Prunus domestica L.). *LWT—Food Science and Technology* 85: 506–509.

Versari, A., Biesenbruch, S., Barbanti, D., Farnell, P.J., and Galassi, S. 1997. Effects of Pectolytic Enzymes on Selected Phenolic Compounds in Strawberry and Raspberry Juices. *Food Research International* 30: 811–817.

Vlaic, R.A. et al. 2018. The Changes of Polyphenols, Flavonoids, Anthocyanins and Chlorophyll Content in Plum Peels During Growth Phases: From Fructification to Ripening. *Notulae Botanicae Horti Agrobotanici Cluj-Napoca* 46: 148–155.

Walkowiak-Tomczak, D., Reguła, J., and Łysiak, G. 2008. Physico-Chemical Properties and Antioxidant Activity of Selected Plum Cultivars Fruit. *Acta Scientiarum Polonorum Technologia Alimentaria* 7.

Wang, C., Duan, H., Liu, L., Luo, Y., and Dai, J. 2014. Effect of Juicing on Nutrition Qualities of “Sanhua” Plum (Prunus Salicina Lindl.) Juice from 4 Cultivars. *Food Science and Technology Research* 20: 1153–1164.

Wang, W., Jung, J., Tomasino, E., and Zhao, Y. 2016. Optimization of Solvent and Ultrasound-Assisted Extraction for Different Anthocyanin Rich Fruit and Their Effects on Anthocyanin Compositions. *LWT—Food Science and Technology* 72: 229–238.

Wu, C.H., Pan, C.H., and Wu, C.H. 2019. *Composition of Wood Ear, Shiitake, Hawthorn Fruit, Roselle, Celery and Fruit of Chinese Plum for Treatment and/or Prevention of Hyperlipidemia, Atherogenesis and Obesity*. Google Patents.

Xue, H., Shen, L., Wang, X., Liu, C., Liu, C., Liu, H., and Zheng, X. 2019. Isolation and Purification of Anthocyanin from Blueberry Using Macroporous Resin Combined Sephadex Lh-20 Techniques. *Food Science and Technology Research* 25: 29–38.

Yao, G.L., Ma, X.H., Cao, X.Y., and Chen, J. 2016. Effects of Power Ultrasound on Stability of Cyanidin-3-Glucoside Obtained from Blueberry. *Molecules* 21.

Yu, S., Hongkun, X., Chenghai, L., Chai, L., Xiaolin, S., and Xianzhe, Z. 2016. Comparison of Microwave Assisted Extraction with Hot Reflux Extraction in Acquirement and Degradation of Anthocyanin from Powdered Blueberry. *International Journal of Agricultural and Biological Engineering* 9: 186–199.

Yuan, J., Li, H., Tao, W., Han, Q., Dong, H., Zhang, J., Jing, Y., Wang, Y., Xiong, Q., and Xu, T. 2020. An Effective Method for Extracting Anthocyanins from Blueberry Based on Freeze-Ultrasonic Thawing Technology. *Ultrason Sonochem* 68: 105192.

Zhu, N., Zhu, Y., Yu, N., Wei, Y., Zhang, J., Hou, Y., and Sun, A.D. 2019. Evaluation of Microbial, Physicochemical Parameters and Flavor of Blueberry Juice After Microchip-Pulsed Electric Field. *Food Chemistry* 274: 146–155.

Index

C

D

E

G

H

I

J

K

L

Q

R

S

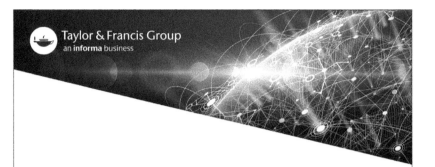

Ingram Content Group UK Ltd.
Milton Keynes UK
UKHW020614130723
425041UK00001B/10

9 781032 062426